207
Topics in Current Chemistry

Springer-Verlag Berlin Heidelberg GmbH

Small Ring Compounds in Organic Synthesis VI

Volume Editor: Armin de Meijere

With contributions by
A. de Meijere, L. P. Hadjiarapoglou,
N. Iwasawa, A. F. Khlebnikov, S. I. Kozhushkov,
K. Narasaka, J. Salaün

Springer

This series presents critical reviews of the present position and future trends in modern chemical research. It is addressed to all research and industrial chemists who wish to keep abreast of advances in the topics covered.

As a rule, contributions are specially commissioned. The editors and publishers will, however, always be pleased to receive suggestions and supplementary information. Papers are accepted for "Topics in Current Chemistry" in English.

In references Topics in Current Chemistry is abbreviated Top. Curr. Chem. and is cited as a journal.

ISSN 0340-1022
ISBN 978-3-662-15648-3 ISBN 978-3-540-48255-0 (eBook)
DOI 10.1007/978-3-540-48255-0

Library of Congress Catalog Card Number 74-644622

Originally published by Springer-Verlag Berlin Heidelberg New York in 2000
Softcover reprint of the hardcover 1st edition 2000

Cover design: Friedhelm Steinen-Broo, Barcelona; MEDIO, Berlin
Typesetting: Fotosatz-Service Köhler GmbH, 97084 Würzburg

SPIN: 10552172 02/3020 – 5 4 3 2 1 0 – Printed on acid-free paper

Volume Editor

Prof. Armin de Meijere
Institut für Organische Chemie
der Georg-August-Universität
Tammannstraße 2
D-37077 Göttingen, Germany
E-mail: ameijer1@uni-goettingen.de

Editorial Board

Preface

Why yet another volume in the series "Small Ring Compounds in Organic Synthesis" after five previous ones (Vol. 133, 135, 144, 155, 178)? Well, small ring chemistry is still flourishing, and probably even more than ever, at least as far as the applications of small rings in Organic Synthesis are concerned. Such applications range from total syntheses of cyclopropyl group containing natural products and non-natural biologically active compounds or compounds with other important properties (e.g. liquid crystalline) via syntheses of cyclopropyl analogues of natural and non-natural physiologically active compounds, e.g. peptidomimetics, to the use of cyclopropyl groups as reactive subunits in so-called composite functional groups.

Thus, the first contribution in this current volume by *J. Salaün* is to provide an insight into the wide range of biological activities of cyclopropyl containing compounds. Two articles which originate from the editor's own writing, are to present up-to-date comprehensive reviews on the multiple applicabilities of two multifunctional small ring building blocks (methyl 2-chloro-2-cyclopropylideneacetate and bicyclopropylidene – which have been developed in the authors' own laboratory) in Organic Synthesis. The fourth contribution by *K. Narasaka* and *N. Iwasawa* for the first time summarizes a striking new synthetic approach to cyclopentenones from easily accessible 1-ethynylcyclopropanols, which nicely complements other cyclopentenone syntheses. This report provides one more example of a synthetic method which derives from the reactivity of the carbocyclic three-membered ring.

The authors, the editor and the publisher express the hope that this latest volume "Small Ring Compounds in Organic Synthesis" just like the previous five will help to stress the ever-increasing importance of cyclopropane chemistry and stir new interest in its multifold applications.

Göttingen, September 1999 Armin de Meijere

Contents

Contents of Volume 178

Small Ring Compounds in Organic Synthesis V

Volume Editor: Armin de Meijere
ISBN 3-540-60495-2

Cyclopropane Derivatives and their Diverse Biological Activities

Jacques Salaün

Laboratoire des Carbocycles (Associé au CNRS), Institut de Chimie Moléculaire d'Orsay, Bât. 420, Université de Paris-Sud, F-91405 Orsay, France. *E-mail: jasalaun@icmo.u-psud.fr*

Natural and synthetic cyclopropanes bearing simple functionalities are endowed with a large spectrum of biological properties ranging from enzyme inhibitions to insecticidal, antifungal, herbicidal, antimicrobial, antibiotic, antibacterial, antitumor and antiviral activities.

The simple 2-substituted 1-aminocyclopropanecarboxylic acids (ACCs) are currently attracting special attention because of their potential use in conformationally restricted peptides, providing biosynthetic and mechanistic probes. As the immediate biosynthetic precursor of ethylene, the phytohormone that initiates and regulates many aspects of plant growth (germination, inhibition, senescence, fruit ripening, etc.), the parent ACC, structurally related to glycine, is a potent and selective ligand of the glycine modulation site coupled to the *N*-methyl-D-aspartate (NMDA) receptor, one of the four different receptors that mediate the action of the excitatory amino acids (EAA) in the brain transmitter systems; thus, such amino acids have also been proven to be useful in neurochemical studies. The mechanisms responsible for the diverse specific biological activities of compounds containing three-membered carbocyclic moieties are also being discussed.

Keywords: Cyclopropanes, Methanoamino acids, Enzyme inhibition, Drug design, Biomechanisms

Topics in Current Chemistry, Vol. 207

1 Introduction

The cyclopropane ring, due to its unusual bonding and inherent ring strain (27.5 kcal/mol) is unique among carbocycles in both its properties and reactions [1]. Therefore, cyclopropane-containing compounds are of great general interest, particularly to synthetic organic chemists and to bioorganic chemists.

Thus, cyclopropane derivatives provide building blocks of unprecedented synthetic potential [2]. Moreover natural and synthetic cyclopropanes bearing simple functionalities are endowed with a large spectrum of biological properties ranging from enzyme inhibitions to antibiotic, antiviral, antitumor and neurochemical properties [3]. Present in animals, plants and microorganisms, or generated transiently in primary and secondary metabolisms, they provide convenient biological probes for mechanistic studies and allow the design of new drugs [4]. The aim of this article is to review the diverse biological activities of compounds containing three-membered carbocyclic moieties, and to indicate when the introduction of a cyclopropane ring will improve the overall biological potency.

2 Mechanisms Responsible for the Bioactivities of Cyclopropane Derivatives

2.1 Addition to the Cyclopropane Bond

Because the reactivity of a cyclopropane closely resembles that of an olefinic double bond [2], haloperoxidases (chloroperoxidase from *Cadariomyces fumago*, bromoperoxidase from *Penicillus capitalus*) add readily to the ring of cyclopropanes **1a,b** in the presence of halide ions and hydrogen peroxide to

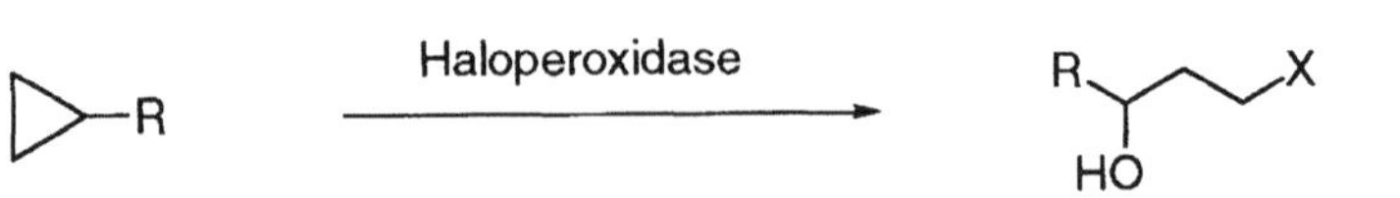

(1)

1a R = Me
b Ph

2a, b (X = Cl, Br)

provide α, γ-halohydrins **2a, b.** The enzyme-mediated cyclopropane ring-opening follows the Markovnikov rule, with the halogen going to the least substituted carbon and the hydroxyl group going to the carbon with substituents best able to stabilize a positive charge, Eq. (1) [5].

2.2 Enzymatic Oxidation

Cyclopropanes have been used to test the reactivity of oxidizing enzymes. Thus, while the monooxygenase enzyme from *Methylococcus capsulatus* oxidizes cyclopropane **3** to cyclopropanol **4**, on the other hand methylcyclopropane **1a** is oxidized to cyclopropylmethanol **5**, also without ring opening, Eq. (2) [6].

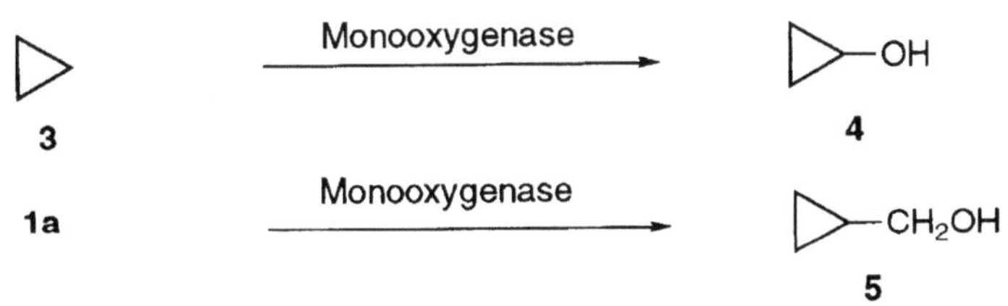

(2)

2.3 One-Electron Oxidation Processes

Pyrroloquinoline quinone (PQQ) (or methoxatin) **6** is a coenzyme, responsible for the oxidation of methanol [7]. It has been found that cyclopropanol **4** inactivates the enzyme from *M. methanica* [8], the dimeric methanol dehydrogenase and the monomeric enzyme from a *Pseudomonas* PQQ-dependent methanol dehydrogenase [9] by forming adducts such as **7**, through a one-electron oxidation process and the ready ring opening of a *cyclopropyloxonium radical*, Eq. (3) [8, 9].

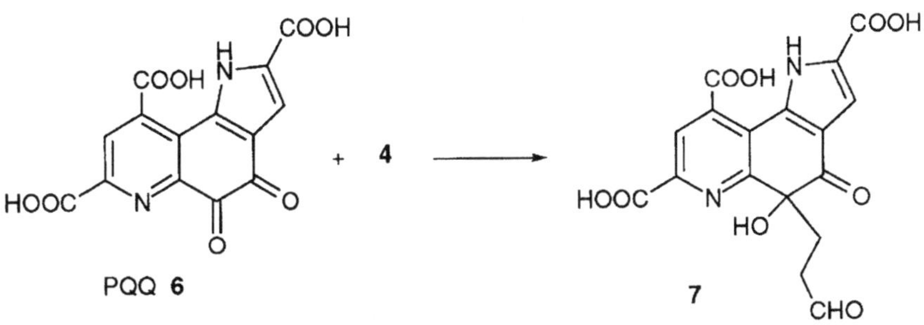

(3)

A flavoprotein oxidase, which is also a methanol oxidizing enzyme, was inhibited by cyclopropanol **4** through the formation of a *N*-5 flavin adduct with a ring opened cyclopropyloxy radical [10].

The *N*-benzyl-1-methylcyclopropylamine **7** is an irreversible inhibitor of the mitochondrial flavoenzyme monoamine oxidase (MAO). It was suggested that

MAO oxidizes amine substrates also by a one-electron route via the *cyclopropylamine radical cation* **8** which undergoes ready ring opening to the iminium radical cation **9** [11]. Then capture by a flavin radical, may cause the enzyme inactivation [12]. This mechanism was established by labeling experiments, Eq. (4) [13].

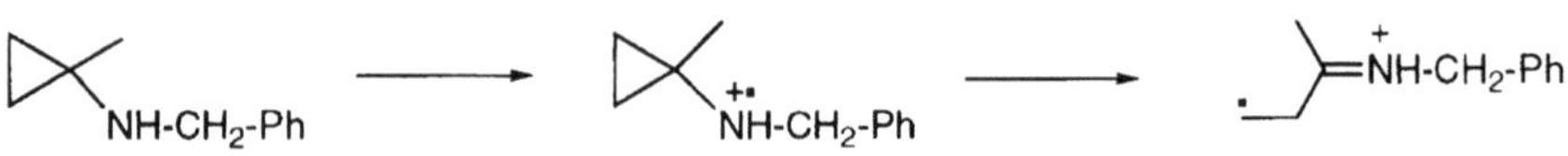

(4)

7 **8** **9**

2.4
Two-Electron Oxidation Processes

The bioactivity of the *trans*-2-phenylcyclopropylamine **10** (*tranylcypromine*) which is also a potent inhibitor of MAO and an efficient, albeit dangerous tranquillizing drug, would result of a net two-electron oxidation process leading to the *cyclopropyliminium ion* **11**, which then undergoes nucleophilic substitution by a thiol group of cysteine yielding **12**, Eq. (5) [14].

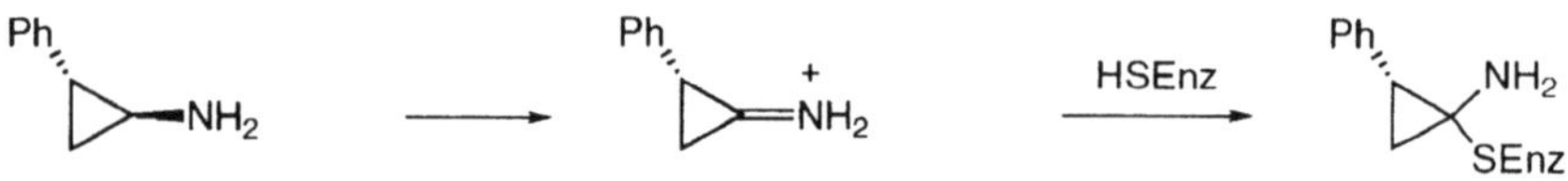

(5)

10 **11** **12**

Likewise, inactivation of MAO by *N*-cyclopropyl-*N*-(arylakyl)amines **13a, b** has been shown to involve the iminium ions **14a, b** which could accumulate to form the flavin adducts **15a, b** Eq. (6) [15].

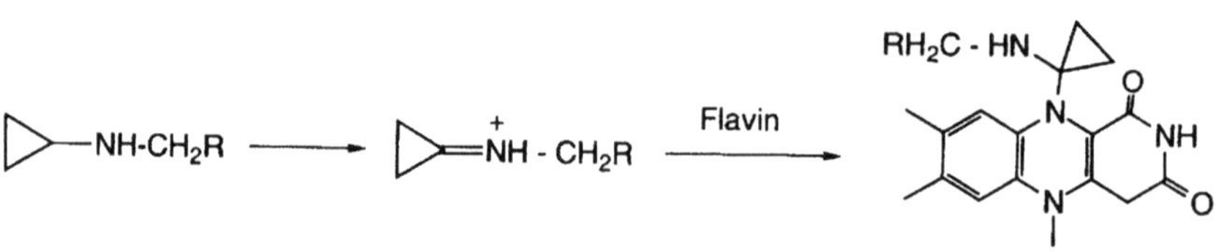

(6)

13a R = Phenyl
b Indolylmethyl

14a,b

15a,b

2.5
Nucleophilic Substitutions

The fruiting body of the inky cap mushroom *Coprinus atramentarius*, Bull. (Basidiomycetes) is apparently non-toxic when eaten alone, but induces in humans and in experimental animals an over-sensitivity to ethanol [16].

The effect is mainly due to the inhibition of NAD^+-dependent aldehyde dehydrogenase (ALDH) which causes an accumulation of acetaldehyde in the body after ethanol ingestion [17]. The compound responsible for the physiological activity of *C. atramentarius* is coprine **16**, a N^5-(1-hydroxycyclopropyl)-L-glutamic acid amide which has been isolated and synthesized, [16a, b]. Thus, when fed with a combination of mushroom and ethanol, rabbits exhibit a drop in blood pressure [18] and mice show a significant increase in their blood acetaldehyde level [19], while ethanol alone has a negligible effect. 1-Aminocyclopropanol ACP-**17**, the hydrolysis product of coprine **16** formed under acidic conditions or by glutaminase enzymes in mammals and in bacteria, is in fact the actual inhibitor in vivo and in vitro of ALDH, Eq. (7) [17].

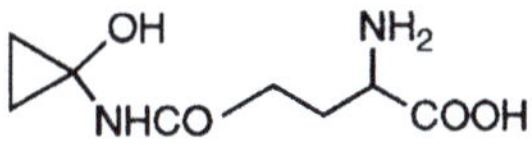

16

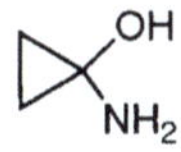

ACP **17** (7)

However, the hemiaminal **17** is unstable as a free base and readily undergoes exchange reactions. Since the hydroxy moiety of **17** is more easily displaced than the amine moiety, a highly reactive *cyclopropyliminium salt* **18** is formed, which then reacts with weak nucleophiles such as ethanol, to give e.g., **19**. Otherwise in water solution **17** can also probably eliminate ammonia to form the highly reactive cyclopropanone **20**, which is in equilibrium with its hydrate **21** and hemiacetal **22**, Eq. (8) [20]. It has been reported that hydrate **21** is also a potent inhibitor of ALDH [20,21].

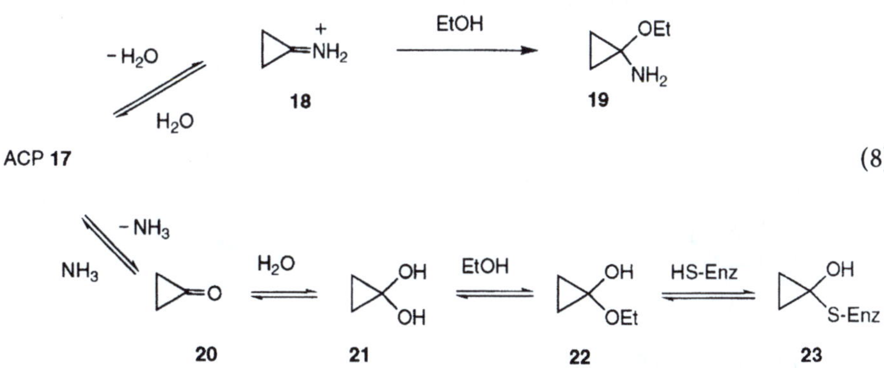

(8)

It has been suggested that the inhibition of ALDH by ACP **17** starts with an interaction between the amino group of **17** and the cysteinyl thiolate side chain of the enzyme to form a modified holoenzyme, or that the enzymic reaction may either proceed through the cyclopropanone **20** yielding **24a** or through the iminium ion **18**, yielding the modified enzyme **24b** [17]. Thus, the covalent hemithioacetal **24a** or hemithioaminal enzyme derivatives **24b** rapidly accumulate in the enzyme microenvironment and lead to the observed activity loss, Eq. (9) [21].

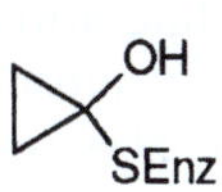

24a

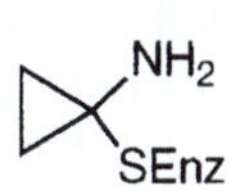

24b (9)

Other enzymes may also be similarly inactivated by such cyclopropanone adducts generated in situ by catalytic unravelling of some latent precursors. NAD^+ as coenzyme favours the electrophilic attack of ACP **17** on the enzymic thiol group, and considerably increases the rate of inhibition [17]. ALDH in brain was also inhibited in rats pretreated with coprine, and aldehyde reductase was slightly inhibited by ACP 17, in vitro [22].

2.6 Electrophilic Ring Opening

The bracken fern, *Pteridium aquilinum* is widely distributed throughout the world and is consumed as a human food in Japan and some other countries. Its toxic effects on livestock have been known since the end of the 19th century; cattle which consume bracken fern exhibit the syndrome known as "*cattle bracken poisoning*". The features include hemorrhage, anorexia, extensive intestine damage, ulceration, pyrexia and bladder carcinomas in animals [23], and enhance esophagal cancer risk in humans [24]. After intensive investigation, ptaquiloside **25** was isolated as the carcinogenic principle of bracken fern [25]. Under weakly alkaline conditions ptaquiloside **25** was converted, with D-(+)-glucose liberation, into the unstable dienone **26** which is responsible for the bracken fern carcinogenicity, Eq. (10) [25].

Ptaquiloside **25** **26** (10)

In fact, the cyclopropyl group of **26** is strongly electrophilic and reacts readily with amino acids, nucleosides and nucleotides under mild conditions [25d]. Thus dienone **26** forms covalent adducts with DNA and causes DNA strand breaks; DNA is the principal biological target of ptaquiloside **25** [26].

It has been recently shown that the selective alkylation and strand scission of deoxytetranucleotide d(GTAC)-**27** chosen as DNA model, results from the formation of covalent adducts **28** and **29** on the *N*-7 of guanine and *N*-3 of adenine with opening of the cyclopropane ring, respectively. Thermal treatment of **28** (90 °C, 5 min) afforded the d(deoxyribose-TAC) **30** with liberation of *N*-7 alkylguanine **31**, while treatment of **29** provided the d(GT-deoxyribose-C) **32** and the *N*-3 alkyladenine **33** [27]. The stabilities of adducts **28** and **29** were $t_{1/2}$ = 31 h and 3.2 h, respectively; therefore, the cleavage reaction of adduct **29** proceeds much faster than that of **28**, Eq. (11) [27].

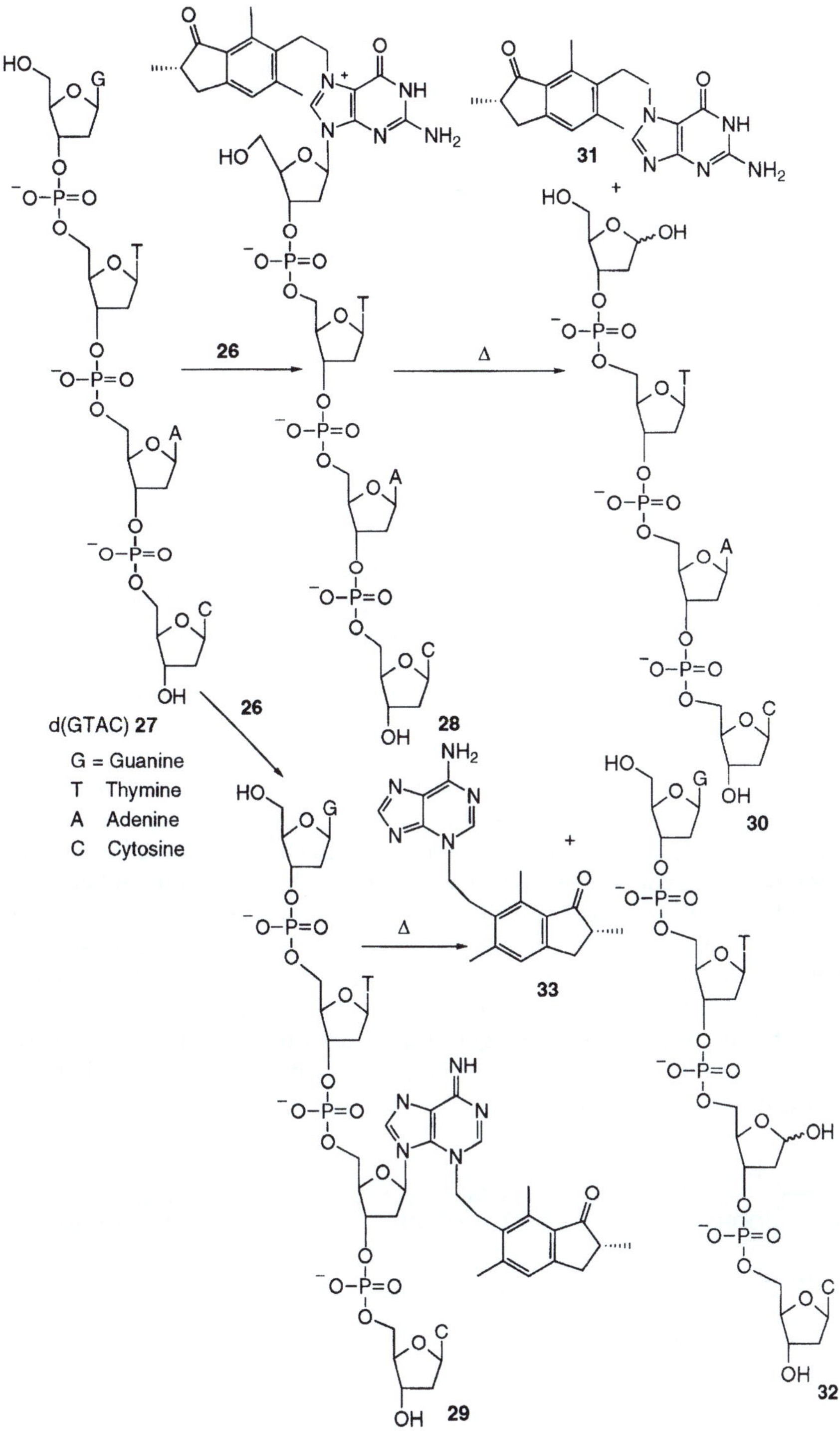
26
Δ
26
Δ
d(GTAC) 27
G = Guanine
T Thymine
A Adenine
C Cytosine
28
29
30
31
32
33

(11)

2.7
Geometric and Electronic Potency

1-Hydroxycyclopropanecarboxylic acid phosphate HCP **34** is an analogue of phosphoenolpyruvate (PEP) **35** which is metabolized by various enzymes. HCP **34** is a potent competitive inhibitor of enzymes utilizing PEP **35**, such as PEP carboxylase, enolase, pyruvate kinase, and probably other enzymes. It is a substantially better inhibitor than phospholactate **36** or phosphoglycolate **37**, presumably because of the similarity of its geometric and electronic structures with phosphoenol pyruvate, Eq. 12 [28].

(12)

HCP **34**   **35**   **36**   **37**

Microorganisms and plants, unlike mammals and other higher organisms, have the ability to biosynthesize amino acids from structurally simple precursors [29], so inhibitors of amino acid biosynthesis may be useful as selectively toxic herbicides and antimicrobial agents [29]. The enzyme dehydratase (2,3-dihydroxy acid hydrolase EC 4.2.1.9) catalyses the metal ion dependent conversion of 2,3-dihydroxy acids to the corresponding α-keto acids, which are the immediate precursors of the amino acids such as valine and isoleucine [30]. Enzymatic dehydration has been shown to proceed through enzyme bound enol intermediates **38a–d** [31]. As stable compounds that resemble the transition state for dehydration should be potent inhibitors of the enzyme [32], the 1-hydroxycyclopropanecarboxylic acids **39a–d**, whose electronic properties of the C-C bond closely resemble that of an olefinic double bond [2] have been tested as inhibitors of this enol producing enzyme. Effectively, the three-membered hydroxy acids **39b–d**, which contain alkyl substituents analogous to the substrates **38b–d** are all more potent competitive inhibitors for the dehydratase from yeast. The 2,2-dimethyl-1-hydroxycyclopropanecarboxylic acid **39b**, is a moderately effective inhibitor of dihydroxy acid dehydratase from yeast and *Escherichia Coli*, but its ring-opened analogue **40** does not inhibit enzymatic dehydration, indicating that the cyclopropane ring does contribute significantly to inhibitory potency, Eq. (13) [33].

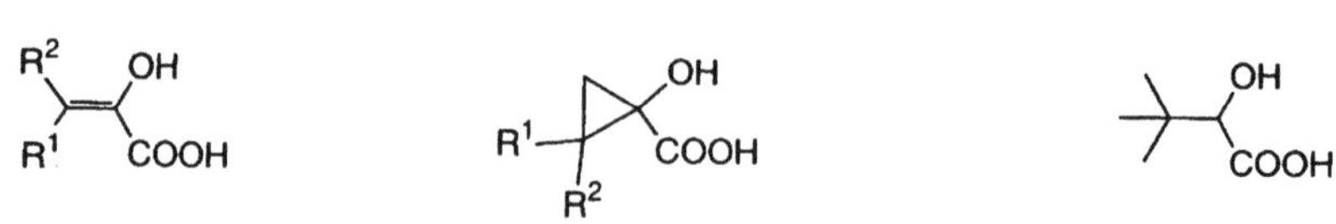

38a $R^1, R^2 = H$ **39a - d**   **40**   (13)

b Me

c $R^1 = Et, R^2 = Me$

d $R^1 = Me, R^2 = Et$

2.8 Conformational Affinity and Potency

The biological activity can be correlated with the conformation of molecules and partial constraint can be obtained by incorporation of a cyclopropane ring. L-Glutamic acid functions at many synapses in the mammalian central nervous system (CNS) as an excitatory neurotransmitter [34] and is implicated in the construction of memory and early learning [35] as well as in the pathogenesis of neuron damage to cause various neuronal diseases [36–38]. It is suggested that glutamate neurotransmission in different synapses is mediated through distinct receptors and combinations of different receptors. Therefore, development of selective and powerful agonists and antagonists appears to be essential for the investigation of molecular mechanisms of glutamate receptors and their physiological functions.

Glutamate receptors have been classified into two types: the ionotropic (iGluRs) and metabotropic (mGluRs) types. The former are further subdivided into *N*-methyl-D-aspartic acid (NMDA), *α*-amino-3-hydroxy-5-methyl-4-isoxazolepropionate (AMPA), and kainic acid (KA) receptors according to their selective action as agonists [34]. Starting from the hypothesis that each receptor subtype would require a particular conformation of glutamate for its selective activation, i.e. conformational requirements for activating receptors, the four diastereomers of L-2-(carboxycyclopropyl)glycine CCGs **41**–**44** have been synthesized [39]. They restrict the conformation of glutamate to an extended A or folded form B, Eq. (14).

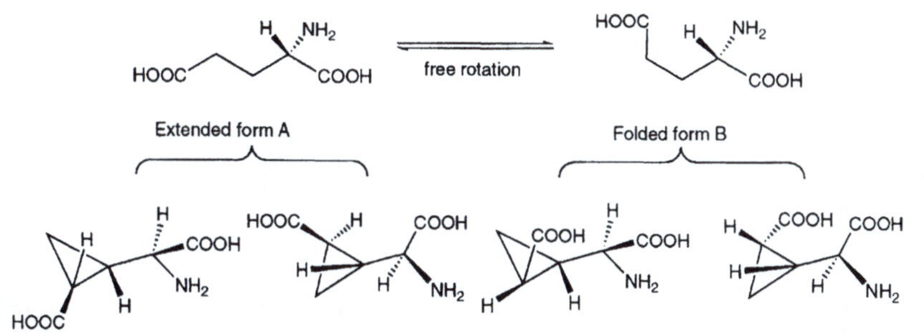

(14)

L-CCG-I (2*S*,1′*S*,2′*S*)-**41** (mGluRs agonist)

L-CCG-II (2*S*,1′*R*,2′*R*)-**42** (uptake inhibitor)

L-CCG-III (2*S*,1′*S*,2′*R*)-**43** (uptake inhibitor)

L-CCG-IV (2*S*,1′*R*,2′*S*)-**44** (NMDA agonist)

Among the four diastereomers of CCG **41**–**44**, one of the extented types CCG **41** was identified as a selective and powerful agonist of mGluRs. On the other hand, CCG **44**, one of the folded types, exhibited potent affinity to NMDA receptors. These results strongly suggested that the conformational requirement of mGluRs was an extended conformation of glutamate, while that of NMDA receptors was a folded conformation. The other isomers, CCG **42** and **43**, were not potent agonists but were inhibitors of glutamate transport systems at the excitatory synapses. Thus, CCGs are not only applied as a useful pharmacological tool in the neuroscience field but also provide proof that a specific conformation of glutamate is one of the most important factors for activation of distinct types of receptors [40].

L-2-[2-Carboxy-3-(methoxymethyl)cyclopropyl]glycines (MCGs) and other related amino acids have been newly synthetized also in a stereoselective manner [41]. Some of them were found not only to be selective and powerful agonists for the receptors but also to provide useful information with regard to the conformational requirements of glutamate receptors. These amino acids are useful as leading compounds for further conformational studies of glutamate receptors as well as for developing effective drugs for various brain diseases [41].

2.9 Conformational Flexibility

Recent efforts have focused on minimizing unwanted biological activities in existing series, while attempting to boost multiple drug resistance (MDR) reversal potency [42]. Due to the lack of understanding of the interaction of MDR reversal agents rational drug design stratagems have so far not led to more potent compounds [43]. MS-073 **45** is a prototypical agent with noteworthy MDR reversal properties containing a dibenzosuberylpiperazine group attached to the 5-position of quinoline *via* a 2-hydroxypropyloxy spacer. The relatively low oral bioavailability reported for this compound is presumably a consequence of its acid lability ($t_{1/2}$ = 15 min at pH 2.0 and 37 °C).

Extensive structural modifications of MS-073 **45** invariably resulted in a loss of potency, usually at least by a factor of ten. However, annelation of a cyclopropyl group to the dibenzosuberane improved or maintained activity ($t_{1/2}$ = 3 h at pH 2.0 and 37 °C). In addition, difluoro or dichloro substitution of the cyclopropane (R=F, Cl) in **46** conferred excellent acid stability to these compounds ($t_{1/2} \gg$ 72 h at pH 2.0 and 37 °C), Eq. (15) [44].

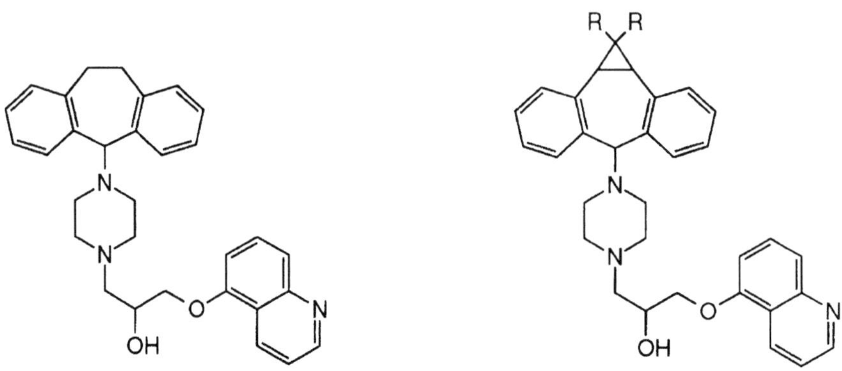

MS 073 **45**   **46** (R = F, Cl)   (15)

The dibenzosuberyl group has in fact been used as an amine protecting group which can be removed under mildly acidic conditions [45]. The *syn* and *anti* diastereomers of methanodibenzosuberylpiperazine **46** have been synthesized and tested on the proliferation and viability of multidrug resistant CH[R]Cs chinese hamster cells [44]. The superior potency of the *anti* derivatives of **46** with an axial piperazine over their *syn* counterparts with an equatorial piperizine, has been interpreted as the result of more efficient π-stacking of the rigid *anti* methanodibenzosuberyl group with overlapping phenylalanine side chains extending outwards form the α-helical backbone of glycoprotein p 170 (P-gp), an energy dependent pump. Such phenylalanine repeat motifs are present in several transmembrane regions of human and murine P-gp [46].

3 Enzyme Inhibitory Activities

In the preceding paragraph, the mechanisms by which cyclopropyl moieties can induce inactivation of specific target enzymes have been described explicitly. The inactivation is not only related to the inherent chemical reactivity of the three-membered ring resulting from its intrinsic ring strain (enzymatic cleavage, oxidation, etc.) but also from heteroatomic (oxygen, nitrogen, etc.) or electron-withdrawing substituents (aldehydes, ketones, esters, hydroxymethyl esters, etc.) which through the formation of *cyclopropyloxy radical*, *cyclopropylamine radical cation*, *cyclopyliminium cation*, or *cyclopropylcarbinyl radical* or *cation* provide suicide substrates for specific enzymes. On the other hand, cyclopropanone equivalents such as the hemiaminal **17** generated in situ by catalytic unraveling of some latent precursors [17], inactivate enzymes such as aldehyde dehydrogenase for instance, by trapping an essential cysteinyl thiolate side chain in its active site [21].

Thus the cyclopropanone hydrate **21** is an inhibitor of yeast aldehyde dehydrogenase (ALDH) through the nucleophilic substitution of a hydroxyl group by an enzymic thiol **47** leading to the cyclopropanone hemithioacetal **24a**, Eq. (16) [21].

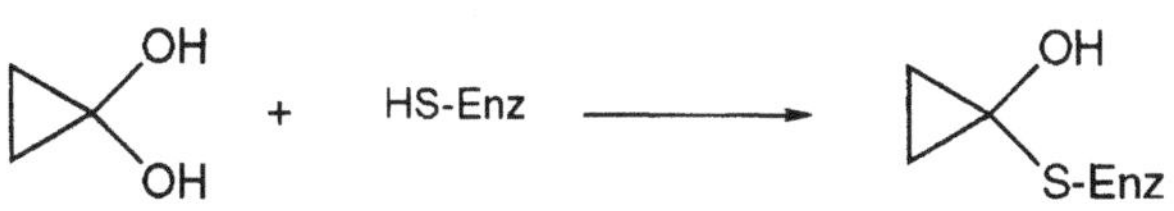

(16)

21 **47** **24a**

Horse radish peroxidase on the other hand, is a hemoprotein which is inhibited by alkylation of the porphyrin ring **48** by a β-propionic acid radical resulting from the ring cleavage of the cyclopropanone hydrate **21**, providing the carboxylate **49**, Eq. (17).

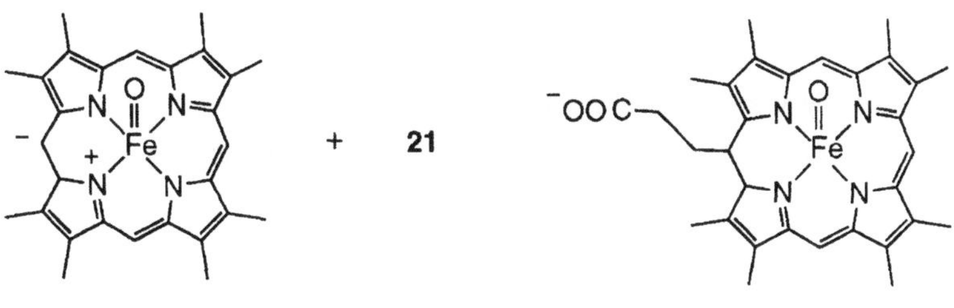

(17)

48 **49**

Cytoplasmic and mitochondrial aldehyde dehydrogenases (from beef liver) have also been inactivated by the hydrate **21** [21].

The enzyme chemistry of cyclopropylmethanols has been studied both as inhibitors and mechanistic probes [4, 47]. Thus, a series of alkylcyclopropylmethanol derivatives have been proved as being inhibitors of horse liver alcohol dehydrogenase. There are two sites in the cyclopropylmethanol inhibitors able of reacting with nucleophiles:

- the methylene carbon bearing the hydroxyl group
- the carbons of the cyclopropane ring

However, mechanistic experiments have involved the apices of the cyclopropane ring as targets for the nucleophilic group of the enzyme. Only the *pro*-R hydrogen was usually removed by the enzyme and transferred to NAD^+, and the stereochemical course of the nucleophilic ring opening of the cyclopropanes was consistent with predictions on the basis of frontier orbital theory prediction [48].

The cyclopropylmethanol ancymidol **50**, applied as soil drench or as a foliar spray, shortened the internodes of ornamental plants. Trials with cultivation of *Chrysanthenum morifolium*, *Euphorbia pulcherrina* and *Tulipa gesneriana* have been described, Eq. (18) [49].

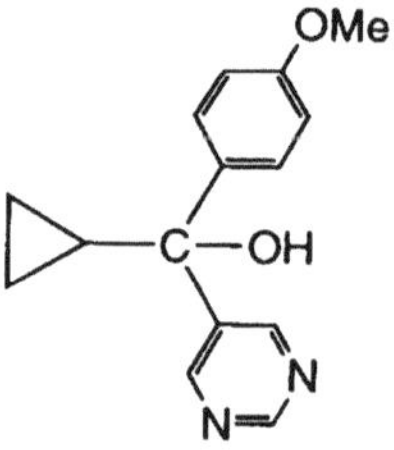

(18)

Ancymidol **50**

(2*S* 4*S*)-3-(2-Methylenecyclopropyl)alanines, so-called hypoglycine A **51** (R = H) and B **52** (R = γ-glutamyl) are the toxic principle of the *isin* (Nigeria) and *ackee* (Jamaica) fruit, *Blighia sapida*, Eq. (19) [50, 51].

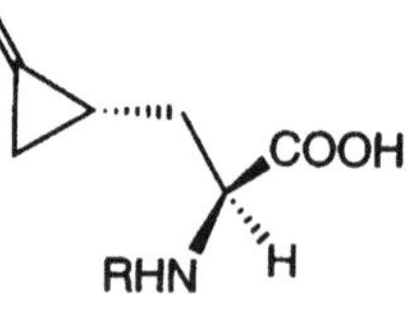

(19)

Hypoglycine A (2*S*,4*S*)-**51** R = H

Hypoglycine B (2*S*,4*S*)-**52** $CO(CH_2)_2CH(NH_2)$-COOH

Eating unripe ackee fruits or seeds causes hypoglycaemia and an organic-acidaemia and may have caused 5000 deaths in Jamaica [51–53]. Effectively, administration of hypoglycine A **51** (20–200 mg per kg body-weight) to animals causes the onset of severe hypoglycaemia after a few hours [54]. Gluconeogenesis is strongly inhibited so that animals run out of glucose when their glycogen reserves are exhausted [55].

The metabolism of **51** generates methylenecyclopropylacetyl-Co A (MCPA-CoA) **53** in the mitochondrial matrix through oxidative deamination by branched-chain oxo acid decarboxylase [56]. This impairs the oxidation of fatty acids of all chain lengths by inactivating the general acyl-CoA and butyryl-CoA dehydrogenases [56] and forming adducts with their FAD prosthetic group, Eq. (20) [57].

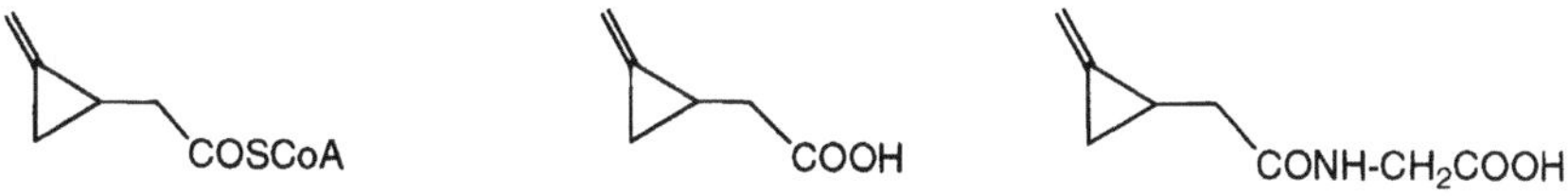

MCPA-CoA **53** MCPA **54** MCPA-glycine **55** (20)

The rate of oxidation is slowed down and allowed to proceed only as far as butyryl-CoA. There is a marked organicacidaemia in rats and people poisoned with hypoglycine with high plasma concentrations of isovalerate and 2-methylbutyrate, together with lower concentrations of butyrate [54b, 58]. Isovaleryl-CoA and 2-methylbutyryl-CoA are metabolites of leucine and isoleucine, respectively, and isovaleryl-CoA and 2-methylbutyryl-CoA dehydrogenases are also inhibited by (MCPA-CoA) **53**. This is then hydrolyzed to give the (2-methylenecyclopropyl) acetic acid **54**, or conjugated with glycine to form MCPA-glycine **55** catalyzed by glycine *N*-acylase, with CoASH release. A massive excretion of dicarboxylic acids: *cis*-dec-4-ene-1,10-dioic, *cis,cis*-dec-4,7-diene-1,10-dioic, *cis*-oct-4-ene-1,8-dioic, glutamic and adipic acids, M-CPA-glycine **55**, *N*-isovalerylglycine has been found in urine from rats treated with hypoglycine A **51** [59]. Concerning the inactivation of acyl-CoA dehydrogenase from pig kidney by a metabolite of hypoglycin A see, Ref. 57 and for the antagonism of hypoglycin toxicity by glycine, see Ref. 60.

2-Amino-4-cyclopropylidenebutanoic acid (2*S*)-**56**, is a methylenecyclopropane substituted alanine which can be considered as a non-natural isomer of hypoglycine A **51**. It has recently been synthesized racemic [61] and enantiomerically pure [62]. Biological assays have shown that at relatively high concentration the 5,6-methanoamino acid **56** inhibits the metabolism of pyruvate into glucose, but **56** is not active in inducing the mitochondrial oxidation of fatty acids, Eq. (21) [63].

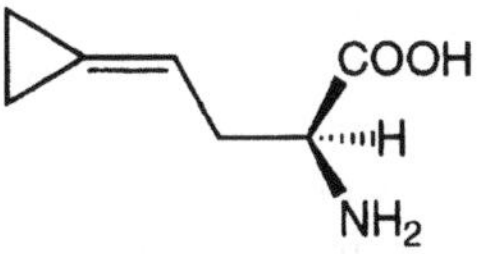

(2*S*)-**56** (21)

A number of 2,3-methanophenylalanine derivatives are efficient inhibitors of DOPA carboxylase [64]. For instance, 2-(3,4-dihydroxyphenyl) ACC **57**, due to its structural analogy with α-methyl DOPA **58**, is a reversible time-dependent inhibitor of DOPA carboxylase and of tyrosine amino transferase, Eq. (22) [65].

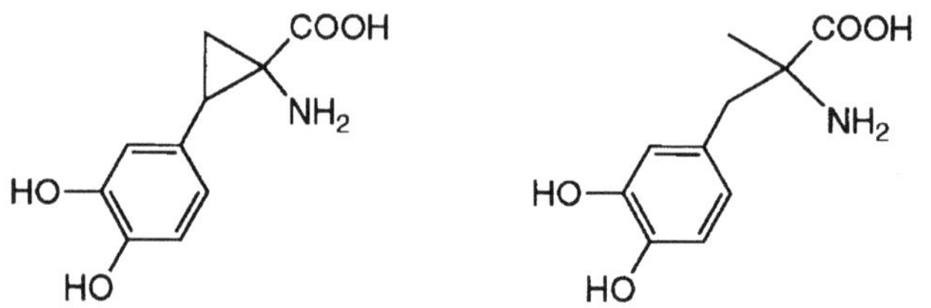

57

α-Methyl DOPA **58**

(22)

(*Z*)-2,3-Methanothyronine **59** and its dibromo derivative **60** have comparable activity with the thyroxine **61**, a thyroid hormone [66], which exhibited thyromimetic activities in basal metabolism and antigoiter tests (comparison of oxygen consumption and heart rate in normal and thyroidectomized rats) but did not have an inhibitory action on the metabolism developed by triiodothyronine [66]. (*Z*)-2,3-Methanohistidine **62**, tested on rat liver, is an effective inhibitor of histidine decarboxylase, Eq. (23) [67].

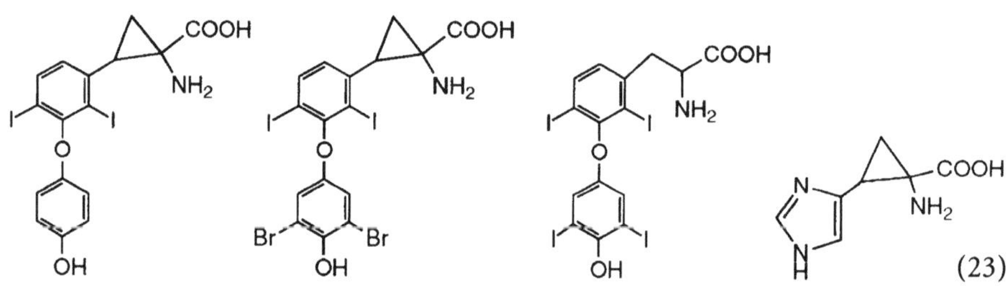

59 **60** Thyroxine **61** **62**

(23)

Prolyl endopeptidase (PEP, EC 3.4.21.26) is the only serine protease which is known to cleave a peptide substrate in the C terminal side of a proline residue [68]. This enzyme was first isolated from the human uterus, then purified from lamb kidney, and subsequently named post proline cleaving enzyme (PPCE) [69]. It is widely distributed in various mammalian tissues such as the brain, liver, and kidney [70]. In the central nervous system, PEP degrades proline-containing neuropeptides involved in the processes of learning and memory such as vasopressin, substance P (SP), and thyrotropin-releasing hormone (TRH) [71]. Moreover, cognitive deficits in Alzheimer's disease patients is reported to show improvement with TRH, and one can postulate that PEP inhibitors could prevent memory loss and increase attention span in patients suffering from senilic dementia.

Among numerous low molecular weight inhibitors of PEP described, two prolinal derivatives have been reported as PEP inhibitors: ONO 1603 **63** [72] and *Z* Pro-prolinal **64** [73], in which the formyl group is able to react with the active site of the enzyme, Eq. (24) [74].

Also a new family of inhibitors exemplified by SUAM 1221 **65** [75] has been described in which a pyrrolidinylcarbonyl function at the P1 site is the crucial entity for enzyme recognition, giving rise to the transition state analog of the enzyme-substrate interaction.

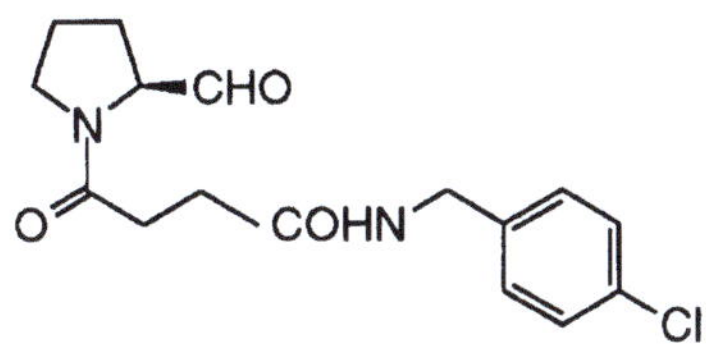

ONO 1603 **63**

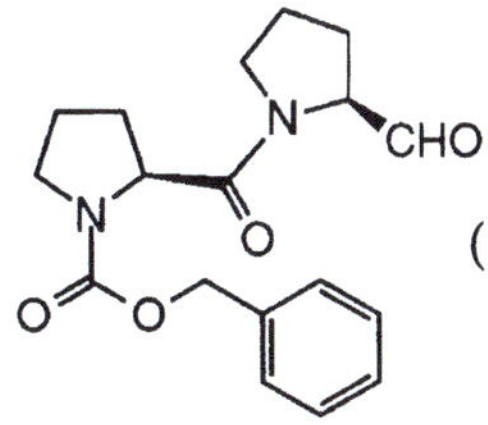

Z Pro-prolinal **64**

(24)

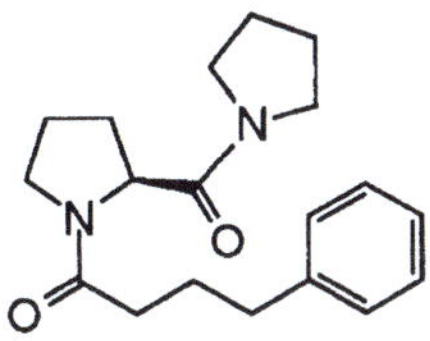

SUAM 1221 **65**

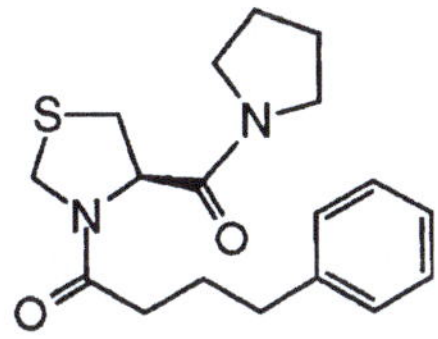

66

(25)

Important improvement in the in vitro activity was obtained when the central proline was replaced by thiaproline (thiaPRO) to give compound **66**, suggesting that modification of the central amino acid (of the proline type) could be of importance in modulating the PEP inhibitory activity, Eq. (25) [76]. In fact, replacement of the proline moiety of **63**–**66** by non-natural amino acids derived from 2-perhydroindole or from 2-azabicyclo[2.2.2]octane and modulation of the side chain by replacement of the terminal phenyl ring by a dicyclopropylcarbinyl moiety afforded derivatives such as **67** and **68** with improved activities (IC_{50} between 10 and 20 nM), Eq. (26).

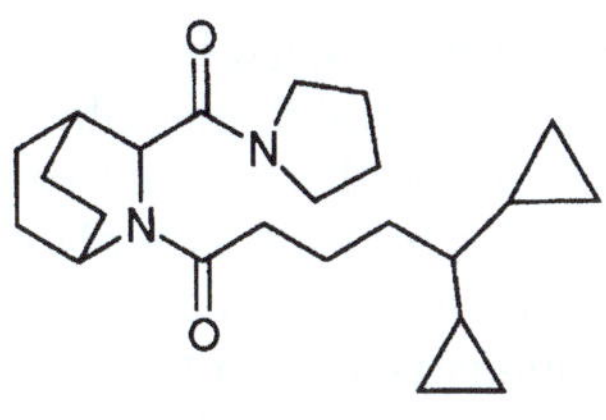

67

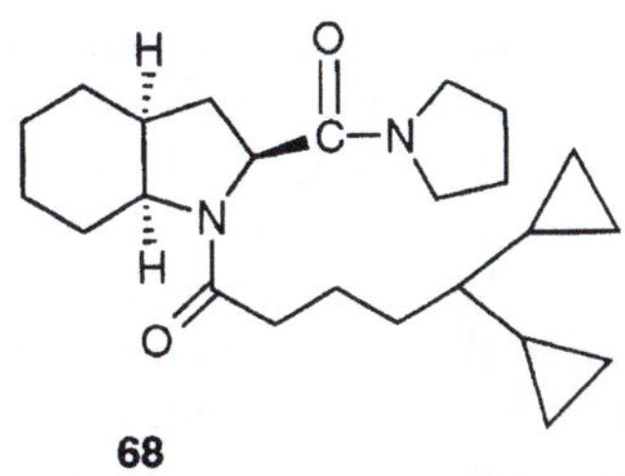

68

(26)

Furthermore, replacing the linear 4-phenylbutanoyl side chain by the (2-phenylcyclopropyl)carbonyl entity as in compound **69**, provided potent inhibitors with IC_{50} culminating at 0.9 nM on a rat cortex enzymatic preparation [77]. Moreover, the configuration on the cyclopropane ring is of prime importance in order to obtain a strong enzymatic inhibitor capacity; it has to be *R*,*R* in order to obtain not only a strong PEP inhibitor in vitro but also a good activity in vivo, as exemplified by inhibitor **70a**, which gives IC_{50} ip and po of 0.3 and 1 mg per kg, respectively. Most important, this original side chain seems to confer a potent and long lasting in vivo PEP inhibitory activity to the central amino acid it is appended to, as exemplified by compounds **69**, **70a**, **b**, Eq. (27) [77].

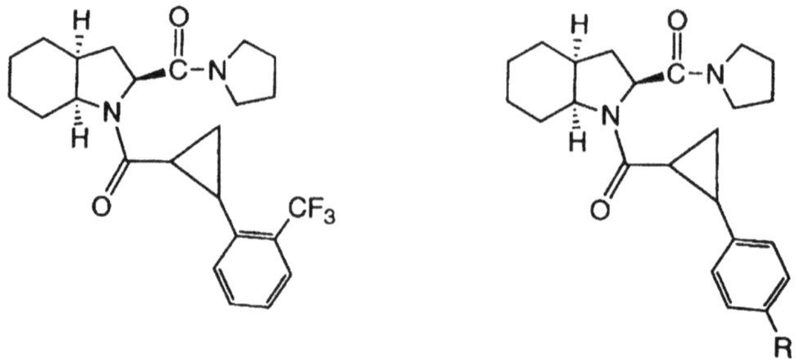

(27)

69 70a R=F
b Me

Some of these compounds are currently undergoing pharmacological studies on models of attention, learning and memory, as well as extensive preclinical evaluation [77].

4 Phytohormones, Phytotoxicity, Plant Growth Regulatory Activities

1-Amino-1-cyclopropanecarboxylic acid ACC **71** and its derivatives are currently attracting special attention because of their outstanding biological activity and potential use in conformationally restricted peptides, providing biosynthetic and mechanistic probes [64]. In fact, they constitute a unique form of constrained amino acids, naturally occurring either unbound or simply linked in dipeptides [53]. However the structure of BZR-cotoxin, the major component of BZR-toxin, which causes leaf spot disease in corn and induced high pathogenicity in rice plants, has been recently determined to be a cyclic nondepsipeptide involving the ACC moiety, Eq. (28) [78].

Present in the tissue of many plants [79], ACC **71** is biosynthesized from (*S*)-adenosylmethionine under the catalytic influence of a pyridoxal 5′-phosphate linked enzyme (ACC synthase) [80]. It is the immediate biosynthetic precursor of ethylene, the phytohormone that initiates and regulates many aspects of plant growth, including germination, inhibition, senescence, ripening of fruits, and is engaged in the metabolism of plants [81]. Rigorous demonstrations of this biosynthetic pathway have been established by ACC formation upon incubation of (*S*)-adenosylmethionine with the crude synthase obtained from ripe tomatoes [82] and by ethylene production on addition of ACC to soybean leaves [83] or to apple slices [84]. Another biological transformation of ACC is its cleavage into α-ketobutyrate and ammonia carried out by certain bacteria [85]; thus, ACC deaminase has been isolated from *Pseudomonas* growing on ACC as their sole nitrogen source [86]. Due to its physiological importance ACC **71** and its derivatives have motivated the initiation of several research programmes aimed at the synthesis of these challenging three-membered ring amino acids [87].

(1*S*,2*S*)-2-Methyl-1-aminocyclopropanecarboxylic acid (*norcoronamic acid*) **72a** was isolated from norcoronatine, a component of the phytotoxic fraction of *Pseudomonas syringae* pv *glycinea* [88]. Its *N*-methyl derivative (1*S*,2*S*)-**72b**, isolated from streptomyces *braegensis* subsp. *japonicus*, has been found to be a constituent of the cyclic peptide portion of the recently discovered DNA-intercalating antibiotics of the quinomycin family, Eq. (28) [89].

(1*S*,2*S*)-2-Ethyl-1-aminocyclopropanecarboxylic acid (*coronamic acid*) **73** was obtained from the hydrolysis of coronatine, a phytotoxin isolated from liquid cultures of plant pathogens *Pseudomonas syringae* pv *atropurpurea*, *Pseudomonas syringae* pv *glycinea*, and *Pseudomonas caronafacience* var. *atropurpurea* [90]. Infection of host plants by these bacteria induces chlorosis on the leaves due to the production of coronatine [90]; this plant toxin also induces hypertrophy of potato cells and inhibition of corn root growth [91]. Plant defense against herbivores involves the release of volatile substances which act as SOS signals, attracting predators that prey on herbivores. It has been shown that coronatine is superior to jasmonic acid in inducing the biosynthesis and emission of volatiles [92]. The biosynthesis of **73** has been demonstrated to occur from isoleucine [93]. Its (1*S*,2*R*) non-natural diastereomer, known as *allo-coronamic acid* is converted into 1-butene by plant tissues; it promises the control of enzymatic processes for plant growth and fruit ripening Eq. (28) [94].

The two other naturally occurring ACC derivatives, (1*S*,2*S*)- and (1*R*,2*S*)-2-(guanidinomethyl)ACC (*carnosadine*) **74**, were isolated from a red alga, *grateloupia carnosa*, Eq. (28) [95].

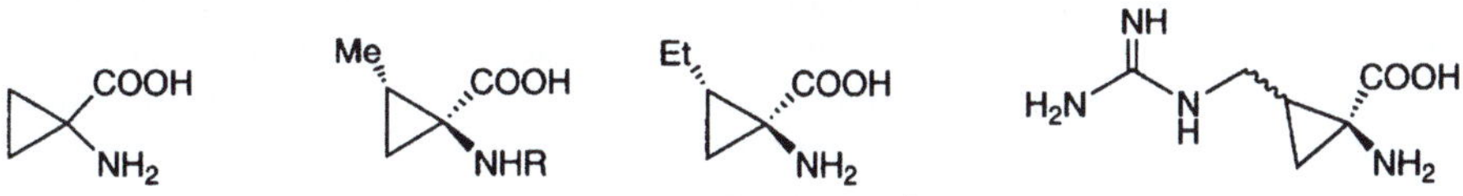

ACC **71** (1*S*,2*S*)-**72a** R = H, **b** Me (1*S*,2*S*)-**73** (1*S*,2*S*)- and (1*R*,2*S*)-**74** (28)

Due to their physiological importance, considerable efforts are currently devoted towards the total synthesis of 2,3-methanoamino acids (ACCs). The parent compound ACC **71** has been readily prepared from acrolein, through the base-induced (K_2CO_3) cyclization of 2-amino-4-chlorobutyronitrile [96] or from one-pot Strecker reaction of cyclopropanone hemiacetal [97].

The total asymmetric syntheses of natural (1*S*,2*S*)-norcoronamic **72a** and of (1*S*,2*S*)-coronamic acid **73** have been obtained from the diastereoselective cyclization of chiral non-racemic 2-(*N*-benzylideneamino)-4-chlorobutyronitriles [98]; but one of the shortest syntheses of these attractive amino acids was based on the diastereoselective palladium(0)-catalyzed alkylation and $S_{N'}$ cyclization of 1,4-dichlorobut-2-ene by the anion of 2-aminoacetonitrile derivatives [99]. On the other hand, diastereoselective palladium(0)-catalyzed azidation of chiral non-racemic 1-alkenylcyclopropyl esters provide non-natural (1*R*,2*S*)-norcoronamic acid, enantiomerically pure [100].

The asymmetric syntheses of carnosadine (1*S*,2*S*)-**74** [101] and of its protected derivatives as conformationally constrained surrogates for arginine have also been reported [102]. Different 2-substituted 1-aminocyclopropanecarboxylic acids have also been prepared by azidation of optically active 2-chloro-2-cyclopropylideneacetates [103] and from the cyclopropanation of chiral bicyclic lactams [104].

Selectively deuterated 1-aminocyclopropanecarboxylic acid ACC **71** was prepared to investigate the biosynthesis of ethylene in plants [105a] and of ammonia and 2-ketobutyrate in *Pseudomonas* [105b].

When subjected to drought stress, excised wheat (*Triticum aestivum L.*) leaves increase ethylene production as a result of an increased synthesis of ACC **71** and an increased activity of the ethylene-forming enzyme (EFE) which catalyzes the conversion of ACC **71** to ethylene. Rehydratation to relieve water stress reduces EFE activity to levels similar to those in non-stressed tissue. Pretreatment of the leaves with *N*-benzyladenine (BA) **75** or indole-3-acetic acid IAA **76** prior to drought stress caused further increase in ethylene production. Conversely, pretreatment of wheat leaves with abscisic acid ABA **77** reduced ethylene production to levels of non-stressed leaves, accompanied by a decrease in ACC **71** content, Eq. (29).

ACC **71** was also found to be converted into a non-volatile product, i.e., into 1-(malonylamino)cyclopropanecarboxylic acid **78** (R, R^1 = H) in both nonstressed and water-stressed tissues [106]. The formation of this major metabolite of ACC **71** has also been identified in etiolated buckwheat leaves fed with exogenous ACC or in etiolated soybean (*Glycine soja*) seedlings treated with IAA **76** [107]. Derivatives of **78** (R, R^1 = H, NH_4, alkali or alkali earth metals, Mn) were prepared as plant growth regulators. Thus amino diacid **78** (R, R^1 = H) was a more effective soybean growth regulator than 1-formamidocyclopropanecarboxylic acid [108]. Analogously, treating etiolated cotton seedlings with IAA **76** increased ACC synthase activity, ACC **71** content and the rate of ethylene formation [109]. A direct relationship between the content of IAA **76** in the hypocotyl tissue and the rates of ACC **71** and ethylene formation was observed, Eq. (29) [109].

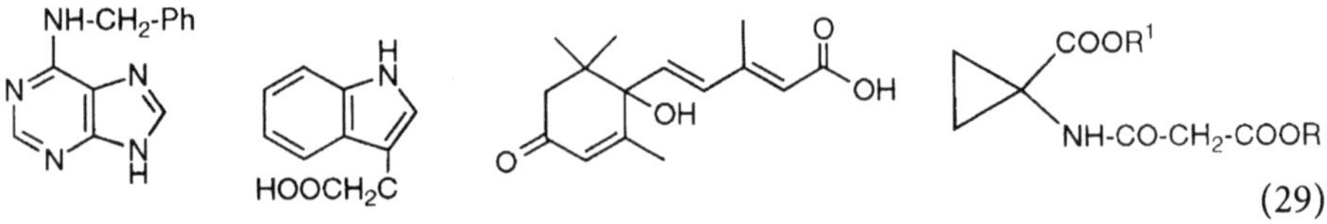

BA **75** IAA **76** ABA **77** **78** (29)

1-Aminocyclopropanecarboxylic acid-metal complexes **79** (M = transition metal with tetracoordination) also provide useful plant growth regulators, Eq. (30) [110].

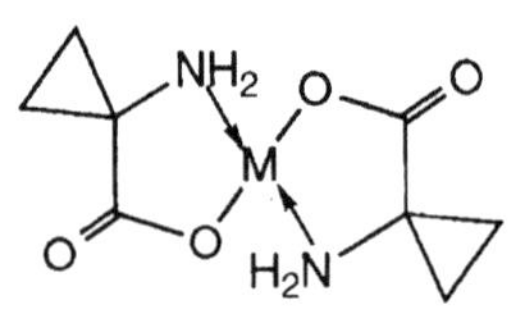

79 (30)

Non-natural ACC derivatives have been synthesized in order to test their eventual biological activity. Thus, in alkyl-ACC derivatives the alkyl and carboxyl groups must be in *trans* positions to be metabolized by plant tissue [111a]. For instance, the *trans*-methyl-ACC **80** serves as a good inhibitor of ethylene production and is a substrate for propylene production in mungbean hypocotyl segments [111 a]. 1-Amino-2-cyclopropylcyclopropanecarboxylic acid **81**, is also a good inhibitor of ethylene biosynthesis probably due to the bulky cyclopropyl group which makes it difficult for **81** to penetrate into the bioactive center; incubation of mungbean hypocotyl segments with various concentrations of **81** provided instead, 1,4-pentadiene by a mechanism involving a 1,2-hydrogen migration in a diradical intermediate, and this is responsible for the suicide inactivation [111b–d]. 1-Aminocyclopropenecarboxylic acid **82** has been shown to be an inhibitor of the ethylene-forming enzyme, an extremely poor substrate for acetylene production and an inhibitor of senescence (carnation antisenescence assay of flowers). The inactivation was postulated to result from the ready formation of the stabilized cyclopropenone imine **83** (aromatic stabilization) which can then undergo nucleophilic attack by an enzymic residue, (vide supra, Sect. 2.5) Eq. (31) [112].

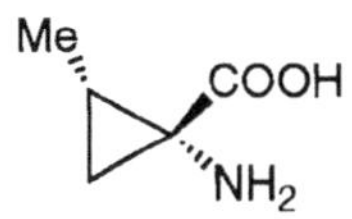

80

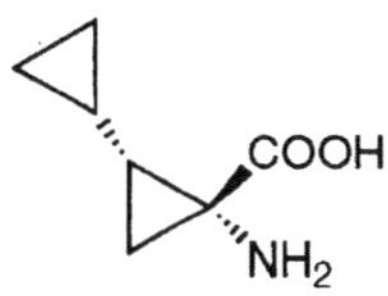

81

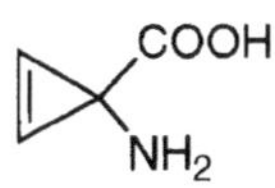

82

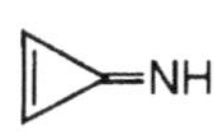

83 (31)

Racemic (*Z*)-2,3-methanoglutamic acid **84** and racemic 2,3-methanopyroglutamic acid derivatives **85a–d** have been prepared by cyclopropanation of a (*Z*)-dehydroglutamic acid derivative [95b,113]. The *β*-naphthylamide **85d** was shown to be stable to enzymatic hydrolysis by pyroglutamate aminopeptidase in vitro [113b]. Racemic 2,3-methanoproline **86** was found to be a weak inhibitor of ethylene-forming enzyme in cucumber cotyledons strips and germinating squash seeds [114]; however 2,3-methanohomoserine **87**, which has a strong ability to inhibit ethylene production from ACC in mungbean hypocotyl segments, will provide a useful tool to deliver reagents to ACC-binding proteins and to remove them from protein mixtures (affinity purification techniques) or elicit an immune response (generation of antibiotics), Eq. (32) [115].

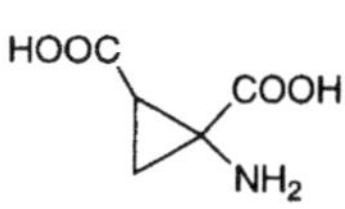

84

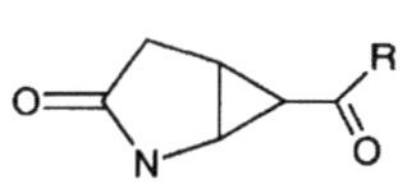

85 a R = OMe
b OH
d NH-β-Naphth
c NHMe

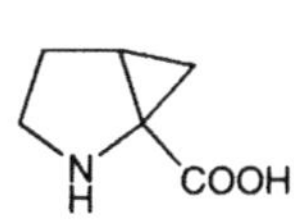

86

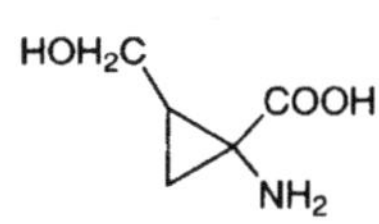

87 (32)

ACC **71** synthase, i.e. (*S*)-adenosylmethionine methylthioadenosine lyase (EC 4.4.1.14), has been purified from several plant tissues [116]. Recently, ACC synthase cDNA clones have been isolated and sequenced from wounded fruit tissues of tomato, winter squash, zucchini, ripening apple and tomato fruit. Using the polymerase chain reaction (PCR), four different ACC synthase gene fragments were obtained by amplification of cDNA derived from mRNA of tomato fruit and tomato cell culture and used to examine the expression of different ACC synthase transcripts of enhanced ethylene production. It has been proven that tomato ACC synthase is encoded by a multigene family and that the expression of each gene is differentially activated by different developmental, environmental and hormonal factors [117].

Auxin-induced ACC **71** synthase (*acc A*) was purified from slices of immature cucumber fruits and partial amino acid sequences were determined. By using oligonucleotides a cDNA for auxin-induced ACC **71** synthase from winter squash (*Cucurbita maxima* Duch. cv Ebisu) was cloned and its sequence determined. This sequence was markedly different from that for the wound-induced enzyme (*acc W*) from the same plant. The results showed that the gene for *acc A* is different from that for *acc W*; therefore ACC **71** synthase is encoded in two different genes differentially expressed by auxin and wounding. The biological significance of the presence of two different genes for the enzyme has been discussed [118].

Three divergent members (OS-ACS_1, OS-ACS_2 and OS-ACS_3) of a multigene family encoding ACC **71** synthase in rice, which grows in the deepwater regions of South-East Asia, have been cloned. It has been shown that OS-ACS_1 is induced in the shoots whereas OS-ACS_3 is induced in the roots. The protein contains all eleven invariant amino acid residues that are conserved between aminotransferases and ACC synthases cloned from various dicotyledonons plants. The amino acid sequence shares significant identity to other ACC synthases. The extraordinary degree of divergence among ACC synthase isoenzymes within each species arose early in plant evolution and before the divergence of monocotyledonons and dicotyledonons plants [119].

Polymeric forms of ACC **71** were synthesized and their biological activities evaluated. Synergistic interactions between ACC polymers and the cytokinin 6-*N*-benzylaminopurine have been indicated [120].

Germination of witchweed (*S. asiatica*), an important parasitic weed on several poaceous crops, is stimulated by several synthetic compounds. The role of ethylene biosynthesis and action in cytokinin-induced germination was investigated. Whereas conditioned striga seeds treated with ACC **71** produced little ethylene, on the other hand treatments with cytokinin-ACC combinations enhanced ethylene production. Seeds treated with cytokinin-ACC combinations have displayed higher rates of germination. Addition of ACC **71** overcame the effect of aminoethoxyvinylglycine (AVG), which is a potent ACC-synthase inhibitor. A model in which striga germination and embryo growth are limited by low capacity of the seeds to oxidize ACC **71** were consistent with the results. The cytokinin promotes ACC **71** conversion into ethylene and consequent striga germination by enhancing ACC oxidase activity and/or synthesis [121].

5 Insecticidal, Antifungal and Herbicidal Activities

The *trans*-chrysanthemic acid **88** is an essential component of naturally occurring pyrethrin esters which are present in the flower of *Chrysanthenum cinerariaefolium* and has a defense function in these plants [122]. Very effective as an antifeedant for herbivores, it presents a broad spectrum as an insect repellent.

Synthetic derivatives such as the (*S*)-bioallethrin **89** or bioresmethrin **90** are known for their high insecticidal activity with low mammalian toxicity, Eq. (33, 34) [123].

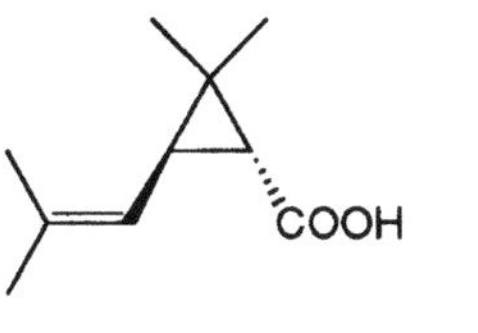

trans-Chrysanthemic acid **88**

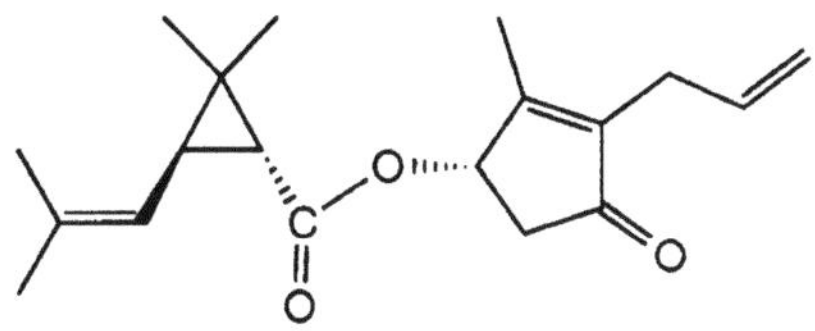

(*S*)-Bioallethrin **89**

(33)

Various structural modifications particularly involving the chrysanthemic acid moiety led to compounds with enhanced stability and insecticidal activity for a wide range of insect pests. It has been reported that plant growth was also stimulated by the photostable insecticidal biopermethrin **91**, Eq. (34) [123c].

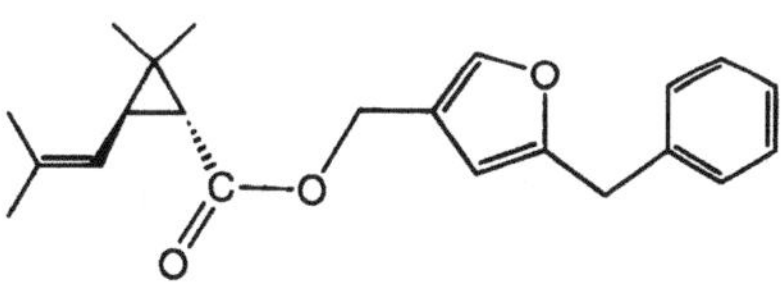

Bioresmethrin **90**

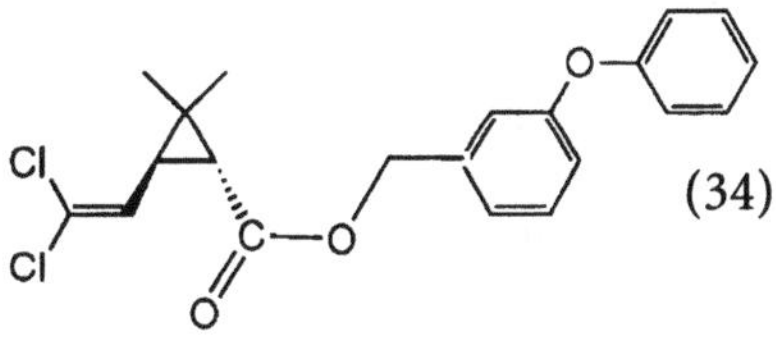

Biopermethrin **91**

(34)

The highest insecticidal activity was reached with the deltamethrin **92**, by introduction of a benzylic cyano group and of a dibromovinyl substituent with the *cis* configuration, Eq. (35) [123c].

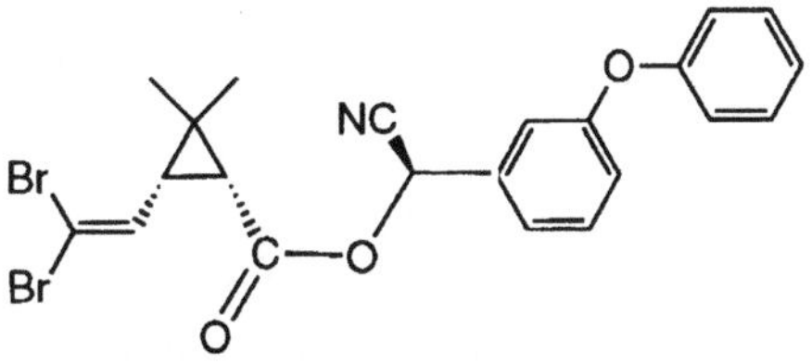

Deltamethrin **92**

(35)

The mechanism of insecticidal action has been attributed to blocking of the sodium channel in target cell membranes and consequent blocking of ion transport [124]. Following these observations a considerable effort was developed for the preparation of synthetic pyrethroids as commercial insecticides.

Triazolylcyclopropane derivatives are endowed with antimycotic properties [125]. They are also prepared as plant growth regulators and fungicides; for instance **93** (R = unsubstituted and substituted aryl, heteroaryl) markedly inhibited the growth of rice, cotton and soybeans in hot tests [126]. 1-(1,2,4-Triazolyl)-2-(2,4-dichlorophenyl)cyclopropane **94**, is a more effective fungicide against *Podosphaera leucotricha* and a better growth retardant in rice and soybeans than the (1,2,4-triazolyl)pentenone **95**, Eq. (36) [127].

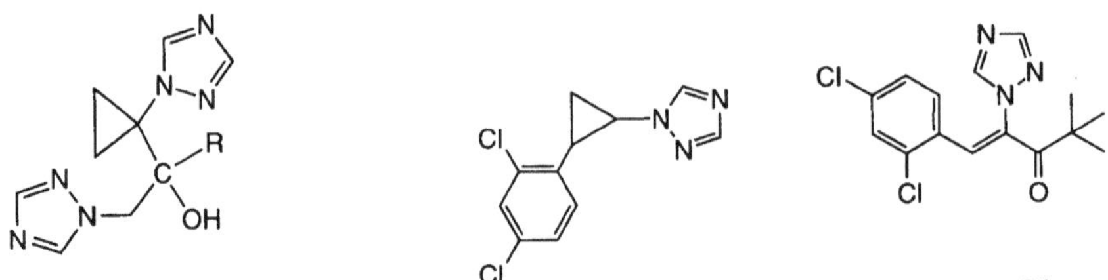

93

94

95

(36)

The *N*-(3-cyclopropyl-2,4-pyrimidyl)- **96a** and *N*-(3-cyclopropyl-2,4,6-triazinyl)-*N*-aryl sulfonyl ureas **96b** (R = halo-, amino-, alkoxyalkyl and cycloalkyl: X = CH, N) are useful as plant growth regulators and herbicides. At 500 g per ha the cyclopropyltriazine derivative **96b** (R = OMe, Ar = 2-carbomethoxyphenyl) killed *Abutilon* species and *Sivapis alba* or prevented germination without affecting wheat, Eq. (37) [128].

Unsubstituted and substituted aryl- and heteroarylcyclopropyl (methyl-1,2,4-triazolyl)-carbinols have been prepared and used as fungicides and plant growth regulators. Thus the cyclopropylcarbinol **97** ($R^1 = p$-ClC_6H_4, $R^2 = SEt$) inhibited the growth of rice, cotton and soybean, and protected apple against *Venturia* infection [129]. The 3-(1,2,4-triazolyl)-3-(1-methylcyclopropylketone)-1-trichloromethylethanol **98** was as effective against *Venturia inaequalis* on apples, Eq. (37) [130].

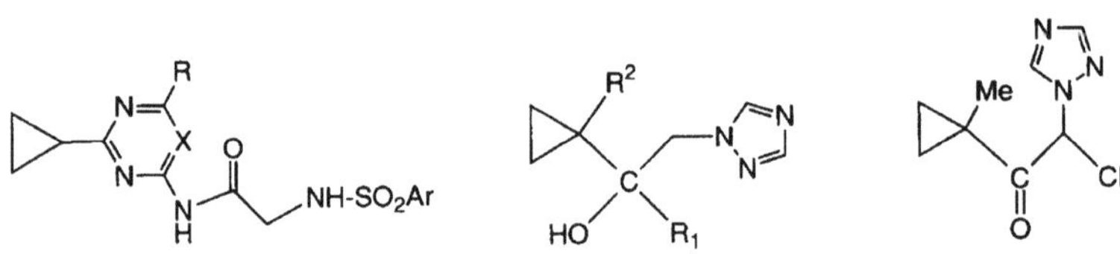

96a X = CH
b N

97

98

(37)

The acid salts and metal complexes of **99** (e.g. R = CCl_3, CCl_2F, $CClF_2$, C_{1-4} perfluoroalkyl) and of **100** (e.g. X = F, Cl, Br, CN, SCN, alkylcarbonyloxy, alkyl carbonylthio, amines) were used as agrochemical fungicides and plant growth regulators, Eq. (38) [131].

99 (n = 0–2)

100

(38)

Several derivatives of **101** were effective against *Pyrenophorateres, Vertusia indequolis* and *Pyricularia oryzae*; while derivatives of **102** (X = CH_2CH_2, CH = CH, OCH_2, SCH_2) were superior to known similar compounds as plant growth regulators in soybeans, rye, wheat and cotton, and as fungicides against *Pyricularia oryzae*, *Leptosphaeria rodorum* and *Sphaerotheca fuliginece*. In a post-emergence test, the 1-(2,4-dichlorophenyl)-2-(triazolylethyl)cyclopropane **103** (0.05 wt. %) showed 95% control of *Botrysis cinerea* on paprika, compared to 40% control by a similar triazolylcyclopropane; **103** was also an effective plant growth regulator for barley, Eq. (39) [131].

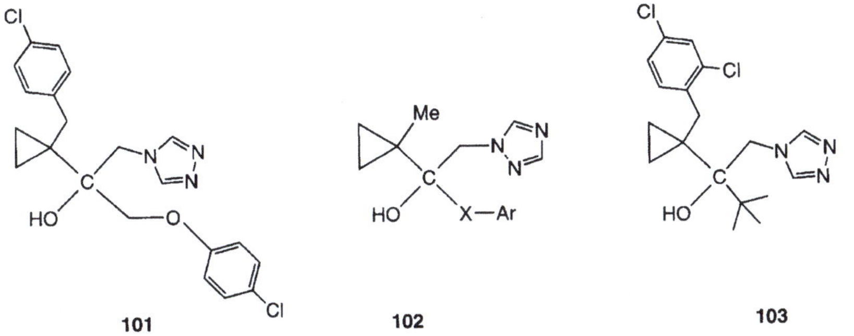

101 **102** **103** (39)

FR 900848 **104** is a natural product isolated from the fermentation broth of *Streptoverticillium fervens*. It shows potent, in vitro selective activity against filamentous fungi such as *Aspergilus niger* and *Mucor rouxianus* with MIC values of 0.05 mg/ml, but is essentially inactive against non-filamentous fungi such as *Candida albicans*, yeasts and gram-positive and gram-negative bacteria [132]. Structurally the molecule is remarkable since it is endowed with five cyclopropanes, four of which are contiguous Eq. (40).

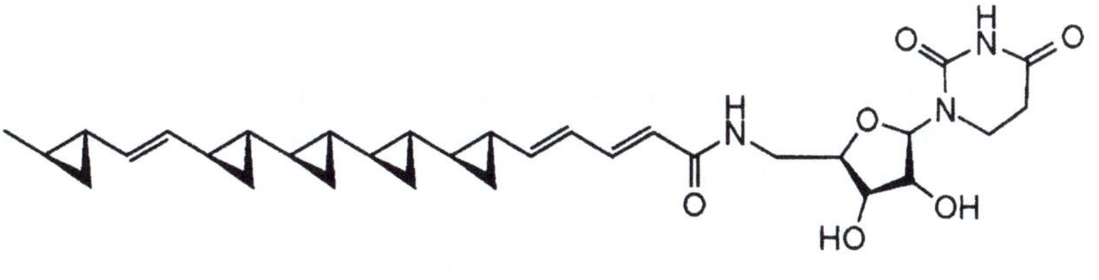

(40)

FR 900848 **104**

The full structure and absolute configuration of FR 900848 **104** has been determined to be (6*R*,8*S*,9*R*,11*S*,12*S*,14*R*,15*S*,17*R*) from X-ray crystallographic study [133]. Strategies for its enantioselective total synthesis are based on an iterative cyclopropanation [134], and on the use of chiral auxiliaries [135]. It has also been prepared by fermentation and isolated from cultures of *Streptoverticillium fervens* to be considered as an agrochemical microbicide [136].

6 Antibiotic, Antimicrobial and Antitumoral Activities

L-2-(1-Methylcyclopropyl)glycine **105** was isolated from the culture broth of *M. miyakanonensis*, and this 3,4-methanoamino acid exhibits an antimicrobial activity against *E. coli* on a synthetic medium, Eq. (41) [137].

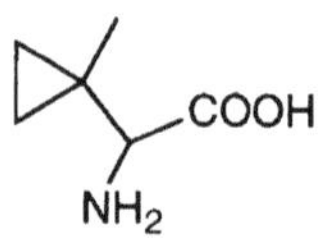
105

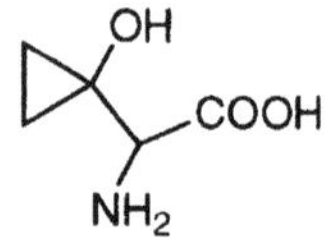
106

(41)

3,4-Methanoserine or cleonine **106** has been isolated from cleomycin **107 a**, an antibiotic from the bleomycin-phleomycin group, which is different from bleomycin **107 b** only in its threonine moiety [38]. This amino-(1-hydroxycyclopropyl)acetic acid is located in the place of the threonine, Eq. (42).

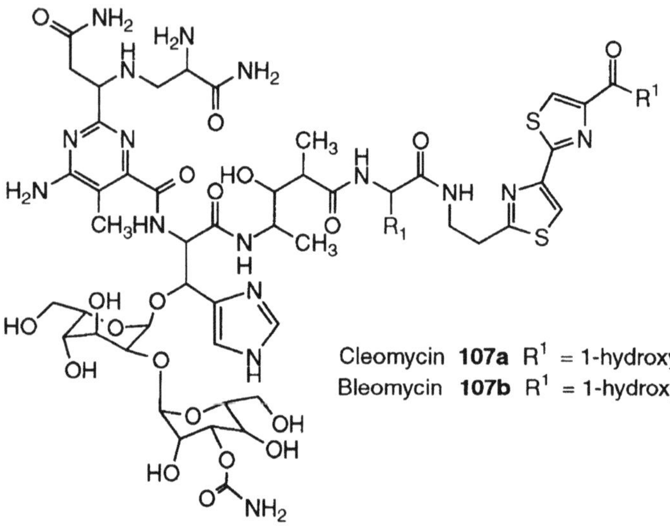

Cleomycin **107a** R^1 = 1-hydroxycyclopropyl
Bleomycin **107b** R^1 = 1-hydroxyethyl

(42)

The configuration of cleonine **106** appears to be *S* from a biosynthetic viewpoint [139]; its synthesis from the readily available cyclopropanone cyanohydrine has been reported [140, 141].

CC 1065 **108** is a highly toxic antibiotic isolated from *Streptomyces zelensis* containing a reactive spirocyclopropane ring. It cleaves DNA through a mechanism similar to the cleavage which occurs upon treatment of DNA with the carcinogenic ptaquiloside **25** (vide supra, Sect. 2.6), namely by depurination of alkyladenine adducts, Eq. (43) [142].

However, significant differences have been observed:

a) The dienone **26**, arising from alkaline hydrolysis of **25**, forms adducts at both guanine and adenine residues
b) The dienone **26** induces the spontaneous cleavage at adenine base sites under physiological conditions, whereas CC 1065 **108** causes cleavage at higher temperature (>70 °C) [143]. The covalent adenine-adduct of CC 1065 **108** has been reported to undergo at 37 °C a retrohomologous Michael reaction to regenerate the initial cyclopropylpyrroloindole structure and likely intact DNA; in contrast, the *N*-3 adenine adduct of dienone **26** depurinate spontaneously at 37 °C [144]. For these reasons ptaquiloside **25** is a typical carcinogen, whereas CC 1065 **108** is an antitumor agent [27].

Adozelesin **109**, a synthetic analogue of CC 1065 **108**, is also a potent antitumoral agent, Eq. (44) [145]. Both the natural and synthetic compounds containing a cyclopropyl pyrroloindole (CPI) unit were also shown to alkylate the *N*-3 atom of adenine in a certain sequence of DNA, a reaction which mediates the potent biological effects of these drugs [146]. They differ from traditional chemotherapeutic alkylating agents, such as the nitrogen mustards and nitrosoureas [147] by their selectivity for target nucleophilic sites in DNA and their lack of reactivity with other biological nucleophiles [146].

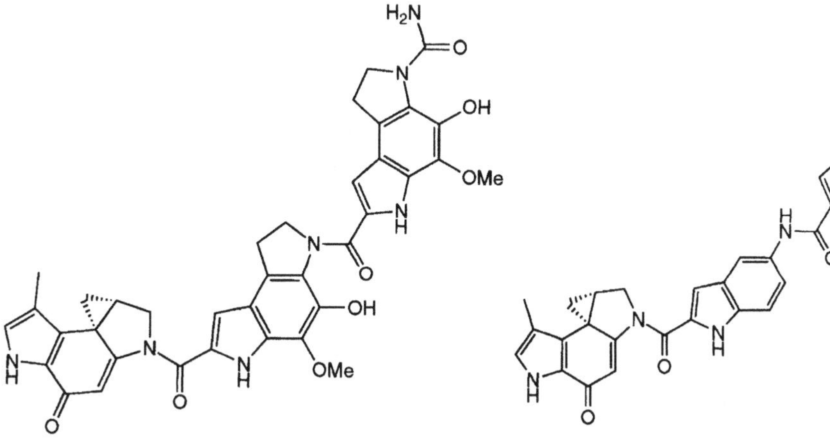

CC 1065 **108** Adozelesin **109** (43)

Comparison with the reported reactivity of the spiro[2.5]octa-1,4-dien-3-one **110** and 4-(hexafluoroisopropylidene)cyclohexa-2,4-dien-1-one **111** have suggested that the conjugated cyclopropane ring of compounds **108** and **109** imparts a strong acid dependence to its reactivity with nucleophiles, Eq. (45). This property is likely to be relevant for the exceptional reactivity of these antibiotics toward DNA [148].

110 **111** (44)

Illudins M **112** and S **113** are sesquiterpenes produced by *Omphalotus illudens*, the jack-o'-lantern mushroom [149]. These compounds demonstrated in vitro selective toxicity for a wide range of tumor cells compared to normal cells, but poor therapeutic indices were found when tested in vivo [150]. The spirocyclopropane and α,β-unsaturated ketone moities present in the illudin skeleton constitute a bis-electrophile that is responsible for the DNA damage (vide supra Sect. 2.6) [150, 151]. Illudin derivatives with greatly improved therapeutic indices have been prepared; first of all dehydroilludin M **114** has shown better activity against metastatic MV 522 lung carcinoma xenografts than nine known anticancer agents including cisplatin, cytoxan and paclitaxel and comparable efficacy to that of mitomycin C [152]. Then the analogue acylfulvene **115** exceeded the efficacy of dehydroilludin M **114** and that of mitomycin C [150]. The third generation hydroxymethylacylfulvene (HMAF) **116**, caused complete tumour regression in all animals at the maximum tolerated dose of 10 mg kg^{-1} and has been found to exhibit outstanding activity against breast, colon and skin cancer cell lines derived from human tumours [153]. HMAF **116** can be prepared readily from illudin S **113** or from acylfulvene **115**; its total synthesis has been performed from 4-hydroxy-5-methyl-2-cyclopenten-1-one and 1-acetyl-1-(diazoacetyl)cyclopropane in 14 steps and 15% overall yield [154]. Two recent approaches towards the illudin sesquiterpenes have been based on the 1,3-dipolar cycloaddition reaction of a rhodium(II) carbenoid on 2-cyclopentenones [155].

Mycelial cultures of *Mycena leaiana* produce a bright orange pigment, leaianafulvene **117** which exhibits weak antibacterial activities but pronounced cytotoxic activities; a 50% lysis of Ehrlich ascitic tumour (ECA) cells was observed at 2.5 μg ml^{-1} [156].

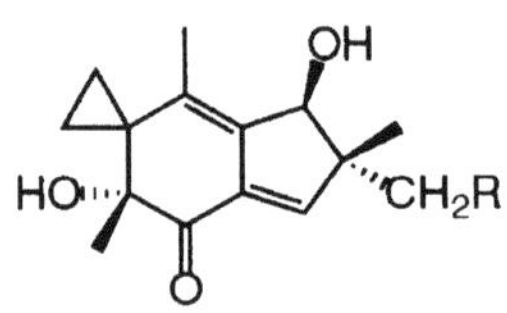

Illudin M **112** (R=H)
Illudin S **113** (R=OH)

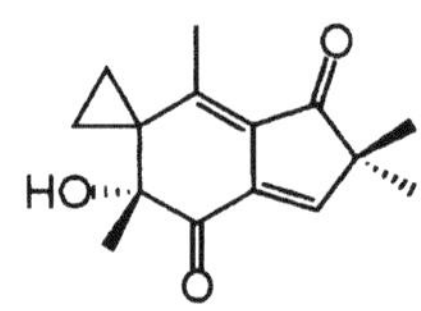

Dehydroilludin M **114**

(45)

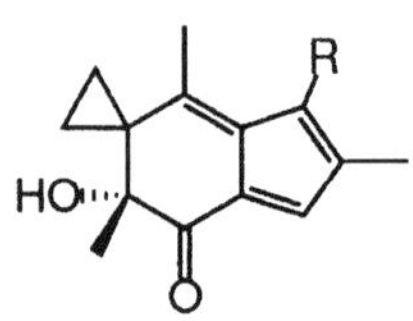

Acylfulvene **115** (R=H)
Hydroxymethylfulvene (HMAF) **116**

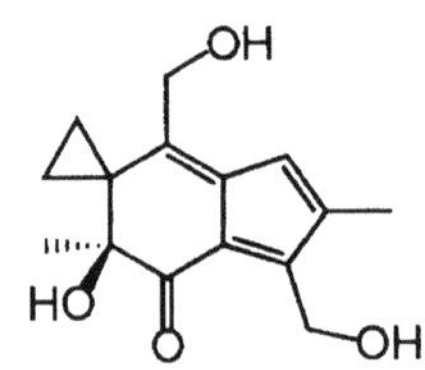

Leaianafulvene **117**

The development of compounds that interfere with the biosynthesis and/or metabolism of (*S*)-adenosylmethionine (AdoMet) (vide supra) as potential chemotherapeutic agents is an area of active investigation [157]. Biological methylation reactions and polyamine biosynthesis, both of which utilize AdoMet, are critically involved in cellular growth and function. Therefore, chemotherapeutic strategies have been based on the antibacterial, antitumor, antiviral and antiparasitic potential of AdoMet derivatives. Among them, has emerged 5′-deoxy-5′-(isobutylthio)adenosine (SIBA) **118** a nucleoside analogue, structurally related to biologically active AdoMet metabolites: (*S*)-adenosylhomocysteine (AdoHcy) (product of AdoMet-mediated methylation reactions) and 5′-deoxy-5′-(methylthio)adenosine (product of spermidine (Spd) and spermine (Spm) biosynthesis, and of ethylene biosynthesis in plants). SIBA **118**, which has potent effects on a number of AdoMet metabolic enzymes (Spd and Spm synthases, MTA phosphorylase, AdoHcy hydrolase) is also an inhibitor of cyclic AMP phosphodiesterase and of cellular nucleoside and sugar transport [158]. 5′-Deoxy-5′-(cyclopropylmethylthio)adenosine (CMPTA) **119**, is a sterically constrained analogue of SIBA **118**, the in vitro and in vivo antitumor activity of which in two murine leukemia cell lines: L1210 (MTA phosphorylase-containing) and L5178Y (MTA phosphorylase-deficient) have been found to be comparable to that of SIBA **118**. These agents are being developed as inhibitors of methylation and/or polyamine synthesis, Eq. (46) [159].

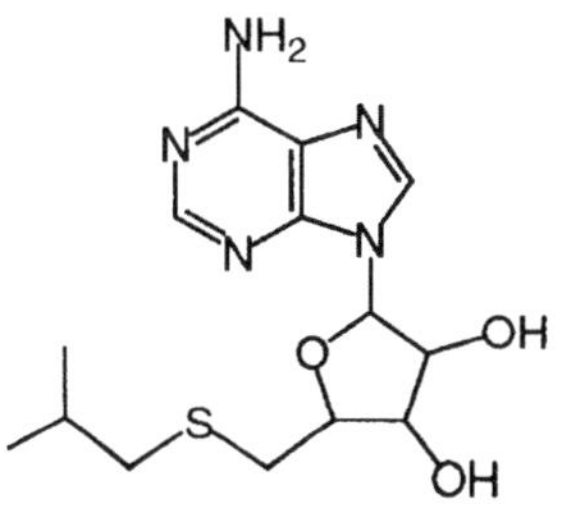

SIBA **118**

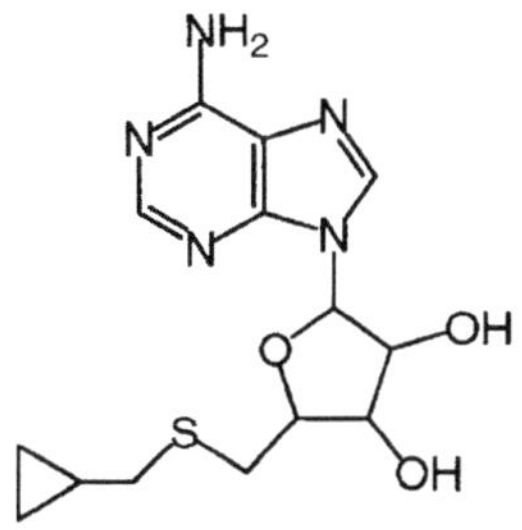

CMPTA **119**

(46)

Marine sponges provide a rich source of uncommon sterols the biogenetic origin of which still rises intriguing questions; indeed steroids with a 26-methyl-26,27-cycloergostane skeleton are very rare [160]. Thus, aragusterol A **120**, isolated from the sponge of the genus *Xestospongia* on the coral reef of Aragusuku Island (Okinawa, Japan), possesses potent antitumor activity [161]. This marine cyclopropyl steroid strongly inhibits the cell proliferation of KB, HeLaS3, P388 and LoVo cells in vitro at IC_{50} 0.042, 0.16, 0.022 and 0.0079 µg/ml, respectively. It also shows potent in vivo antitumor activity toward P388 in mice (T/C 172% at 6.25 mg/kg) and L1210 in mice (T/C 220% at 1.6 mg/kg), Eq. (47) [160].

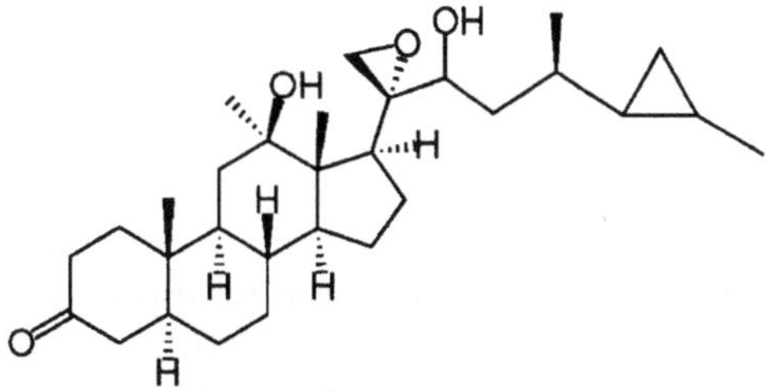

(47)

Aragusterol A **120**

The synthetic cyclopropylandrostenediones **121** (R = H, F; R^1 or R^2 = CH_2) provide useful anticancer agents, Eq. (48) [162].

Several bis(dioxopiperazines) exhibit antitumor activity. Thus, ICRF 159 **122** is an inhibitor of DNA synthesis, blocks the cell cycle in G_2-M phase and inhibits metastases in the Lewis lung tumor (3LL) animal model without impeding the

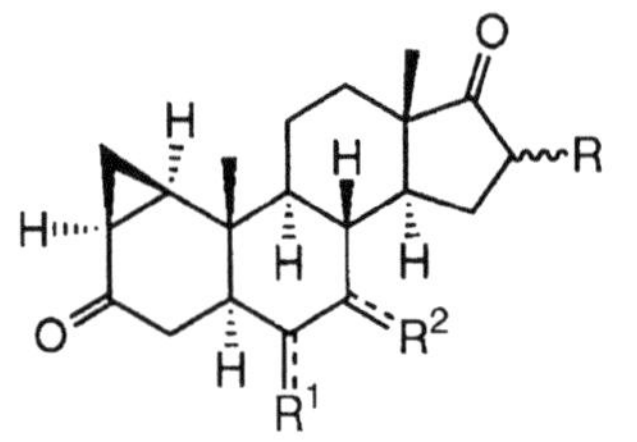

(48)

121 (R = H, F; R^1 or R^2 = CH_2)

growth of the primary implant [163]. Stereoselective effects of *cis*-**123** and *trans*-bis(dioxopiperazinyl)cyclopropane **124** on metastases of a hamster lung adenocarcinoma have been investigated in comparison with conformationally mobile ICRF 159 **122** using a Syrian hamster lung adenocarcinoma (LG 1002). Whereas ICRF 159 **122** and **123** significantly inhibited lung metastases, the *trans*-isomer **124** significantly increased the number of metastatic nodules in the lung. It has been concluded that, at least in one tumor model, antimetastatic activity can be separated from metastatic potentiating activity by controlling the drug geometry [164].

The potentiating effects of **124** may be related to an effect on cell volume and glycosaminoglycan biosynthesis, as previously proposed for the effects of **122** on B16 melanoma cells, whereas the *cis*-isomer **123** may selectively cause normalization of developing blood vessels in the primary tumor and thus inhibits metastases, Eq. (49) [165].

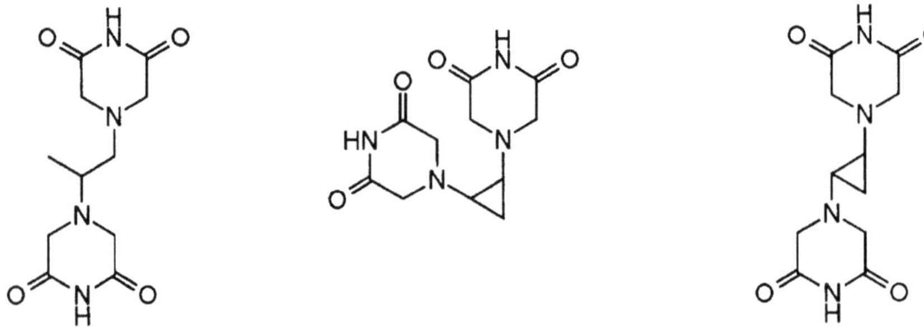

ICRF 159 **122** **123** **124**

(49)

The enzymes involved in the polyamine metabolic pathway have been the subject of intensive study, and a number of specific inhibitors for these enzymes have been designed as potential antitumor or antiparasitic agents [166]. Thus, α-difluoromethylornithine, has become a clinically useful agent [167]. Most of the studies involving inhibitors of polyamine metabolism have focused on enzymes involved in the biosynthetic pathway. Recently, there has been considerable interest generated in the enzyme spermidine/spermine-N^1-acetyltransferase enzyme (SSAT), the rate-limiting step in the back conversion of polyamines. SSAT, in conjunction with polyamine oxidase (PAO), allows for reversal of the biosynthetic pathway and attenuation of the levels of individual polyamines.

The induction level of SSAT in two human lung cancer cell lines which respond differently to treatment with inhibitors of polyamine biosynthesis appears to correlate inversely with the degree of resistance to cytotoxicity following treatment with the polyamine analogue, N^1,N^{12}-bis(ethyl)spermine (BESpm) **125** [168]. Rate of growth and cellular polyamine content in the human small cell lung carcinoma (SCLC) line NCI H82 are minimally affected by BESpm **125**, which appears to down regulate polyamine biosynthesis by the same mechanism as the natural polyamines [161]. By contrast, BESpm **125** was found to be markedly cytotoxic at a concentration of 10 mM in the large cell lung carcinoma (LCC) line NCI H157, accompanied by nearly complete depletion of all intracellular polyamines and a decrease in ornithine decarboxylase (ODC) activity to undetectable levels [170].

In light of the potential value of terminally alkylated polyamines as therapeutic agents, the unsymmetrically substituted polyamine analogues N^1-ethyl, N^{11}-propargyl-4,8-diazaundecane **126** and N^1-ethyl-N^{11}-(cyclopropylmethyl)-4,8-diazaundecane **127** have been synthesized and tested as inhibitors of human SSAT using a crude lysate from H157 cells [168]. Both compounds **126** and **127** were found to be effective inhibitors in this assay system, exhibiting similar potency to the known inhibitor BESpm **125**, and produce a differential superinduction of SSAT in situ which appeared to be associated with a cell-specific cytotoxic response in two human lung cancer cell lines. These BESpm **125** analogues exhibit promising antitumor activity against cultured human lung cancer cells, and should provide additional tools to facilitate the understanding of the regulation of SSAT gene expression, Eq. (50) [1171].

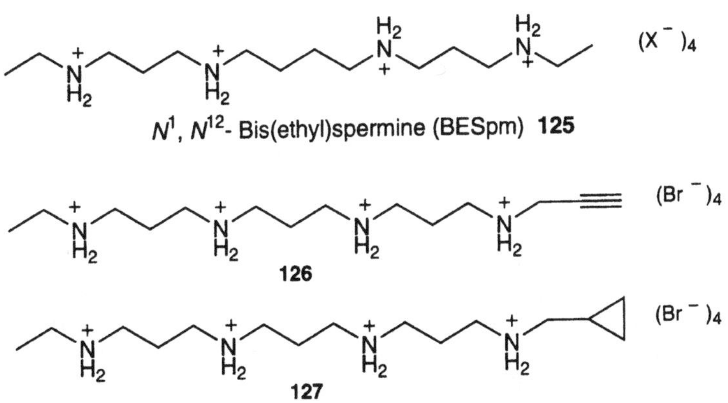

(50)

Antiestrogens block uterine growth and the growth of estrogen-dependent mammary tumors and are effective in the control of other diverse neoplastic diseases, as well as controlling and correcting various endocrine disorders. Their mode of action, although not completely understood, is, however, known to include competitive inhibition at the estrogen receptor (ER) [172], as well as having an estrogen irreversible cytotoxic action linked to their antagonism of calmodulin-activated cellular processes [173]. Tamoxifen (TAM) **128**, a clinically useful triarylethylene (TAE) antiestrogen, elicits varied estrogenic effects, including an increase in the incidence of hepatocellular carcinoma in rats at high doses

[166] and a possible increased risk of endometrial carcinoma [175]. Besides TAM **128**, other TAE antiestrogens also elicit mixed estrogen agonist-antagonist responses [176]. Incomplete remission of estrogen-dependent mammary tumors during treatment with the TAEs appears to be associated, at least in part, with the uterotrophic activity of these compounds [177].

Inhibition of estrogen is a potentially useful strategy for the treatment of hormone-dependent breast tumors in postmenopausal females and possible tumor prevention in premenopausal women. It has been reported that the introduction of a cyclopropyl or dichlorocyclopropyl moiety in place of the olefinic link in estrogenic stilbenes greatly reduces or abolishes their estrogenic activity. Thus, the 1,1-dichloro-*cis*-2,3-diphenylcyclopropane **129**, has antiestrogenic properties with no estrogen agonist activity in the mouse, and is comparable to TAM **128** against the hormone-dependent 7,12-dimethylbenz[a]anthracene-induced rat mammary tumor model [178].

A series of 1,1-dichloro-2,2,3-triarylcyclopropanes (DTACs) **130** (R = H, OH, OMe, OCH_2Ph, $OCH_2CH_2NMe_2$) have been synthesized and tested as antiestrogens, Eq. (51) [179].

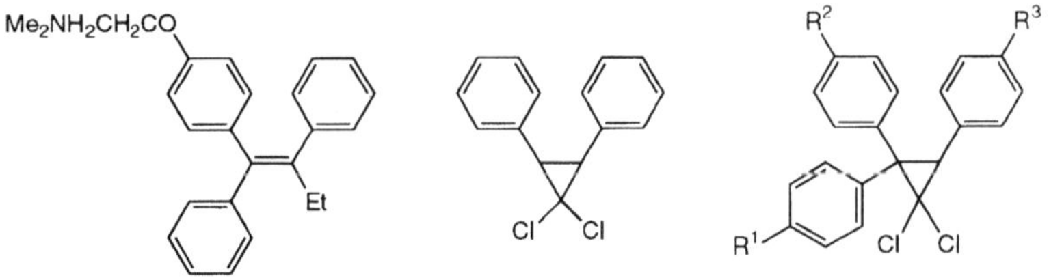

(51)

Tamoxifen (TAM) **128** **129** DTACs **130**
(R^{1-3} = H, OH, OMe, OCH_2Ph, $OCH_2CH_2NMe_2$)

All DTACs **130** were competitive inhibitors of [^{3}H]-estradiol binding in the immature rat uterine cytosol receptor assay, with relative binding affinities of 0.1 – 3.6% of estradiol. None of these compounds were estrogenic in the 3-day immature mouse uterotrophic assay at doses up to 750 µg; moreover DTACs **130** with either a methoxy-, benzyloxy- or (dimethylamino)ethoxy- side chain on the ring produced significant decreases in uterine weight. One compound, (*Z*)-1,1-dichloro-2-[4-[2-(dimethylamino)ethoxy]-phenyl]-2-(4-methoxyphenyl) 3-phenylcyclopropane **130** (R^1=$OCH_2CH_2NMe_2$, R^2=OMe, R^3=H), elicited a dose-dependent decrease in vivo comparable to MER 25 (a triphenylethanol derivative with pure antiestrogen property but precluded for use in humans due to its clinical side effects [180]). These compounds, as well as the parent compound **122**, were active in vitro against the estrogen-dependent MCF-7 human breast tumor cell line in a dose-dependent fashion.

Twenty-four cyclopropyl compounds were screened for their antiproliferative activities, eighteen were found to be active and five appeared to be promising pure antiestrogens, superior to tamoxifen **128** not only in the treatment of the estrogen-dependent tamoxifen-responsive breast cancer patients, but also in the treatment of the estrogen-independent, tamoxifen non-responsive, breast cancer patients.

The studies were conducted in the absence or presence of exogenous estradiol to reveal the estrogen-dependent nature of the cyclopropyl compounds [179, 181].

Ciprofibrate **131** is a potent, long-acting hypolipidemic agent. It is effective in type IIa, IIb, III and IV hyperlipoproteinemias and produces a beneficial elevation of the anti-atherogenic high density lipoprotein, Eq. (52) [182].

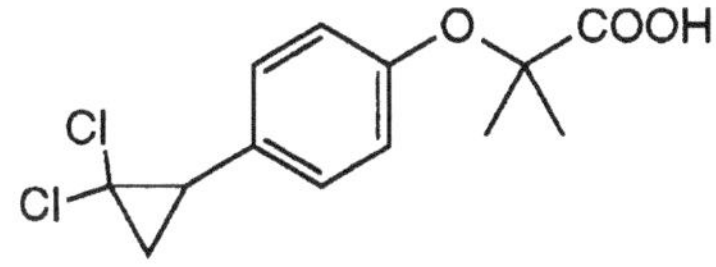

(52)

Ciprofibrate **131**

Enantiomerically pure trifluoromethyl- and trifluoroethylcyclopropylcarbinols **132** and derivatives (R=C_{1-2} fluoroalkyl, alkyl, aryl) appeared useful as intermediates for enzyme inhibitors, physiological active substances and antitumoral agents, Eq. (53), [183].

Esters of 1-(4-pentyloxybenzoyl)cyclopropane-2,2,3-tricarboxylic acid **133** have antineoplastic activity. Thus oral application of 100 mg/kg per day to mice for eight days suppressed the growth of transplanted S 37 tumor by 26%, Eq. (53) [184].

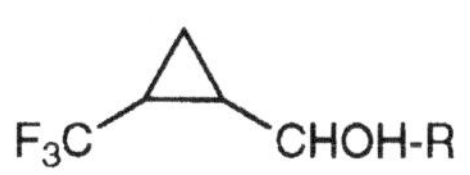

132 (R = C_{1-2} fluoroalkyl, alkyl, aryl)

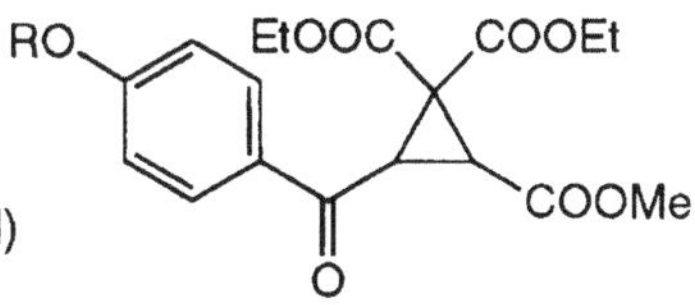

(53)

133

New types of biologically active substances having a spirocyclopropane ring and related to the antibiotic quadrone **134** [185], have been synthesized [186]. The bioassay of 2-methylenetricyclo[4.3.2.$0^{1,5}$]undecan-3-ones **135** and **136** was undertaken against tumor cells of mice in vitro; their cytotoxicity has been observed at almost the same level as that of **134**. Interestingly, **135** has exhibited antimicrobial activity against *Staphylococcus aureus*, *Candida albicans*, *Trichoyhyton foetus* (minimum inhibitory concentration: MIC, 2.5–5 μg/ml) and the activity of **136** was somewhat lower (MIC, 20 μg/ml), while quadrone **134** has no antibacterial or antifungal activity at level of 100 μg/ml, Eq. (54) [187].

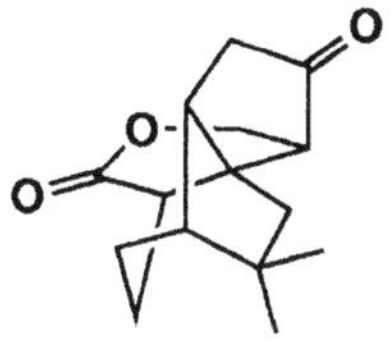

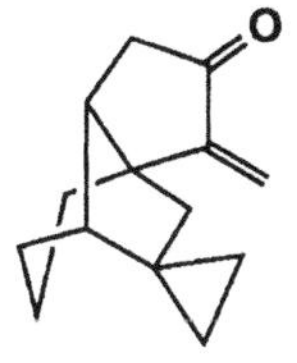

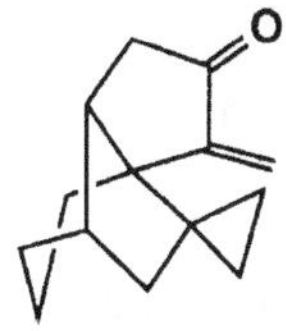

(54)

Quadrone **134** **135** **136**

The mammalian topoisomerase II enzyme catalyzes the double-strand breakage of DNA to allow the second strand passage and thereby control the topology and conformation of DNA [188]. There are many topoisomerase II inhibitors that demonstrate useful antitumor activity (e.g. *m*-AMSA, VP-16 and VM-26) [189], and it has been suggested that enhanced topo-II-mediated DNA cleavage is an important mechanism for these antitumor agents [190]. While quinolone-based inhibitors of bacterial topo-II (DNA gyrase) have long been used successfully as antibacterial agents [191] (vide infra), recent studies have identified some quinolones which also inhibit mammalian topoisomerase II and thus may have a potential as antitumor agents [192].

WIN 57294 **137** is a potent inhibitor of DNA gyrase; it is both clastogenic and mutagenic, which precluded its development as a human anti-infective agent [193]. This compound was subsequently found to possess moderate topo II inhibitory activity (EC_{50}=7.6 µM). Structure activity relationship studies of WIN 57294 **137** resulted in the discovery that the 3-CO_2H group was not a requisite for topo II potency for **138**, (EC_{50}=17 µM) [194]. A conformationally rigid quinolone derivative **139** has been reported to display better topo II potency (EC_{50}=2.77 µM) [195].

A series of novel 4-substituted-1,4-dihydroquinolines **140** were prepared and found to exhibit moderate to excellent mammalian topo II inhibitory activity. Among the compounds prepared, in general, the nitrogen analogues are the most active compounds and the sulfur analogue is the least active one. The most potent analogue **140** (X=NH-2-pyridinyl), had a topo II potency nearly equivalent to VP-16, a clinically useful topo II interactive antitumor agent, Eq. (55) [197].

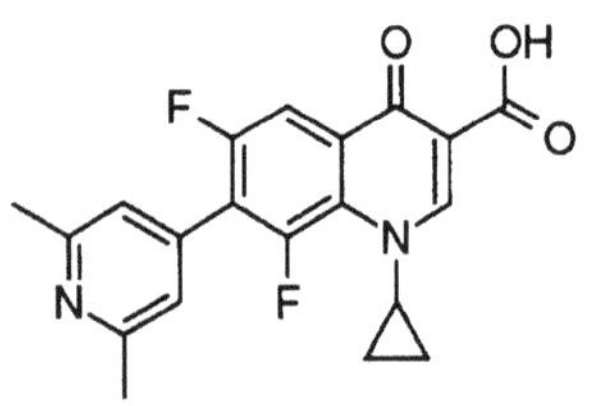

WIN 57294 **137**

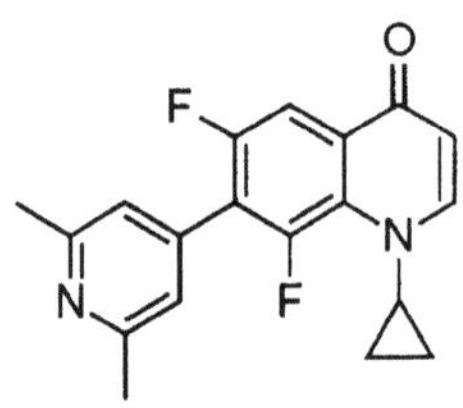

138

(55)

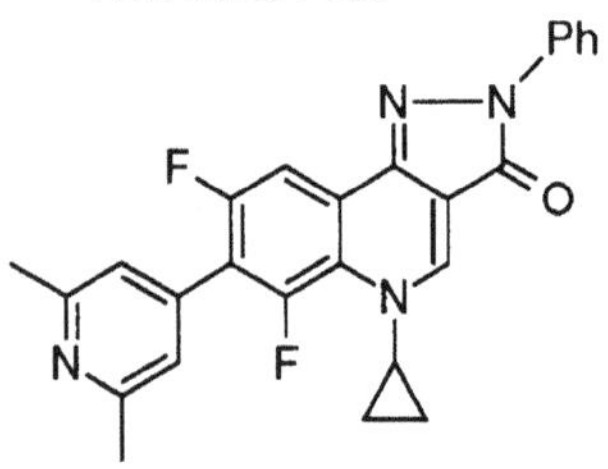

139

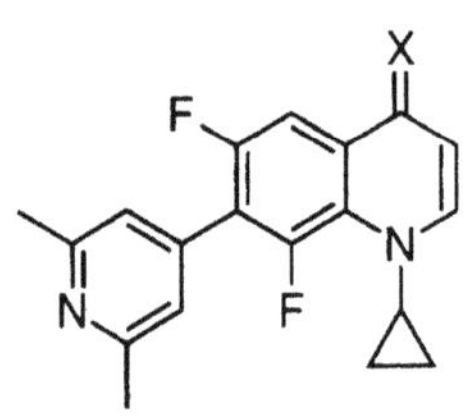

140

7 Antibacterial Activities

Cyclopropane containing fatty acids are found in bacterial membranes; thus lactobacillic acid **141** has been isolated from *Lactobacillus arabinosus, Brucella abortus* and *B. melitensis*, Eq. (56) [188].

141

(56)

Cyclopropyl fatty acids with C_{12}, C_{19} and C_{21} have been formed in many gram-negative and gram-positive bacteria [198]. The mycolic acid **142** (C_{80}) is a major component of the mycobacterial cell wall of the human strain of *M. tuberculosis*, Eq. (57) [199].

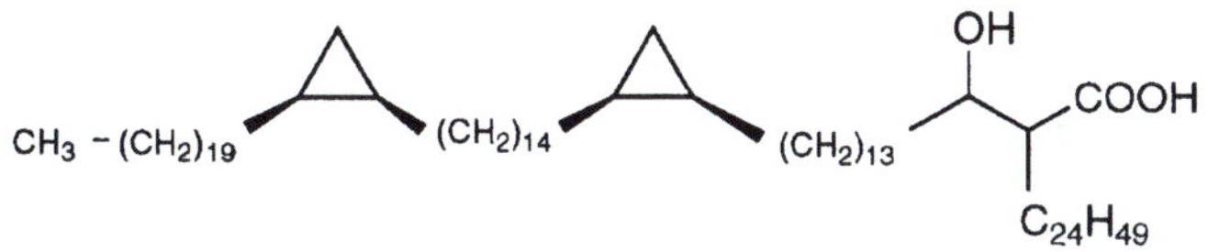

142

(57)

The re-emergence of tuberculosis infections by strains which are resistant to conventional drug therapy has demonstrated the need for alternative chemotherapy against *Mycobacterium tuberculosis.* The fluoroquinolones represent a major class of antibacterials with great therapeutic potential specially against *M. tuberculosis.* Over the years, several structure-activity and side-effect relationships have been developed, covering thousands of analogues, in an effort to improve overall antimicrobial activity while reducing undesirable side effects. The various structural features of the quinolones which govern antibacterial activity and influence the side-effect profile have been reviewed [200]. Those features which most remarkably enhance antimicrobial effectiveness are:

- a halogen (F or Cl) at the 8-position which improves oral absorption and activity against anaerobes
- an alkylated pyrrolidine or piperazine at C^7 which increases serum half-life and potency versus gram-positive bacteria
- and a cyclopropyl group at N^1 and an amino substituent at C^5, both of which improve overall potency [200].

Thus quinolone antibacterial agents, such as ciprofloxacin CPFX **143** [201], ofloxacin OFLX **144** [202], sparfloxacin SPFX **145** [203] and trovaflaxin **146** [204] are members of a major class of antibacterial drugs. These fluoroquinolones show broad-spectrum antibacterial activity and are widely used to treat patients with infections, Eq. (58).

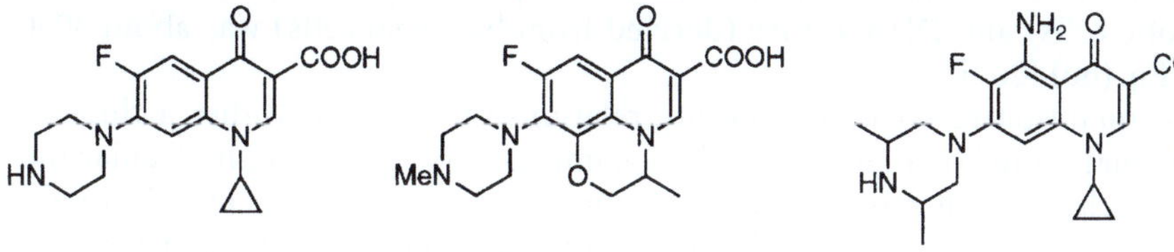

Ciprofloxacin **143** Ofloxacin **144** Sparfloxacin **145**

(58)

A series of substituted 1-cyclopropyl-6-fluoro-1,4-dihydro-5-methyl-4-oxo-3-quinoline carboxylic acids was synthesized and tested for their in vitro and in vivo antibacterial activity [205a–c]. Among them, 1-cyclopropyl-6-fluoro-1,4-dihydro-5-methyl-7-(3-methyl-1-piperazinyl)-4-oxo-3-quinoline carboxylic acid **147** (grepafloxacin) exhibited potent in vitro antibacterial activity gram-positive bacteria such as *Streptococcus pneumoniae* and high in vivo activity on the experimental systemic infections caused by the gram-positive and -negative bacteria tested. It also showed a high distribution to the lung and bronchoalveolar lavage fluid in comparison to reference drugs and is now undergoing clinical evaluation, Eq. (59) [205b].

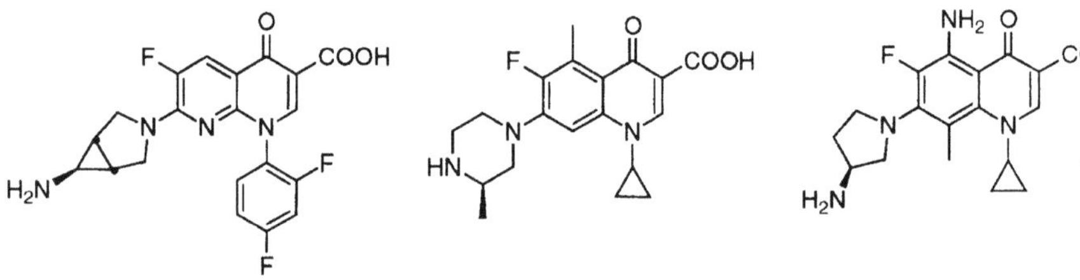

(59)

Trovafloxacin **146** Grepafloxacin **147** **148**

It has been recently found that the 5-amino-8-methyl compound **148** showed strong antibacterial activity (in vitro antibacterial activity of **148** is four times more potent than that of CPFX **149** against both gram-positive and gram-negative bacteria), reduced injury to the chromosome, and reduced quinolone-type toxicity (free from both phototoxicity at a dosage of 30 mg/kg in guinea pigs (i.v.) and convulsion-inducing activity when coadministered with fenbufen at a dosage of 100 mg/kg in mice (i.p.)).

Optimization of the 3-aminopyrrolidine moiety of **148** Eq. (59) [206] was obtained by introduction of C-alkyl (Me, Et, Pr, di-Me, cyclopropyl) and *N*-alkyl groups (Me, di-Me). C-alkylation at the 4-position of the 3-aminopyrrolidine moiety enhanced in vitro and in vivo antibacterial activity. (*S*)-5-Amino-7-(7-amino-5-azaspiro[2.4]hept-5-yl)-1-cyclopropyl-6-fluoro-1,4-dihydro-8-methyl-4-oxoquinoline-3-carboxyclic acid **149** and (3*S*,4*S*)-5-amino-7-(3-amino-4-methyl-1-pyrrolidinyl)-1-cyclopropyl-6-fluoro-1,4-dihydro-8-methyl-4-oxoquinoline-3-carboxylic acid **150** showed strong antibacterial activity (in vitro antibacterial activity including quinoline-resistant bacteria is four times more potent than that of ciprofloxacin CPFX **143**; in vivo antibacterial activity is 1.5 to 20 times more potent than that of CPFX **143** and quinolone toxicity is reduced (free from both phototoxicity at a dosage of 30 mg/kg in guinea pigs (i.v.) and convulsion when coadministered with 4-biphenylacetic acid at a dosage of 20 mg in rats (i.c.v.)). Their selectivity between DNA topoisomerase II (derived from eukaryotic cells) and DNA gyrase (derived from bacterial cells) was about 3000-fold, Eq. (60) [207].

For quantitative structure-activity relationship (QSAR) studies a three-dimensional model of a DNA-quinolone complex was built using molecular modeling techniques. It was based on the intercalation of quinolone into the double helix of DNA. It was concluded that the intercalation model is consistent with most available data on the structure of the quinolone complex. This predicted

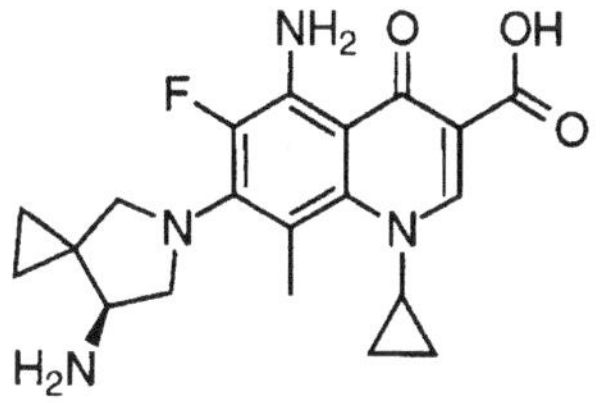

149

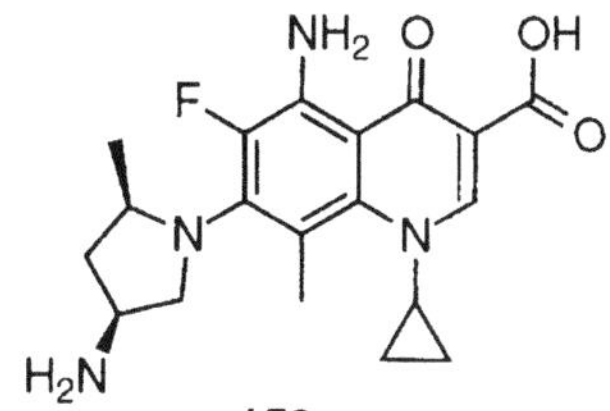

150

(60)

structure is stabilized by the binding of magnesium ion with the sp^2 oxygens present in quinolone, a phosphate and a purine base of the DNA. Substituents of **143**–**144** are predicted to make hydrophobic interactions in the major and minor groove of DNA, respectively. The piperazinyl substituent could also form hydrogen bonds with amino groups of guanines and the aspartic acid residue at position 87 in DNA gyrase subunit A [208].

The 5-methoxyimidazoquinolones **151** have been synthetized and have been shown to be superior to the corresponding ofloxacin type analogues **152** in in vitro antibacterial activity. The activity of **151** was equipotent against *S. aureus*, but 2 to 16 times less potent againtst *E. coli* and *P. aeruginosa* compared to that of the 5-fluoro analogues **153a,b**, Eq. (61) [209].

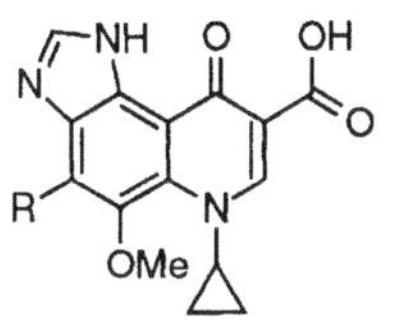

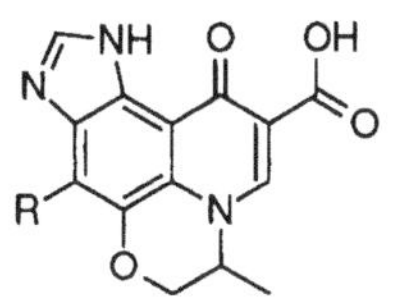

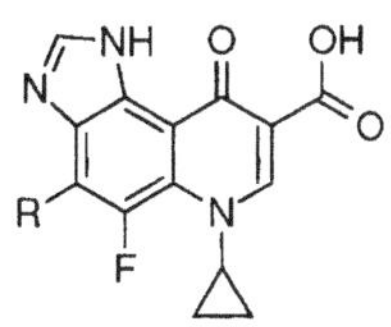

(61)

151 R = cyclic amino group

152 R = cyclic amino group

153a R = 4-N-methylpiperazinyl
b 3-methylpyrrolidinyl

Novel C^7-derivatives of 1-cyclopropyl-6-fluoro-4-quinolonecarboxylic acid have been synthesized and evaluated for in vitro antibacterial activity. Compounds **154** (3-aminocyclobutyl), **155** (1-aminocyclopropyl), **156** ((2-aminomethyl)vinyl), and **157** (1-aminomethyl)vinyl) showed significant inhibitory activity, comparable to that of cyprofloxacin **143**, against gram-negative bacteria including *P. aeruginosa*. A good pharmacokinetic profile (serum and brain concentrations and urinary recovery) was obtained for the two cyclic compounds (**154** and **155**), but that of the vinylic compounds (**156** and **157**) was less favorable. Compound **156** was less toxic than **155**, ciprofloxacin **143** or or ofloxacin **144** in terms of acute toxicity and convulsion-induction, Eq. (62) [210].

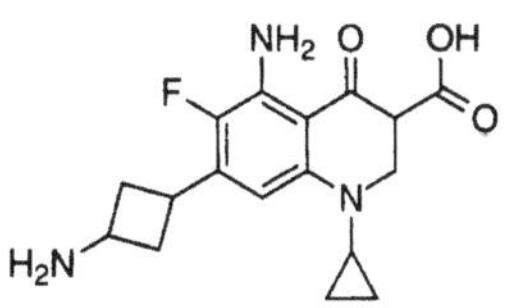

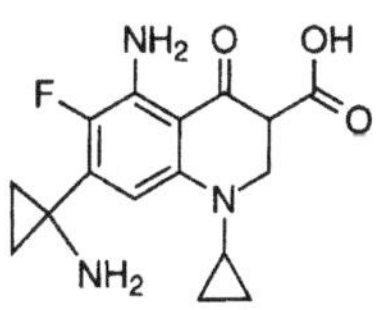

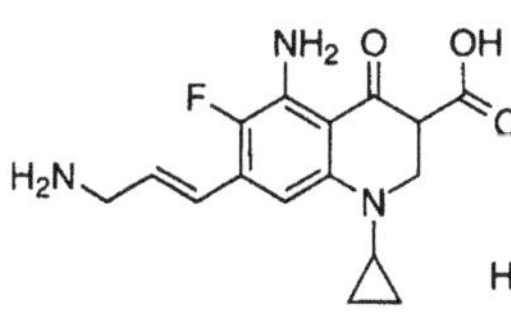

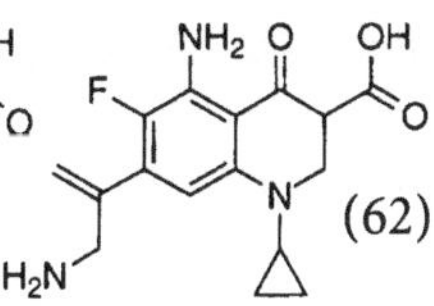

(62)

154 155 156 157

The three quinolones **158–160** exhibited potent antibacterial activities against both gram-positive and gram-negative bacteria, which are comparable to those of ciprofloxacin CPFX **143** and ofloxacin OFLX **144.** Among the three compounds, the best pharmacological and pharmacokinetic profile was obtained with **160**, an OFLX analogue, which was considerably less toxic than the three reference quinolones **155**, CPFX **143** and OFLX **144**, Eq. (63) [211].

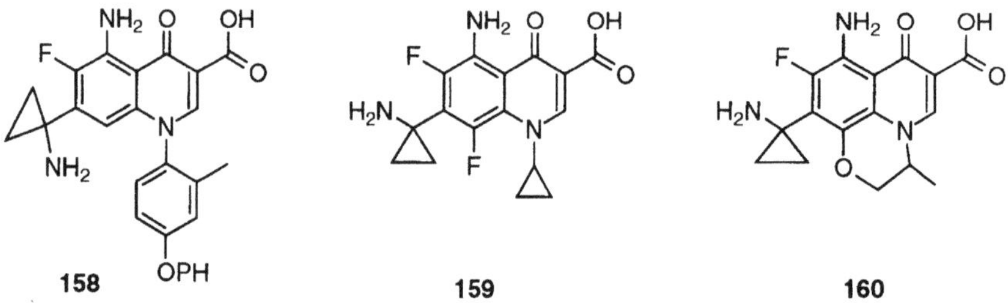

158 **159** **160** (63)

For other structure-activity relationship of the quinolone antibacterials improved by the presence of cyclopropyl substituents see Ref. [212].

Intensive efforts to expand the antibacterial spectra of existing oral β-lactams [213] have led to a new type of an orally absorbable cephalosporin, ceftibuten **161**, which shows a broad and potent antibacterial activity against most gram-negative bacteria with limited activity against gram-positive ones [214]. Antibacterial activity against gram-positive bacteria was potentiated by increasing the alkyl moiety in the 7β-side chain with decreasing the activity against gram-negative ones. Compounds **162a–c** carrying alkyl groups such as isopropyl, cyclopropyl and cyclopentyl connected directly to the vinyl carbon were slightly less active against gram-positive bacteria and more active against gram-negative bacteria than **162d–e** substituted *via* a methylene group such as propyl, cyclopropylmethyl and cyclopentylmethyl, Eq. (64) [215].

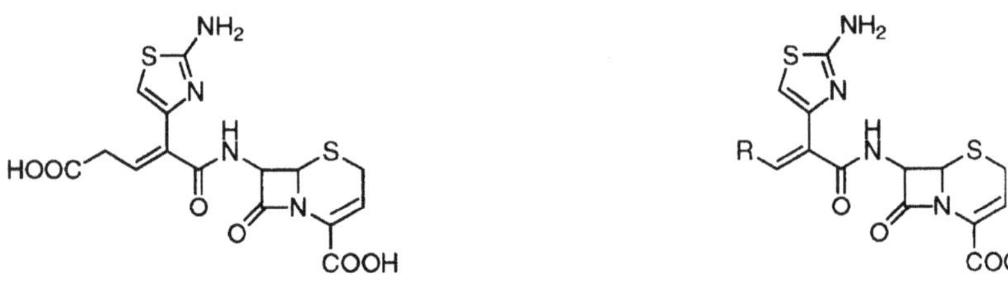

(64)

Ceftibuten **161**

162a R = iPr **d** cyclopropylmethyl
b cyclopropyl **e** cyclopentylmethyl
c cyclopentyl

8
Antiviral Activities

More than 60 per cent of all diseases in Europe, North America and Japan are caused by the action of viruses, amongst them bronchitis, hepatitis, influenza, infections by several strains of herpes as well as by human immunodeficiency viruses (HIV) [216]. Modified nucleosides that inhibit the replication of the viruses have been used in the chemotherapy of these infections [217]. These analogues, which act through similar mechanisms, can be divided into three categories: 1) phosphate modified, 2) base modified, and 3) sugar modified. Most of the known active compounds belong to the two latter groups [218]. Of main interest are compounds the ribose unit of which has been subject to major changes, either by replacement by a cyclopentane or cyclopentene (carbasugars), oxetane or cyclobutane ring, or by an acyclic chain. Thus Penciclovir **163** had emerged as a potent and selective anti herpes-virus agent, particularly active against herpes simplex types 1 and 2 (HSV-1 and HSV-2) and varicella zoster virus (VZV) [219]. In order to clarify the relationship of side chain conformation and flexibility to biological activity, cyclopropane-containing guanine, purine and pyrimidine nucleosides, such as the 9-[2,2-bis(hydroxymethyl)cyclopropyl]guanine **164** for instance, have been synthesized in order to test their potential anti-retroviral activity. The approach overlaps the area of carbocyclic nucleoside analogues, such as the isomeric compound SQ 32,829 **165** [220], which have recently received considerable attention due to their potent antiviral activity, Eq. (65) [221].

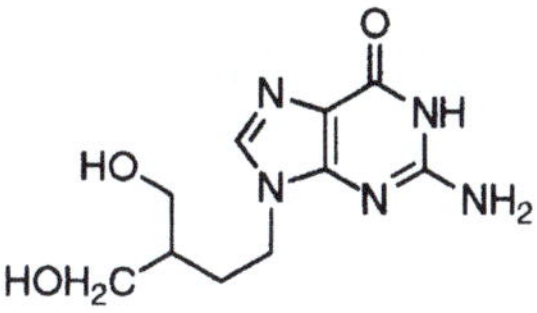

Penciclovir **163**

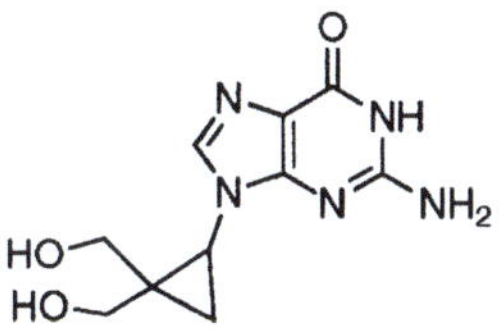

164

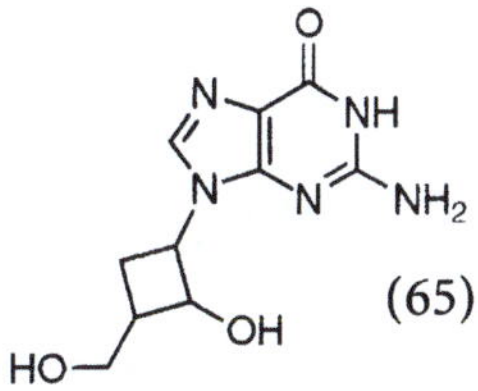

SQ 32,829 **165**

(65)

However, the nucleoside analogue **164** was found to be devoid of activity against HSV-1, HSV-2, VZV and the cytomegalovirus (CMV) in human fibroblast (MRC-5) cells. In this case the decreased conformational flexibility resulting from the introduction of the cyclopropyl group into **164** appeared to be unfavourable for interaction with the enzymes involved (vide supra, Sect. 2.9) [222]. Likewise, the cyclopropylpyrimidine **166c–f** and **167**, the cyclopropylpurine nucleosides **168** showed no antiviral activity against HSV-1, HSV-2, HCMV and HIV-1 in cell culture, Eq. (66) [223].

A new class of purine nucleoside analogs such as **169** comprising a methylenecyclopropane moiety has recently been reported to exhibit potent in vitro activity against a number of herpes viruses including human and murine cytomegalovirus (HCMV and MC-MV), Epstein-Barr virus (EBV), varicella-zoster virus (VZV), human herpes virus 6 (HHV 6), herpes simplex virus type 1 and 2 (HSV-1 and HSV-2). They also inhibit the replication of hepatitis B virus and human immunodeficiency virus type 1 (HIV-1) [214]. The (*Z*)-isomer revealed higher biological activity (EC_{50}=26 μM for **169** and comparatively >100 μM for its trans isomer); and (–)-(*Z*)-**169** was highly effective against HIV-1 with EC_{50} = 13 μM [215]. Efficient syntheses of these purine nucleoside analogs have been recently reported from the coupling reaction of vicinal dibromocyclopropane derivatives with adenine [225, 226].

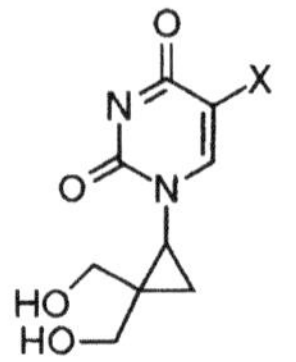

166a X = H **d** I
b Cl **e** Me
c Br **f** (E)-CH =CHBr

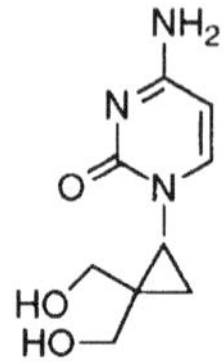

167

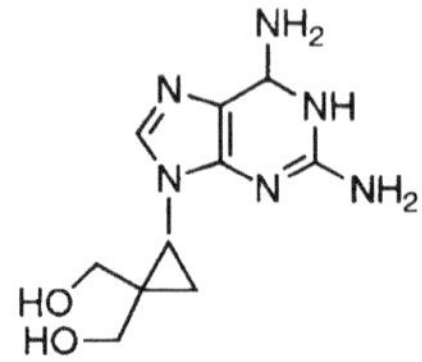

168

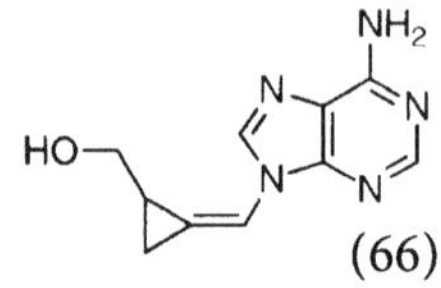

169

(66)

Although the methano homologues of oxetanocin, i.e. the (*Z*) and (*E*) nucleosides **170** (R = adenine, thymine, 5-fluorouracil) were devoid of anti-HSV-1 activity [227], the guanine derivative **170** (R = guanine) showed activity against herpes simplex HSV-1 strain KOS in cell cultures (ID_{50} of 0.023–120 µg/ml) [228]. In vitro, the 9-[2-(1,2-dihydroxyethyl) cyclopropyl]guanine **171** produced an inhibition of virus-induced cytopathogenic effects in E-377 cells challenged with herpes simplex virus (virus rating of 1.8 at 101 mg/ml), Eq. (67) [229].

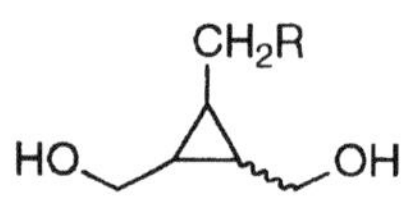

(*E*)- and (*Z*)-**170**
(R = adenine, thymine, 5-fluorouracil, guanine)

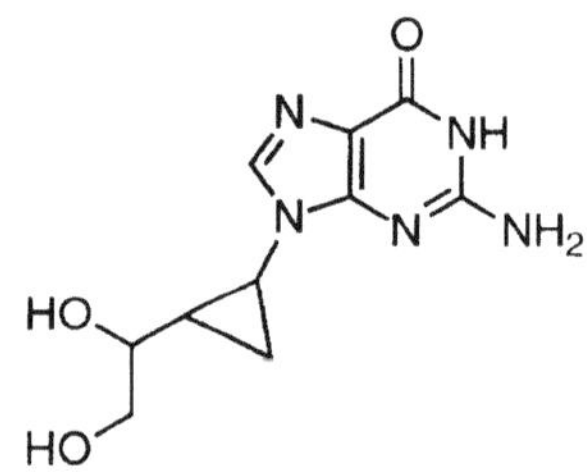

171

(67)

The cyclopropylcarbinol nucleosides **172**–**177** have also been prepared and tested as antiviral, antitumoral, antibacterial agents and neoplasm inhibitors, Eq. 68) [230–232].

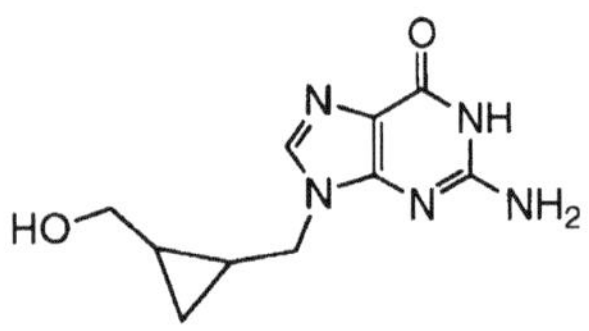

172

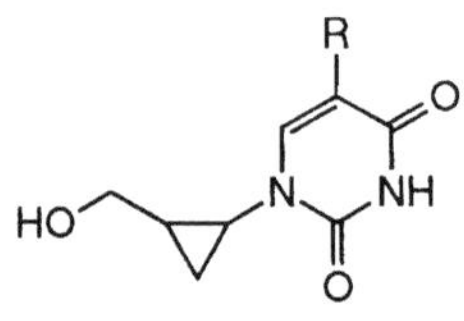

173 (R = Alkyl)

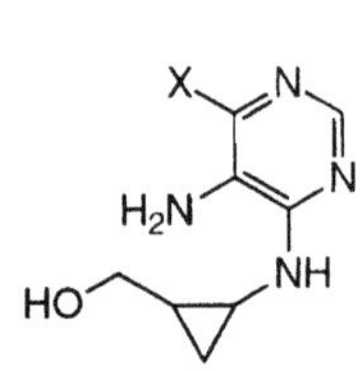

174

(68)

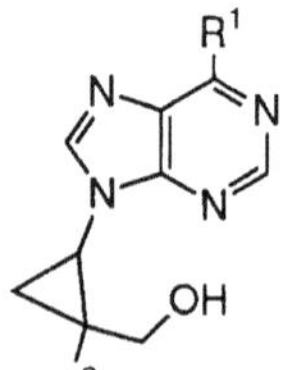

175a R^1 = halo, amino
b R^1 = Cl, R^2 = CH_2OH

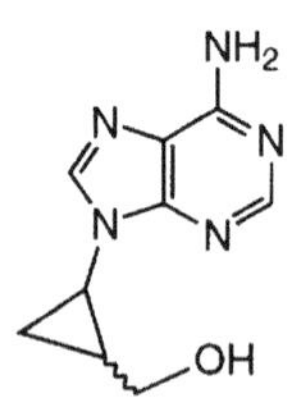

176

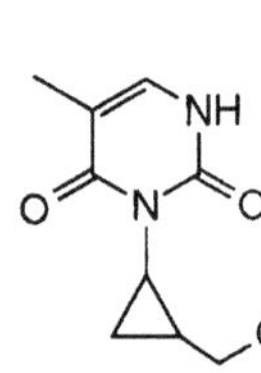

177

Nucleosides such as **178** and **179** containing a cyclopropane ring fused to the sugar portion have shown antiviral activity; their stereocontrolled syntheses have recently been achieved, Eq. (69) [233].

178 **179** (69)

The two diastereomers of 5-(2,2-dichlorocyclopropyl)-**180a** (X^1, X^2=Cl) and the four diastereomers of 5-(2-chlorocyclopropyl)-2′-deoxyuridine **180b** (X^1=Cl, X^2=H), as well as the corresponding brominated **180c,d** and fluorinated derivatives **180e,f** have been prepared and examined for antiviral and cytotoxic activity, in comparison with (*E*)-5-(2-bromovinyl)-2′-deoxyuridine BVDU **181** and 5-fluoro-2′-deoxyuridine FDU **182.** The (1*R*,2*R*)-5-(2-chlorocyclopropyl)-2′-deoxyuridine **180b** was the most active antiviral agent against herpes simplex HSV-1, relatively to BVDU **181** (ED=0.082 µg/ml). Compounds having the *R* configuration at the C-1 and/or C-2 positions of **180a,b** exhibited the most potent antiviral activity. The (1*R*)-difluoro compound **180e** was also more active than BVDU **181** against HSV-1 and a cytotoxic agent in the CCRF-CEM (IC_{50}=230 µM) screen relative to FDU **182**; the (1*S*)-**180e** diastereomer was inactive in both screens. Moreover (1*R*)-**180e** was more resistant to glycosidic bond cleavage by thymidine phosphorylase than its (1*S*) diastereomer, Eq. (70) [234].

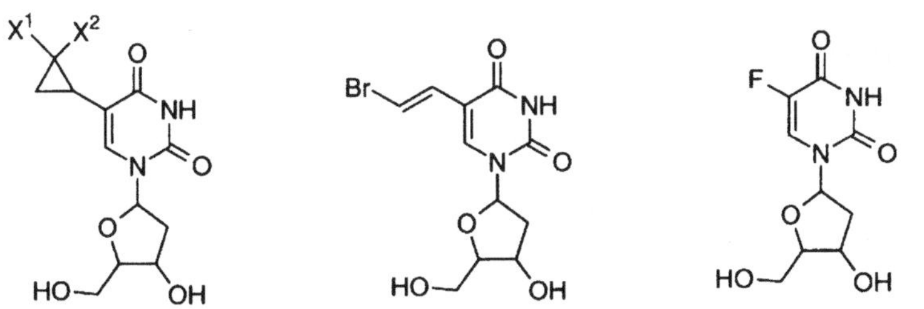

(70)

180a X^1, X^2 = Cl
b X^1 = Cl, X^2 = H
c X^1, X^2 = Br
d X_1 =Br, X_2 = H
e X_1, X_2 = F
f X_1 =F, X_2 = H

BVDU **181**

FDU **182**

The rapid spread of acquired immune deficiency syndrome (AIDS) has prompted numerous efforts to develop therapeutic agents against the human immunodeficiency virus type 1 (HIV-1) [235]. Efforts have focused on inhibition of the virally encoded reverse transcriptase (RT) enzyme, which is responsible for the conversion of retroviral RNA to proviral DNA. The nucleoside RT inhibitors 3′-azidothymidine (AZT) and dideoxyinosine (ddI) have proven to be clinically useful anti HIV-1 agents [236], but due to their lack of selectivity versus other DNA polymerases, these compounds are flawed by their inherent toxicities [235]. Therefore a considerable number of potent non-nucleoside HIV-1 RT inhibitors which act at an allosteric site unique to HIV-1 RT, providing selectivity versus other DNA polymerases, have been reported [238].

Among them, the cyclopropyldihydroquinazolinones **183a, b** were shown to be potent inhibitors of HIV-1 RT; however their potential therapeutic utility was hampered by metabolic liability. Thus, the 3-methyl group is susceptible to oxidative metabolism, resulting in loss of this substituent; removal of this methyl group results in large losses in inhibitory potency and severe limits in oral bioavailability [239]. The (–)-(4*S*)-enantiomer of compound **184**, where the 3-methyl of **183a** was replaced by a 3-(2-pyridinoethynyl), had exhibited the most favorable overall biological profile to overcome these problems; effectively (4*S*)-**184** was a potent antiviral agent in cultured MT-4 cells infected with HIV-IIIb, possessing a CI C_{95} value of 25 nM (n > 10); its (+)-(4*R*) enantiomer was essentially inactive, Eq. (71).

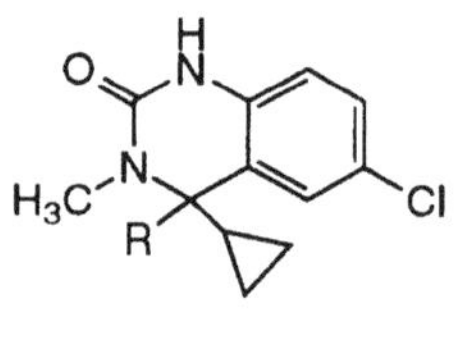

183a R = alkyl
b C≡CH

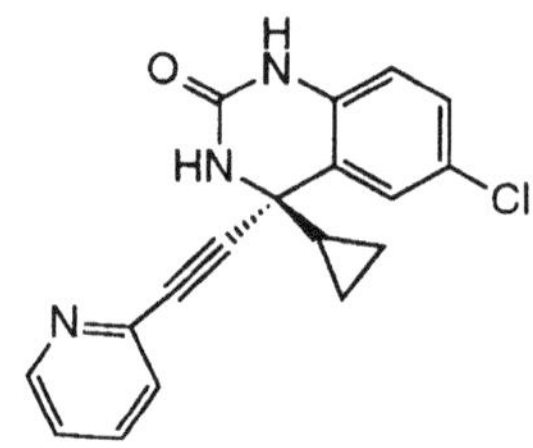

(–)-(4*S*)-**184**

(71)

Oral administration of (4*S*)-**184** to rhesus monkeys at 10 μg/kg provided peak levels of 12 mM at 2–4 h, while levels of parent drug remaining above 4 mM after 24 h. In view of this, (4*S*)-**184** was chosen as a candidate for further preclinical investigations; its potential to induce resistance and its activity against a number of known HIV-1 RT mutants are under current investigation [239]. The low oral bioavailability and rapid biliary excretion of peptide-derived HIV protease inhibitors have limited their utility as potential therapeutic agents. From a broad screening program to discover nonpeptidic HIV protease inhibitors, phenprocoumon (compound 185, K_i = 1 mM) was previously identified as a lead template. Overlay of the crystal structures of HIV protease complexes contain-

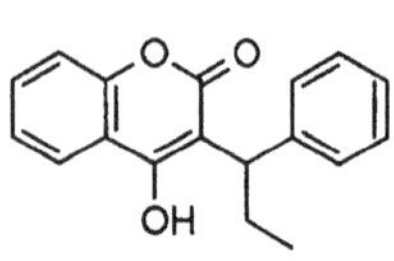

185

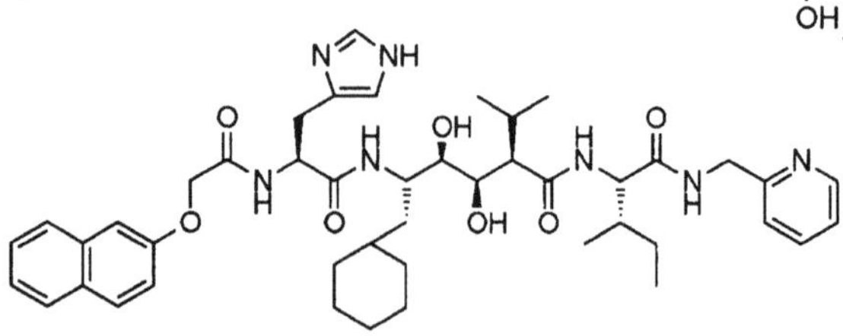

186

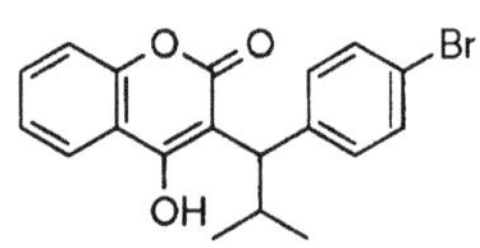

187

(72)

ing the peptide-derived inhibitor **186** and the nonpeptidic inhibitor **187** provided the basis for a molecular-modeling study that suggested the incorporation of a carboxamide functionality at the *meta* position of the benzyl side chain at C-3, Eq. (72) [240].

Compound **188** was prepared and shown to have improved inhibitory activity over the reference compound **185**. A crystal structure of the inhibitor **188** HIV protease complex was then determined, and the conformation of the inhibitor **188** in the enzyme active site was compared to that of the conformation expected from the modeling study. On the basis of this structure-based design of compound **188**, additional sets of analogues in the 4-hydroxy coumarin and the 4-hydroxy-2-pyrone series were prepared to evaluate the structure-activity relationship and to discover inhibitors with improved inhibitory potency. The inhibitor **189**, in the 4-hydroxycoumarin series, exhibited high HIV protease inhibitory activity with a K_i value of 1.4 nM. This finding of a specifically added carboxamide functionality to the inhibitory template, which resulted in inhibitors with improved enzyme-binding affinity, provides a new direction for the preparation of new promising series of potent and nonpeptidic HIV protease inhibitors. Although the inhibitors **189** and **190**, contain amino acid residues, compounds **191** and **192**, without amino acids, also showed high inhibitory activity. These latter two compounds provided a basis for the further exploration of more structure-based design experiments with non-amino acid-containing 4-hydroxycoumarin and 4-hydroxy-2-pyrone analogues which are expected to result in potent HIV protease inhibitors, Eq. (73) [240].

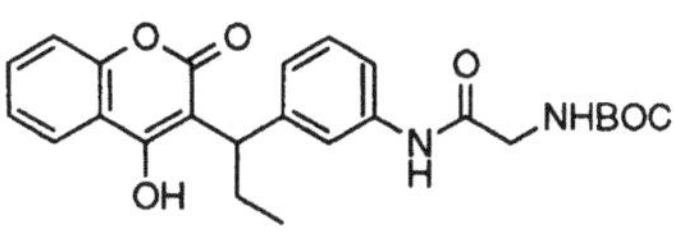

188

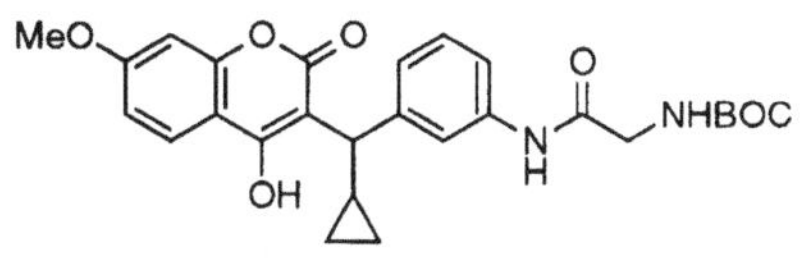

189

(73)

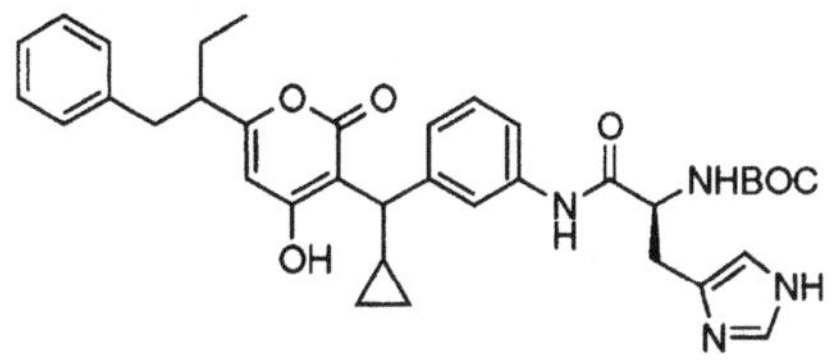

190

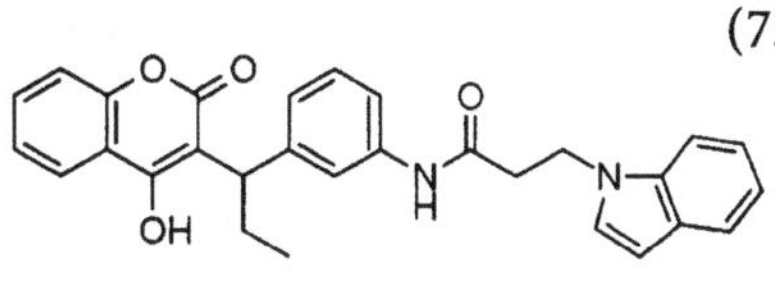

191

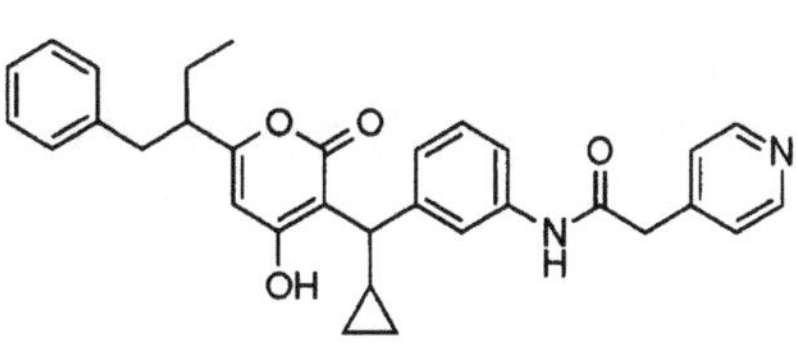

192

Recently, the novel cyclooctylpyranone HIV protease inhibitor **193a** was identified and an X-ray structure analysis of this inhibitor complexed with HIV-2 protease was obtained. This crystal structure was used to develop two strategies for creating derivatives of **193a** with enhanced enzyme inhibitory activity. The first strategy, substitution on the cyclooctyl ring, met with limited success, but provided some interesting information about the conformationally flexible cyclooctyl ring on the inhibitors. The second strategy, substitution at the *meta* postion of the aromatic ring, was far more successful and generated compounds, such as the carboxamide derivatives **194a** ($K_i = 3.0 \pm 0.4$ nM) and **194b** ($K_i = 4.0 \pm 0.8$ nM), which were significantly more active than the corresponding unsubstituted cyclooctylpyranone **193b** ($K_i = 11.7 \pm 4.7$ nM). An X-ray crystal structure of **194b** complexed with HIV-1 protease indicated the increase in binding affinity is most likely due to the additional interactions between the amide substituent and the S3 region of the protease, Eq. (74) [241].

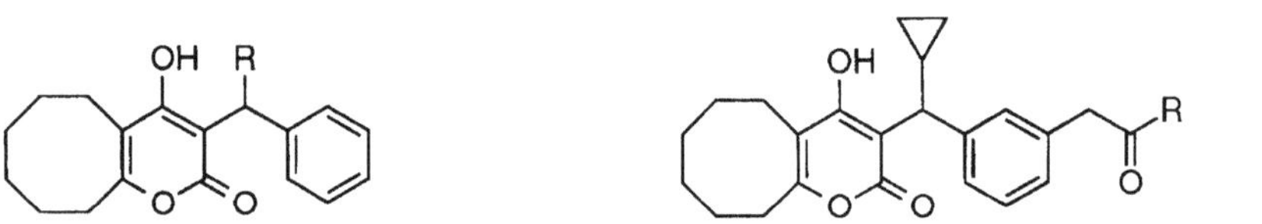

(74)

193a R = Et
b cyclopropyl

194a R = N-BOC- 2-(3-imidazol)ethylamine
b N-BOC -ethylamine

The only agents currently approved for the treatment of the acquired immune deficiency syndrome (AIDS) exert their therapeutic effects at the level of the human immunodeficiency virus type 1 (HIV-1) reverse transcriptase (RT) enzyme. These therapeutics, the nucleoside analogs AZT [241], DDI [242], DDC [243] and D4T [244], after intracellular transformation to the triphosphates, are incorporated by RT into the nascent proviral DNA and thereby terminate its synthesis. In addition to the nucleoside analogs, there is a second class of RT inhibitors, the non-nucleosides, exemplified by the dipyridodiazepinone nevirapine **195a** [245]. The non-nucleoside RT inhibitors [246–250] bind close to the active site [251–254] inducing conformational changes that affect the catalytic efficiency of the enzyme [255, 256]. Notwithstanding the differing mechanisms of action, the emergence of resistant viruses is a major limitation associated with the use of either nucleoside or non-nucleoside inhibitors of RT [257–259].

The primary cause of viral resistance to nevirapine **195a** is the mutation which substitutes cysteine for tyrosine-181 in RT (Y181 RT) [258]. This Y181C RT is less sensitive to nevirapine **196a** than the wild-type enzyme and also less sensitive to other non-nucleoside inhibitors [259]. Besides improving potency against the wild-type enzyme, a major focus was to achieve significant activity against the Y181C RT. Of the previously reported dipyridodiazepinones [260], only the 2-chloro derivative **195b** displayed significant inhibition of the Y181C RT ($IC_{50} = 0.21$ mM).

The potency of the dipyridodiazepinone class against the wild-type RT has been enhanced, and inhibition has been extended to the Y181C RT and other mutant RT enzymes by substitution at the 2-position of the dipyridodiazepinone ring system. Excellent activity against wild-type RT can be achieved with methyl or methoxy substituents **195c, d**, although in these cases there is only moderate activity against the Y181C mutant enzyme. Potency against both wild-type RT and the Y181C RT can be achieved with chloro **195b**, pyrrolyl **195e**, pyrazolyl **195f**, substituted phenyl **195g**, and substituted pyridyl groups **195h**. In addition, some of these substitutions confer activity against mutant RT enzymes resistant to other classes of non-nucleoside RT inhibitors. It remains to be seen whether or not new mutations in the RT enzyme can confer resistance to these more potent analogs of nevirapine, Eq. (75) [261].

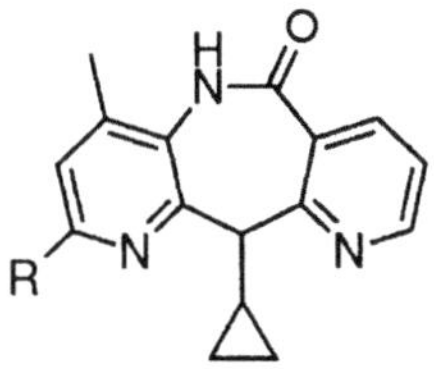

(75)

Nevirapine **195a**	R = H	**e**	pyrrolyl
b	Cl	**f**	pyrazolyl
c	Me	**g**	4-NH2Ph
d	OMe	**h**	4-NH_2Pyridyl

Following the discovery of the antiviral activity of the azidothymidine analog, and the activity of the 3′-azido-2′-dideoxyguanosine analog [262], the synthesis of a series of 2-amino-6-substituted-(3′azido-2′,3′-dideoxy-*b*-D-*erythro*-pento-furanosyl)purine analogs **196** was undertaken to explore the structure-activity relationships.

Among the various substituents at the 6-position of the purine ring, only the *trans*-(2-phenylcyclopropyl)amino derivative **196b** consistently demonstrated inhibition of MT4 cell growth (vide supra **10** in Eq. (5), Sect. 2.4), Eq. (76) [263].

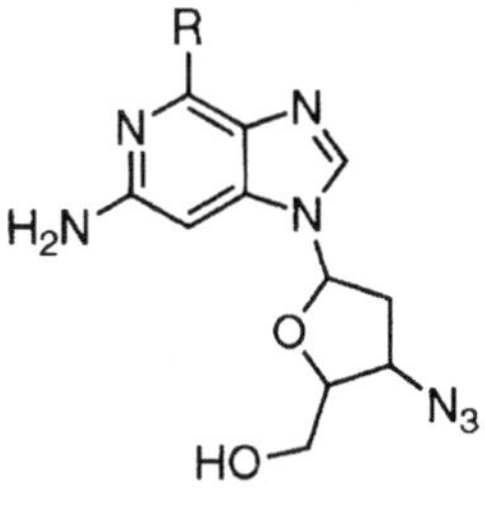

(76)

196a R = OH, OR, NH_2, NR_2, SH, SR, Me, CN
b *trans*-2-phenylcyclopropylamino

9
Neurochemical Activities

Anatomical and pharmacological studies have both indicated that the two major transmitter systems within the brain are the inhibitory GABA ergic and the excitatory amino acid (EAA) pathways [264]. At least four different receptors mediate the action of EAA, they are named according to the most selective ligand used to characterize them: *N*-methyl-D-aspartate (NMDA), quisqualate, kainate and L-2-amino-4-phosphonobutanoic acid (AP4) receptors. The NMDA receptor has received the most investigation based upon recent advances in the availability of pharmacological tools to study this receptor. It rapidly became clear that the NMDA receptor is a macromolecular complex possessing negative modulatory sites which bind phencyclidine (PCP), Zn^{2+} and Mg^{2+} [265]. Electrophysiological and neurochemical investigations have demonstrated that glycine can modulate the activity of NMDA-operated cation channels [266]. The glycine B site was characterized as a modulatory site of the NMDA receptor complex [267] which can influence the binding of PCP ligands [268]. Various types of antagonists [269], partial agonists [270] and agonists have been identified. Among them ACC **71**, which is structurally related to glycine, is a potent and selective ligand of the glycine modulatory site coupled to NMDA receptors, and appeared to be an even more specific ligand for the glycine B receptor than glycine itself. Therefore, ACC **71** (vide supra, Eq. (28)), may prove useful in neurochemical, pharmacological and electrophysiological studies of the NMDA receptor complex [271].

2,3-Di[^{3}H]-ACC **197** has been synthesized as a specific ligand for the glycine-B binding site [272]. The hypothesis of NMDA-mediated cell death in stroke, and the possible involvement of this receptor in neurodegenerative diseases such as Alzheimer's disease and Huntington's chorea have been suggested [273]. Although, the role of the Gly-B site in convulsions (epilepsy) has not been clearly defined, glycine has been shown to be active as an anticonvulsant, Eq. (77) [266].

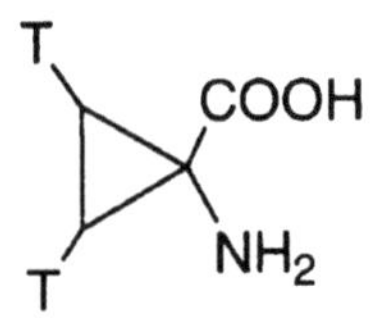

(77)

Both the Gly-B agonist D-cycloserine [275], a positive modulator of the NMDA receptor which enhances performances of learning tasks in rats, and the glycine prodrug, milacernide [276] have demonstrated memory enhancing actions (memory disorders). Therefore, such glycine-B agonists allow the characterization of the possible receptor subtypes and the understanding of the molecular biology of the receptor complex, which are crucial in the design of optimal pharmacological modulators [277].

The syntheses of the four diastereomers of α-(carboxycyclopropyl)glycines (*Z*)-**198a** and (*E*)-**198b** isolated from *Aesculus parviflora* [276], have been reported [277]. Neurobiological assays using a β-hydroxy-L-glutamate (L-BHGA) sensitive neuron from an african giant snail (*Achatina fulica Ferrusae*) [278] have indicated clearly a conformation-activity relationship; thus the diastereomer of **198** with erythro configuration and extended conformation, i.e. with the active conformation of L-BHGA when its interacts with the receptor, was markedly recognized by the neuron receptor, Eq. (78).

(78)

(*Z*)-(2*S*,3*S*,4*R*)-**198a** (*E*)-(2*S*,3*S*,4*S*)-**198b**

As it is now clear that excitatory amino acids (EAA) play a critical role as neurotransmitters in the brain [279], in addition to the putative endogenous agonists, synthetic agonists such as NMDA **199** and (2*R*,3*S*,4*R*)-a-(carboxycyclopropyl)glycine **198c** have been identified, Eq. 79 [280].

(79)

NMDA **199** (*Z*)-(2*R*,3*S*,4*R*)-**198c**

They have provided clues about the conformation of agonists at the receptor site. The cyclopropane ring was used to prepare conformationally restricted glutamic acid analogues which have exhibited affinity and potency at the NMDA receptor similar to those of glutamic acid itself and more selectively, they showed decreased affinity at other EAA receptors. These results, led to the consideration of the synthesis of hybrid molecules such as the (*E*)- and (*Z*)-phosphonomethanoamino acids **200**–**202**, in which the basic framework of the related 2-amino-5-phosphonopentanoic acid AP5-**203a**, has been rigidified by a cyclopropane ring, as competitive antagonists for the NMDA receptors.

Biological evaluation using [^{3}H]-L-glutamate as radioligand, has shown that among the AP5 analogues, the position of the methano bridge becomes less and less favorable when it moves from the 4,5-position to the 2,3-position. Thus the 4,5-methanol-AP5 compound **202** has a higher affinity for the receptor than its 3,4-isomer **201** (n=1), and the most favored configuration for the 3,4-methano AP5 **201** is *trans* (*E*). A surprising agonist-like efficacy was manifested by the AP7 analogue **201b** (n=2), Eq. (80) [281].

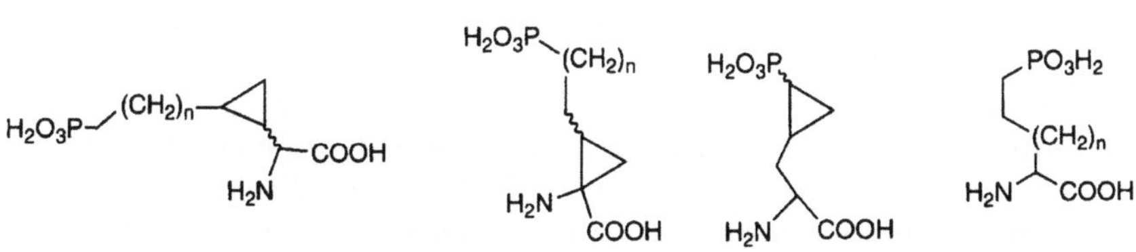

(80)

(*E*)/(*Z*) -**200** (*E*)/(*Z*) **201a** n = 1; **b** 2 **202** **203a** n = 1; **b** 2

Peptidomimetics of the anti-opiate neuropeptide Phe-Met-Arg-Phe-NH_2 have been synthesized by exchanging the Met with each of the four isomers of 2,3-methanomethionine **204**. All these peptides entailed more morphine abstinence in morphine addicted rats, although receptor binding studies in vitro have shown that the methanologues were less strongly bound than the parent peptide [282]. CCK methanologues containing (*Z*) or (*E*)-methanophenylalanine **205** showed different selectivities for the CCK-A or CCK-B receptors, although they were less highly bound than CCK itself, Eq. (81) [283].

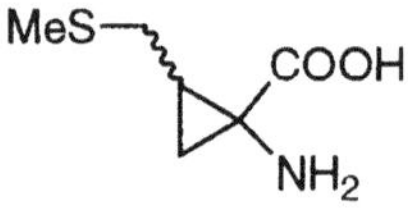

204

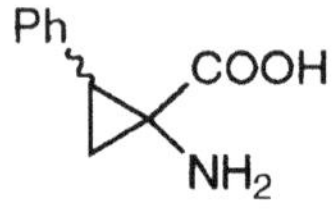

205

(81)

The development of the benzodiazepine class of drugs for the treatment of a variety of neurological indications has proven to be an outstanting success story in the field of chemotherapy. However, these compounds often produce undesirable side effects when used as anti-anxiety or hypnotic agents. These side effects include sedation, physical dependence, amnesia, muscle relaxation, and ethanol potentiation. The development of a benzodiazepine receptor-based anxiolytic agent devoid of these side effects would constitute a major advance in the field and has been the focus of significant research efforts [284].

Benzodiazepines exert their influence by interacting with the benzodiazepine receptor (BzR) located on the α-aminobutyric acid ($GABA_A$) chloride ion channel complex. Associated with the $GABA_A$ ion channel are a variety of recognition sites for small molecules, which can directly influence the ability of this channel to transport chloride ion across neuronal membranes. In addition to the benzodiazepine receptor, there exist binding sites for γ-aminobutyric acid (GABA), barbiturates, picrotoxin (and other convulsant agents), and neurosteroids [285]. When GABA, the major inhibitory neutrotransmitter in the central nervous system (CNS), binds to its receptor, the flow of chloride ion through the channel is increased and the excitability of the neuron is reduced [286]. Of the many types of receptor-ligand interactions that influence this GABA-induced chloride flux, the benzodiazepine receptor and its ligands have been the most widely studied, with many structural classes discovered which span the entire activity spectrum. Full agonists potentiate the GABA-induced chloride flux to further decrease the excitability of the neuron and have found wide-spread use as anxiolytic, hypnotic and anticonvulsant agents. In contrast, inverse agonists which decrease the flow of chloride ion are proconvulsant and anxiogenic in nature. Antagonists which have minimal or no effect on the chloride flux have neutral activity. Presumably, partial agonists lie within this activity continuum [287]. This is especially intriguing in that partial agonists may display anti-anxiety properties but, due to their lowered intrinsic activity, lack the undesirable side effects often associated with full agonists [288].

The search for viable partial agonists or subtype selective ligands has led to the development of a variety of compounds representing diverse structural types including imidazoquinoxalines, benzodiazepines, imidazopyridines and β-carbolines. In an effort to identify replacement candidates for the partial agonist pandiplon U 78875 **206** [289], which was removed from clinical trials due to liver enzyme induction, a variety of analogs were prepared and evaluated. One class of compounds studied consisted of imidazo[1,5-*a*]quinoxaline amides, carbamates, thiocarbamates, and ureas of which U-91571 **207** is representative [290]. Analogs within this series had varying activity; however, like **207**, most were partial agonists. Another related class of compounds that was explored involved a series of tetracycles as represented by U-89267 **208**, in which the carbonyl group was constrained to point toward the arene ring by incorporating a *C*(4)-*N*(5) tether (imidazo[1,5-a]-quinoxaline numbering [291]. Interestingly, derivatives from this subseries were full agonists by in vitro measurement (TBPS shift ratio) and were extremely potent in in vivo assays such as the metrazole antagonism assay. Furthermore, compounds such as **208** had unusually high affinity (13 nM) for the $\alpha_6\beta_2\delta_2$ subtype, whereas most derivatives from the "uncyclized" series (e.g. **208**) did not bind to this subtype, Eq. (82) [290–292].

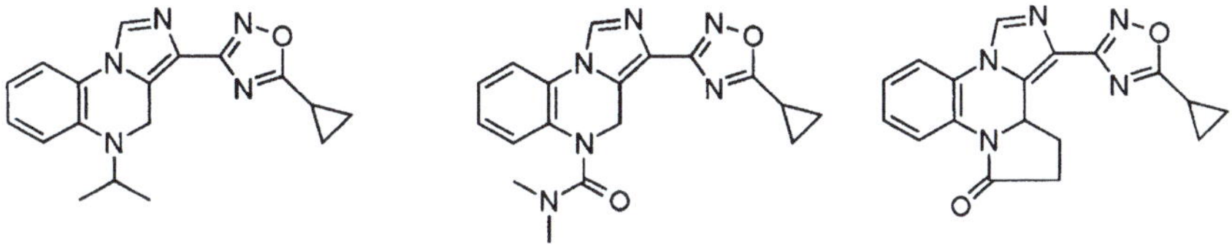

(82)

U 78875 **206**
(Pandiplon)

U 91571 **207**

U 89267 **208**

Constraining the appended substituent of U-91571 **207** into a tetracyclic imidazo[1,5-*a*]quinoxaline ring system provided a series of high-affinity ligands for the $GABA_A$/benzodiazepine receptor complex. In addition, this constraint, which forces the carbonyl group into a relatively planar ring system, provided analogs with an antagonist to partial agonist intrinsic activity profile as indicated by TBPS shift and Cl^- current measurement. Only **209 a, b**, which contain out-of-plane substituents at the 4-position, were nearly full agonists (in vitro). Most analogs were active in a metrazole antagonism assay consistent with ani-convulsant and possible anxiolytic activity. While the most effective analogs in this assay contained out-of-plane 4-substituents, several of the planar derivatives, including **210**, were surprisingly effective, especially considering their intrinsic activity. In contrast to **209**, none of the analogs reported herein had reasonable affinity for the diazepam insensitive $\alpha_6\beta_2\delta_2$ subtype. Clearly, the orientation of the carbonyl group (and added steric bulk) in **210** and related analogs prevents effective interaction with the $\alpha_6\beta_2\delta_2$ subtype in contrast to the nonplanar lactam ring of **208**. In addition, orientating the carbonyl group away from the aryl ring in a planar configuration provides analogs with partial agonist (or antagonist) properties which can, however be overridden by bulky out-of-plane substituents to provide full agonists, Eq. (83) [293].

A series of imidazo[1,5-*a*]quinoxaline amides, carbamates, and ureas which have high affinity for the *γ*-aminobutyric acid A/benzodiazepine receptor complex was developed. Compounds within this class have varying activities ranging from antagonists to full agonists. However, most analogs were found to be partial agonists as indicated by [^{35}S]TBPS and Cl^- current ratios. Many of these

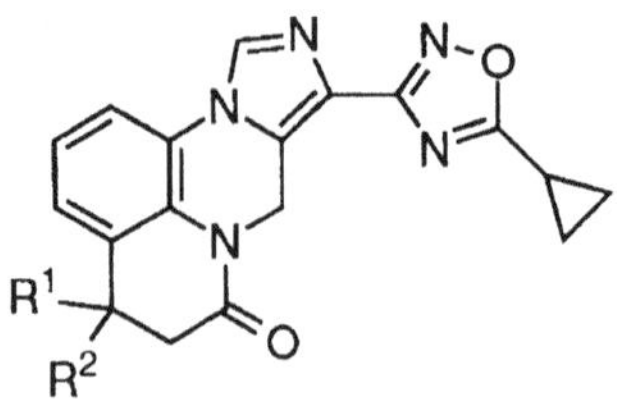

209a R[1], R[2] = Me
b R[1] = H, R[2] = Ph

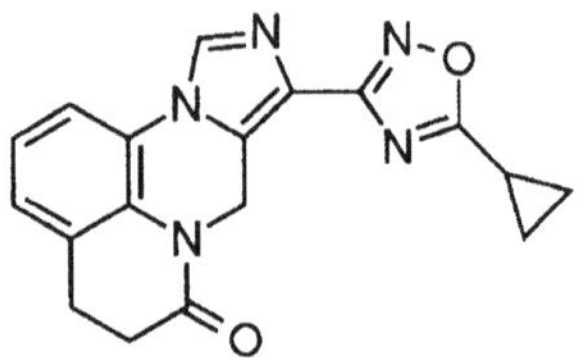

210

(83)

compounds were also effective in antagonizing metrazole-induced seizures in accordance with anticonvulsant and possible anxiolytic activity. Selected quinoxalines displayed limited benzodiazepine-type side effects such as ethanol potentiation and physical dependence in animal models. *N,N*-Dimethyl urea **211** emerged as the most interesting analog, having a partial agonist profile in vitro while possessing useful activity in animal models of anxiety such as the Vogel and Geller assays. In accordance with its partial agonist profile, **211** was devoid of typical benzodiazepine side effects, Eq. (84) [294].

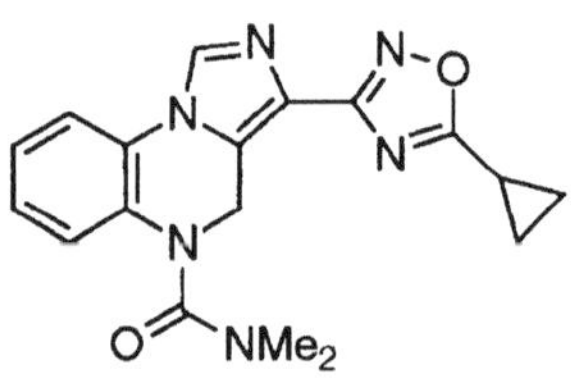

(84)

211

Schizophrenia is one of the most severe phychiatric illnesses and is characterized by hallucinations, delusions, and disorganized thought and behavior which result in major impairment of the patient's social and occupational function. Current medications utilizing typical neuroleptic antipsychotics such as haloperidol **212** show some promising activity in controlling the positive symptoms of schizophrenia. However, their effects are only partial, and they induce a substantial incidence of extrapyramidal symptoms (EPS) as neurological side effects [295]. Traditionally, two dopamine receptor subtypes have been classified on the basis of pharmacological evaluation, namely, the D_1 and D_2 receptors. Existing anti-psychotics are considered to act via the blockade of the classical "D_2 receptor" [296]. Recently, however, molecular biological approaches have led to the discovery of the novel dopamine D_3 and D_4 receptor isoforms [297], which are classified as the D_2-like (D_2, D_3 and D_4) receptor subfamily, and the D_5 receptor [298], which is classified as the D_1-like (D_1 and D_5) subfamily. The D_2-like receptor subfamily isoforms correspond to the classical D_2 receptors. D_3 and D_4 receptors are particularly concentrated in the mesolimbic and mesolimbocortical regions of the central nervous system, respectively [299]; areas which are thought to control emotional and cognitive functions and to be implicated in the pathology of schizophrenia [298]. In contrast, few D_3 and D_4 receptors are found in the nigrostriatal region, which is rich in D_2 receptors and of which blockade of the dopamine system has been suggested to be associated with EPS [299].

A series of *N*-(3-pyrrolidinyl)benzamide derivatives have been synthesized and evaluated for their binding affinity for dopamine D_2-like receptor subtypes. The SAR studies indicate that the 4-substituent on the benzamide nuclei and the *N*-substituent on the pyrrolidine ring play a critical role in improving D_3 and D_4 selectivity over D_2 receptors. Some preferential D_3 and D_4 antagonists also exhibited potent inhibitory activity against apomorphine-induced climbing behavior in mice. Among them the novel [(cyclopropylcarbonyl)amino]benzamide **214** possesses high affinity for D_3 and D_4 receptors (K_i values of 21 and 2.1 nM respectively) and selectivity for D_4 and D_3 receptors (K_{iD2}/K_{iD4} = 110, K_{iD2}/K_{iD4} = 10) with weak or negligible affinity for other neurotransmitter receptors. In vivo, **213** exhibited inhibitory activity against apomorphine-induced climbing behavior with an ED_{50} value of 0.32 mg/Kc (sc), this biological profile is markedly different from those of known antipsychotics [299]. Thus compound **213** may produce unique pharmacological effects, including atypical antipsychotic effects. Further, it is believed that **213** would contribute to the understanding of the physiological and pharmacological functions of D_2-like receptor isoforms, Eq. (85) [301].

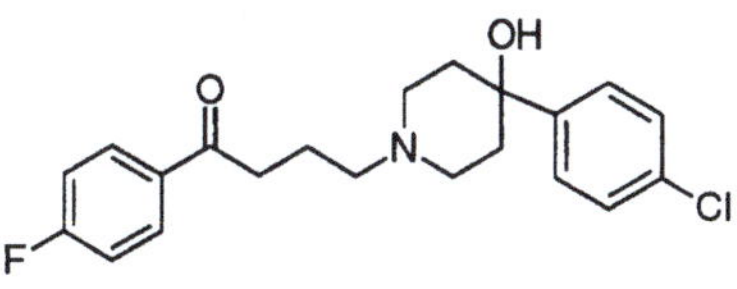

Haloperidol **212**

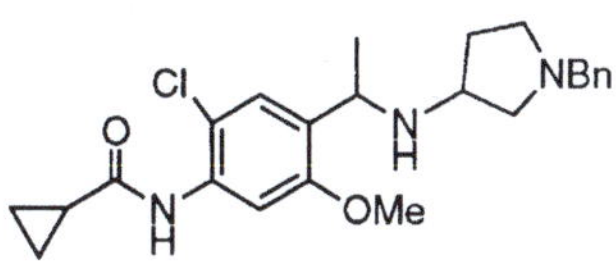

213

(85)

10 Miscellaneous

During work on a series of aspartyl dipeptides containing ACC **71** (vide supra, Eq. (28), Sect. 4) at the carboxyl terminus, it was reported that dispartame Asp-ACC-OMe had a distinct sweet taste [302] and that the corresponding *n*-propyl ester had 250–300 times the sweetness of sucrose [303]. However, replacement of phenylalanine by 2,3-methanophenylalanine gave tasteless analogues of aspartame [293, 304], and some dimethyl-ACC **214** (methanovaline) and trimethyl-ACC **215** aspartame analogues [Asp-(Me)$_n$-ACC-OMe] have a bitter taste. These taste properties, which depend on the number and position of the methyl substituents, have been explained on the basis of topochemical models; thus, a L-shaped conformation of the dipeptide is necessary for sweet taste, Eq. (86) [305].

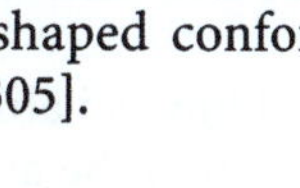

214

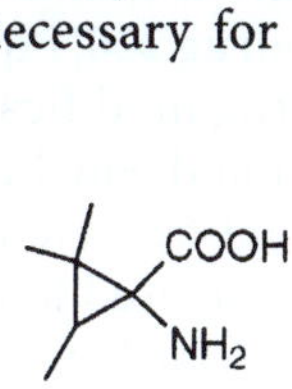

215

(86)

The renin-angiotensin system plays an important role in the pathophysiology of cardiovascular disease and agents such as angiotension converting enzyme (ACE) inhibitors are effective for the treatment of hypertension [306]. Antagonism of angiotension II (AII) AT_1 receptors may also be a useful strategy for the treatment of hypertension and seveal groups have disclosed non-peptidic AII antagonists [307].

Many reported AII antagonists, such as losartan **216** and its more pharmacologically active metabolite EXP 3174 **217**, share structural features believed to be important for activity. These include a heterocycle with one or more basic nitrogen atoms, an alkyl substituent in the 2-position relative to the basic nitrogen, an aryl susbtituent with an acidic group in the 3-position and an acid or its metabolic equivalent in the 4-position. Quinolin-4-carboxylic acids **218** appeared to accomodate all these structural features, Eq. (87). An initial study was the modification of the alkyl group R at the 2-position of the quinoline. The activity of the compound where R is *n*-propyl **218a**, confirmed the hypothesis that the quinoline-4-carboxylic acids can be high affinity ligands for the AII receptor. Reducing the size of the substituent to ethyl **218b** had little effect on binding affinity or functional potency but substitution of a methyl group **218c** reduced functional potency by 10 fold. The most interesting compound in this study was the 2-cyclopropyl compound BMS 183920 **218d** which showed very potent and insurmountable antagonism in the functional assays, Eq. (87) [308].

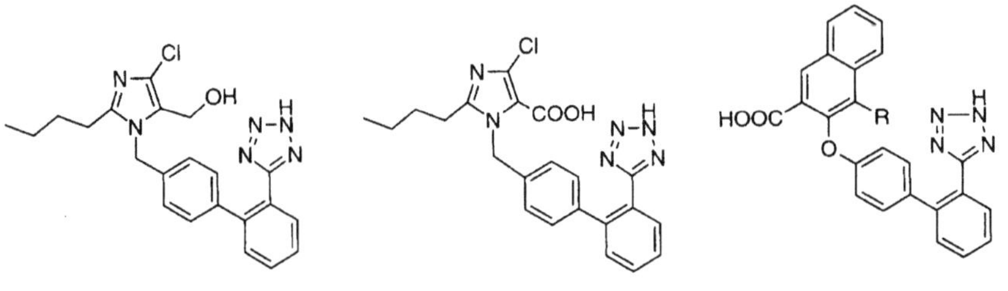

(87)

Losartan **216** EXP 3174 **217** **218a** R = *n*Pr; ***b*** Et; ***c*** Me; ***d*** cyclopropyl

Despite excellent intrinsic potency, weak oral activity limited the in vivo activity of **218d**. Efforts to improve its oral bioavailability by a prodrug approach led to the preparation of two compounds with improved oral activity and bioavailability, the dioxolenonylmethanol ester BMS 184698 **219** and the *N*-alkyl tetrazole prodrug **220**, Eq. (88).

This prodrug modification more than doubled the oral bioavailability of the parent compound. Furthermore, BMS 184698 **219** was shown to be an effective, orally active antihypertensive agent in the sodium-replete spontaneously hypertensive rat. Compared to losartan **216** at 30 mmol/kg *p.o.* once-daily doses, BMS 184698 **219** was the better antihypertensive agent [309].

Retinoids are small molecule hormones that elicit pleiotropic biological responses by activating two families of nuclear receptors which are structurally

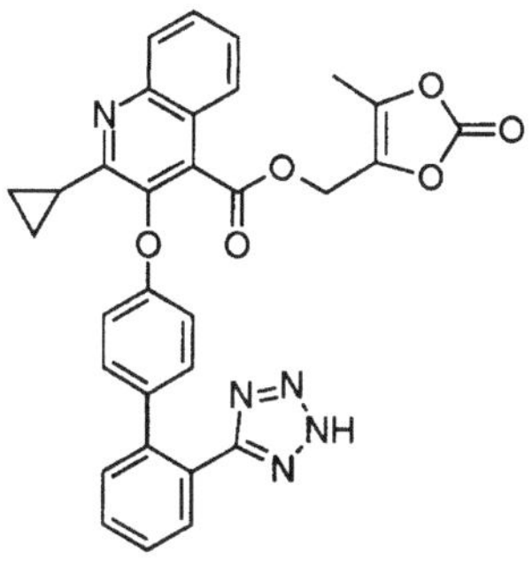

BMS 184698 **219**

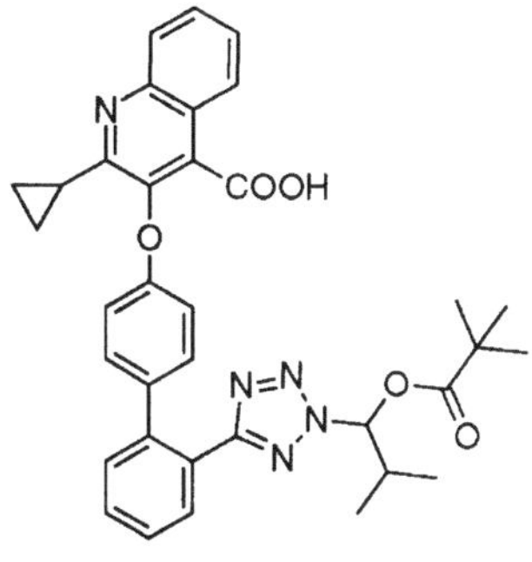

220

(88)

and evolutionarilly related to the steroid/thyroid hormone receptor superfamily [310]. The two families are the retinoic acid receptors (RARs) [311] and the retinoid X receptors (RXRs) [312] and each family consists of three subtypes (α, β, and γ) which are encoded by distinct genes. Physiologically, retinoid hormones regulate a variety of very basic biological functions both in development and in the adult [313]. Disruption of the normal pathways of retinoid homeostasis either by vitamin A deficiency [314] or by alteration of retinoid receptors [315] can lead to disease conditions. Consistent with their broad physiological effects, retinoids are of potential clinical use in a variety of areas including dermatology [316], oncology [317], opthalmology [318] and cardiovascular disease [319].

The physiological hormone for the RARs is all-*trans*-retinoic acid RA **221** and that for the RXRs is its geometric isomer, 9-*cis*-retinoic acid (9-*cis*-RA **222**) [320]. However, these polyolefinic hormones are of only limited use in elucidating the precise biological roles of each receptor family. Synthetic ligands that specifically activate only the RXR or RAR hormonal pathway and which cannot be converted into forms that activate the other pathway would be of much greater use in this regard.

The cyclopropane ring has been used as an isostere for the C9-C10 double bond to obtain locked-9-*cis* and 9-*trans* retinoid analogs. The 9-*cis*-locked analog **223** is the most potent RXR analog described to date. Because of its intrinsic pharmacologic selectivity and because it cannot be converted to an RAR active form, compound **223** is the highest affinity and most potent RXR agonist described to date and would be a very useful tool in defining the biology associated with the RXR hormonal pathways, in vivo (Eq. 89) [321].

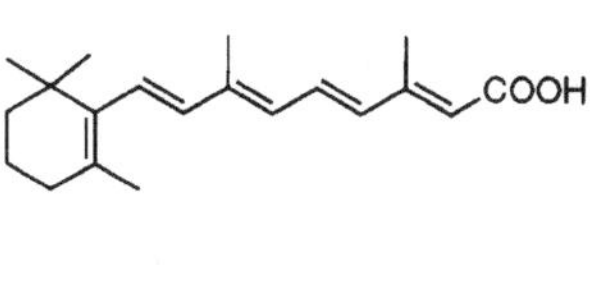

RA **221**

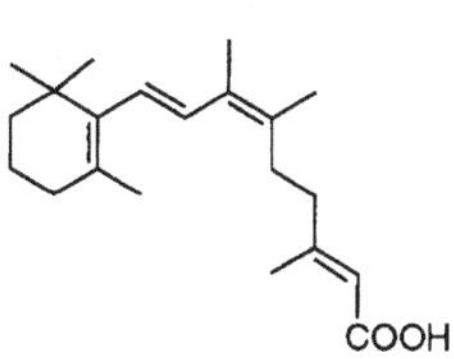

9-*cis*-RA **222**

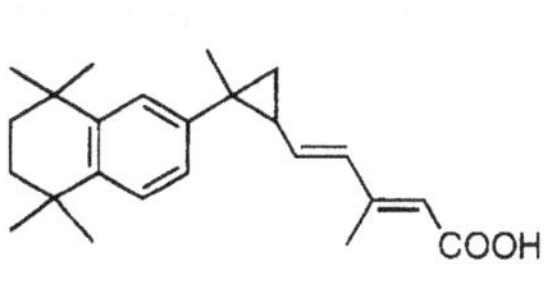

223

(89)

Imidazolines are one of the most studied classes of α-adrenergic drugs. Discrete structural modifications of these ligands reveal agonist or antagonist properties with varying degrees of selectivity for α-adrenergic receptor subtypes. In addition, several of these ligands also exhibit high affinity for a family of membrane proteins termed imidazoline receptors or imidazoline/guanidinium receptive sites (IGRS). Idaxozan-**224**, for example, a selective α_2-adrenergic receptor antagonist [322] belonging to the imidazoline class, has been shown to recognize these sites in a wide range of tissues, with an affinity comparable with that determined at α_2-adrenergic receptors [323]. Cirazoline **225**, a potent α_1-adrenergic receptor agonist and α_2-adrenergic receptor antagonist [324], exhibits high affinity for IGRS in a variety of tissues [323]. Thus, this molecule can serve as a useful starting point to characterize the structure-activity restrictions of this diverse group of ligand-binding pockets.

Removal of the cyclopropyl group to give the unsubstituted compound **226a** results in a 10-fold decrease in potency which parallels the decrease in affinity, while efficacy is virtually unaffected. It is worth noting that, if compound **226a** is considered as the reference, substitution at *ortho* position either with an alkyl or alkoxy group increases both potency, with exception of compound **226b**, and affinity. On the other hand, activity is negatively affected, with the exception of cirazoline **225** and compound **226c**, in which the presence of the cyclopropyl group or of the two methyl groups appears to favor retention of activity. This latter finding supports the hypothesis [324] concerning the role played by the cyclopropyl group in increasing activity, Eq. (90) [325].

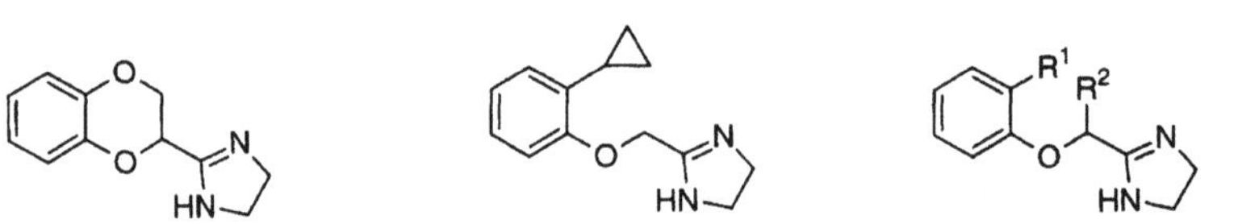

(90)

Idazoxan **224** Cirazoline **225** **226a** $R^1, R^2 = H$; **b** $R^1 = s\text{-Bu}, R^2 = H$; **c** $R^1, R^2 = Me$

Carbetapentane (2-[2-(diethylamino)ethoxy]ethyl-1-phenyl-1-cyclopentane carboxylate) **227** binds with high affinity to σ sites [^{3}H]-(+)-3-PPP ((+)-3-(3-hydroxyphenyl)-*N*-propyl piperidine; $K_i = 11$ nM), [^{3}H]dextromethorphan ($K_i = 11$ nM), [^{3}H)-(+)-pentazocine ($K_i = 32$ nM) [326] and demonstrates anticonvulsant [327], antitussive [328], and spasmolytic [329] actions. In an attempt to determine whether these psychoactivities can be attributed to interaction at σ sites, a series of carbetapentane analogs were prepared. Phenyl ring substitution; contraction, expansion, and replacement of the cyclopentyl ring by a methyl group; replacement of the carboxylate function with an amide, methyl ether, and methylamine; and replacement of the *N,N*-diethylamino substituent with a morpholinyl or piperidinyl moiety were investigated. All of these novel analogs were evaluated for binding, the most selective ligands were found to be compounds **228**–**230**. The chemical modifications including replacing the cyclopentyl ring with a smaller ring system, i.e. cyclopropyl **228** or a methyl group

229, replacing the ester function by an amide function 230 or replacing the diethylamino moiety with a morpholino group resulted in >220 fold selectivity over muscarinic receptor binding. Therefore, several of these novel compounds are potent σ selective ligands which can be investigated as potential antitussive, anticonvulsant, and antiischemic agents Eq. (91) [330].

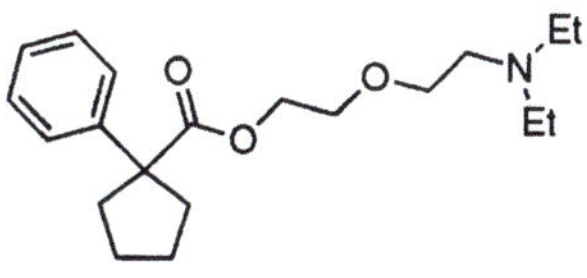

227

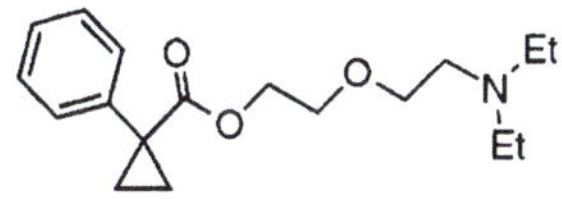

228 (91)

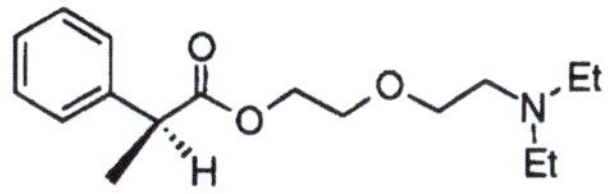

229

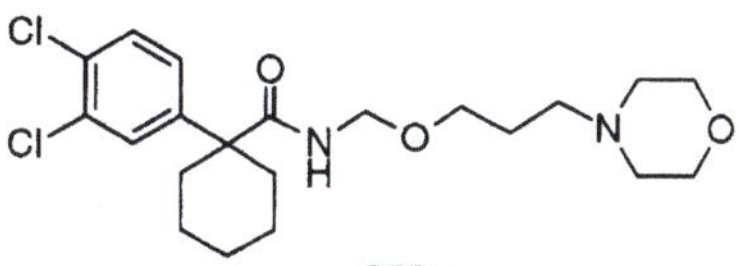

230

Amidinobenzisothiazoles **231** displayed remarkable analgesic action and an interesting antiphlogistic action, which is often dissociated from antipyretic action. New derivatives **232a–e** (R=H, Me) have been synthesized in order to evaluate their antiinflammatory, antipyretic and analgesic activities in vivo as well as their in vitro spasmolytic activity and to improve the understanding of structure-activity relationships. Within this series only the compounds **232c** (R=H and Me), and **232d** (R=Me) display a significant analgesis/antipyretic activity devoided of undesirable side effects. Therefore, the introduction of alicyclic structures, i.e. cyclopropane or cyclopentane rings, in the amidinobenzoisothiazole derivatives afforded effective compounds with improved activity, Eq. (92) [331].

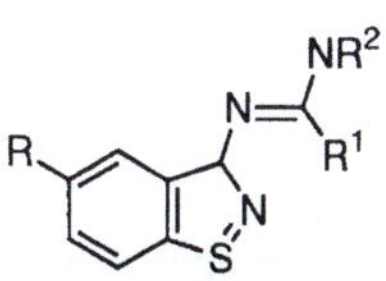

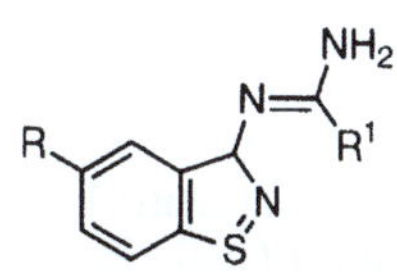

(92)

231 R = H, Me
R^1 = H, Me, Ph
R^2 = H, Me

232a R^1 = CH_2Cl
b CH_2-CH=CH_2
c cyclopropyl
d cyclopentyl
e 2-pyridinyl

Curacin A **233** is a novel antimitotic agent recently isolated from a Caribbean cyanobacterium *Lingbya majuscula* (blue-green algae). It was reported that curacin A inhibited tubulin assembly by binding to the colchicine-binding site [332], which is one of the two distinct drug-binding sites on tubulin. The result is intriguing because curacin A has little structural similarity to known natural and synthetic colchicine-site ligands. Thus, elucidation of the nature of curacin A-binding to tubulin should afford further insight into the molecular mechanism of tubulin-ligand interaction at this site, and could lead to the development of new bioactive agents, Eq. (93) [333]. The relative configuration of the *cis*-disubstituted cyclopropyl ring [332] and its (2*R*,13*R*,19*R*,21*S*) absolute configuration have been determined [333]; its four stereoisomers have been recently synthesized, Eq. (93) [334].

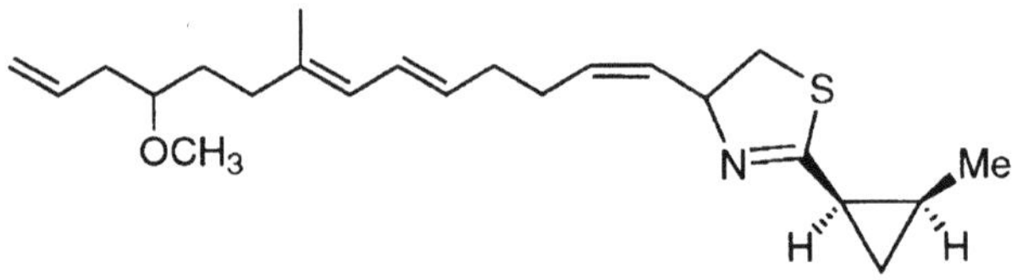

(93)

Curacin **233**

Recently, was reported the isolation of the unusual metabolite U 106305 **234** from the fermentation broth of *Streptomyces sp.* UC 11136 [335]. The compound is structurally remarkable being graced with six cyclopropane rings, five of which are contiguous. U 106305 **234** shows a striking similarity to the potent antifungal agent FR 900848 **104** (vide supra, Sect. 5) which was isolated from the fermentation broth of *Streptoverticillium fervens* [330]. U 106305 **234** is a potent in vitro inhibitor of the cholestery ester transfer protein (CETP) reaction, thus could be of potential application in the prevention of arteriosclerosis [337]. Its enantioselective total synthesis and stereochemical assignment similar to FR 900848 **104** have been reported, Eq. (94) [338].

(94)

U 106305 **234**

References

1. de Meijere A (1979) Angew Chem 91:867; Angew Chem Int Ed Engl 18:809
2. For reviews see: Salaün J (1987) Rearrangements involving the cyclopropyl group. In: Rappoport Z (ed) The Chemistry of the Cyclopropyl Group. Wiley, New York, chap 13; de Meijere A (1986–96) Top Curr Chem 133, 135, 144, 155, 178; Ghosez L (1989) Strain-assisted Syntheses. In: Tetrahedron 45:2875; Michl J (1989) Strained Organic Compounds. In: Chem Rev 89:975; de Meijere A, Wessjohann L (1990) Synlett: 20–32; de Meijere A, In: Ohshiro Y (ed) New Aspects of Organic Chemistry II, Proceedings of the Fifth International Kyoto Conference on New Aspects of Organic Chemistry – IKCOC 5, Kyoto Nov 11 – Nov 15, 1991 (1992) Kodansha, Tokyo, p. 181; de Meijere A (1997) Carbocyclic Three- and Four-membered Ring Systems. In: Houben-Weyl, Methods of Organic Chemistry, vol E 17. Thieme, Stuttgart
3. Salaün J, Baird MS (1995) Curr Med Chem 2:511
4. Liu HW, Walsh CT (1987) Biochemistry of the Cyclopropyl Group. In: Rappoport Z (ed) The Chemistry of the Cyclopropyl Group. Wiley, New York, Chap 16; Kennewell PD, Matharu SS, Taylor JB, Westwood R, Sammes PG (1982) J Chem Soc Perkin Trans 1:2553
5. Perry JJ (1977) CRC Crit. Rev Microbiol 7:387; Geigert J, Neidleman SL, Dalietos DJ (1983) J Biol Chem 258:2273

6. Dalton H, Golding BT, Wates BW, Higgins R, Taylor JA (1981) J Chem Soc Chem Comm: 482
7. Duine JA, Frank Jzn (1981) J Trans Biochem Sci (personal edition) 6:278
8. Mincey T, Bell JA, Mildvan AS, Abeles RH (1981) Biochemistry 20:7502; Parkes C, Abeles RH (1984) Biochemistry 23:6355
9. Dijlstra M, Frank Jzn J, Jongjean JA, Duine JA (1984) Eur J Biochem 140:369
10. Sherry B, Abeles RH (1985) Biochemistry 24:2594
11. Macdonald TL, Zirvi K, Burka LT, Peyman P, Guengerich FP (1982) J Am Chem Soc 104:2050
12. Quin XZ, Williams F (1987) J Am Chem Soc 109:595
13. Silverman RB, Yamasaki RB (1984) Biochemistry 23:1822
14. Paech C, Salach JL, Singer TP (1980) J Biol Chem 255:2700; Silverman RB, Zieske PA (1986) Biochem Biophys Res Commun 135:154
15. Silverman RB, Hoffman SJ (1980) J Am Chem Soc 102:886
16. a) Lindberg P, Bergamn R, Wiskberg B (1975) J Chem Soc Chem Commun: 946; b) Lindberg P, Bergamn R, Wickberg B (1977) J Chem Soc Perkin Trans 1 684; c) Tottmar O, Lindberg P (1977) Acta Pharmac Tox 40:476; d) Tottmar O, Hellström E (1979) Acta Pharmac Tox 45:272; e) Kienzler T, Strazewski P, Tamm C (1992) Helv Chim Acta 75:1078
17. Marchner H, Tottmar O (1983) Biochem Pharm 32:2181 and references cited therein
18. Barkman R, Perman ES (1963) Acta Pharmacol Toxicol 20:43
19. Coldwell BB, Genest K, Hughes DW (1969) J Pharm Pharmacol 21:176
20. For a review on the "Cyclopropanone Hemiacetals" see: Salaün J (1983) Chem Rev 83:619
21. Wiseman JS, Abeles RH (1979) Biochemistry 18:427; Wiseman JS, Tayrien G, Abeles RH (1980) Biochemistry 19:4222; Wiseman JS, Nichols JS, Kolpak M (1982) J Bio Chem 257:6328; Biellmann JF, Tritsch D, Salaün J, unpublished results
22. Hellström E, Tottmar O (1982) Biochem Pharmacol 31:3899
23. Evans IA (1984)Chemical Carcinogens. In: American Chemical Society:Wahsington DC, 2nd (ed) Searle CE Vol 2, p 1171; Hirono I, Yamada K (1987) Naturally Occurring Carcinogens of Plant Origin. In: Hirono I (ed) Kodansha-Elsevier: Tokyo-Amsterdam, p 87); Bryan GT, Pamukcu AM (1982) Carcinogen and Mutagens in the Environment. In: Stich HF (ed) CRC Press: FL, p 75
24. Hirayama T (1979) Naturally Occurring Carcinogens - Mutagens and Modulators of Carcinogenesis. In: Miller EC, Miller JA, Hirono I, Sugimura T, Takayama S (eds) Japan Sci Soc Press-University Park Press: Tokyo-Baltimore, p 359
25. a) Niwa H, Ojika M, Wakamatsu K, Yamada K, Hirono I, Matsushita K (1983) Tetrahedron Lett 24:4117; b) Niwa H, Ojika M, Wakamatsu K, Yamada K, Ohba S, Saito Y, Hirono I, Matsushita K (1983) Tetrahedron Lett 24:5371; c) Ohba S, Saito Y, Hirono I, Niwa H, Ojika M, Wakamatsu K, Yamada K (1984) Acta Crystallogr., Sect. C 40:1877; d) Ojika M, Wakamatsu K, Niwa H, Yamada K (1987) Tetrahedron Lett 43:5261
26. Ojika M, Sugimoto K, Okazaki T, Yamada K (1989) J Chem Soc Chem Commun 1775
27. Kushida T, Uesugi M, Sugiura Y, Higoshi H, Tanaka H, Hirokawa J, Ojika M, Yamada K (1994) J Am Chem Soc 116:479; Yamada K, Ojika M, Higoshi H (1998) Angew Chem Int Ed Engl. 37:1818
28. O'Leary MJ, DeGaoyer WJ, Doughety TM, Andersov V (1981) Biochem and Biophys Res Comm 100:1320; Herrmann KM, Somerville RL (1983) In: Amino Acids: Biosynthesis and Genetic Regulation, Addison-Wesley, Reading, Massachusetts, p 1
29. Ray TB (1986) Trends Biochem Sc 11:180
30. Wixom JL, Wikman JH, Howell GB (1961) J Biol Chem 236:3257; Myers JW (1961) J Biol Chem 236:1414
31. Arfin SM (1969) J Biol Chem 244:2250; Hill RK, Yans S, Arfin SM, (1973) J Am Chem Soc 95:7857
32. Wolfender R (1972) Acc Chem Res 5:10
33. Westkaemper RB, Zenck PC (1988) Eur J Med Chem 23:233
34. a) Monaghan DT, Brudges RJ, Cotman CW (1989) Annu Rev Pharmacol Toxicol 29:365; b) Watkins JC, Krgogsgaard-Larsen P, Honorè T (1990) Trends Pharmacol Sci 11:25

35. a) Collingridge GL, Bliss TVP (1987) Trends Neurosci 10:288; b) Madison DV, Malenka RC, Nicoll RA (1991) Annu Rev Neurosci 14:379; c) Kaba H, Hayashi Y, Higuchi T, Nakanishi S (1994) Science 265:262
36. Medrum B, Garthwaite J (1990) Trends Pharmacol Sci 11:379
37. Olney JW (1990) Annu Rev Pharmacol Toxicol 30:47
38. Choi DW, Rothman SM (1990) Annu Rev Neurosci 13:171
39. Kurokawa N, Ohfune Y (1985) Tetrahedron Lett 26:83; Shimamoto K, Ushida M, Shinozaki H, Ohfune Y (1991) J Org Chem 56:4167
40. Ohfune Y, Shinozaki H (1993) Drug Design for Neuroscience. In: Kozikowski AP (ed) Raven Press. New York, p 261
41. Shimamoto K, Ohfune Y (1996) J Med Chem 39:407
42. Ramuz H, Schmutz W, Huber R (1989) Eur J Med Chem 24:493; Boesch D, Gaveriaux C, Jachez B, Pourtier-Manzanedo A, Bollinger P, Loor F (1991) Cancer Res 51:4226; Dhainaut A, Regnier G, Atassi G, Pierre A, Leone S, Kraus-Berthier L, Prost J-F (1992) J Med Chem 35:2481
43. Hait WN, Aftab DT (1992) Biochem Pharmacol 43:103; Klopman G, Srivastava S, Kolossvary I, Epand RF, Ahmed N, Epand RM (1992) Cancer 52:4121
44. Pfister JR, Makra F, Muehldorf AV, Wu H, Nelson JT, Cheung P, Bruno NA, Casey SM, Zugtshi N, Slate DL (1995) Bioorgan Med Chem Lett 5:2473
45. Pless J (1976) Helv Chim Acta 59:499
46. Chen C-J, Chin JE, Ueda K, Clark DP, Pastan I, Gottesman MM, Roninson IB (1986) Cell 47:381; Gros P, Crrop J, Housman D (1986) Cell 47:371
47. Suckling CJ (1988) Angew Chem Int Ed Engl 27:537
48. Nonhebel DC, Suckling CJ, Walton JC (1982) Tetrahedron Lett 23:4477; Breckenridge RJ, Suckling CJ (1986) Tetrahedron 42:1035
49. Snel M, Eysell E (1974) Proc Br Weed Control Conf 12, Vol. 1:115
50. a) Hassall CH, Reyle K, Feng P (1954) Nature 173:356; b) Hassall CH, Reyle K (1955) Biochem J 60:334; c) Anderson HV, Johnson JL, Nelson JW, Olson EC, Specter ME, Vavra JJ (1958) Chemistry and Ind: 30; d) Eloff JN, Fowden L (1970) Phytochemistry 9:2423
51. These African eatable fruits were introduced to the Carribean Islands and Florida by Captain Bligh. See: Mitchell JC (1974) The posthumous misfortune of Captain Bligh of the Bounty: Hypoglyceamia from Blighia. Diabetes 23:919; Black DK, Landor SR (1963) Tetrahedron Lett: 1065; Black DK, Landor SR (1968) J Chem Soc C: 288
52. Sherrat HSA (1986) Trends Pharmac Sc: 186
53. Fowden L, Lea PJ, Bell EA (1979) The non-protein amino acids of plants. In: Merster A (ed) Advances in Enzymology. Wiley, New York, p 117
54. a) Sherratt HSA, Osmundsen H (1976) Biochem Pharmacol 25:743; b) Billington D, Osmundsen H, Sherratt HSA (1979) Biochem Pharmacol 27:2891
55. Osmundsen H, Billington D, Taylor JR, Sherratt HSA (1978) Biochem J 170:337
56. a) Osmundsen H, Sherratt HSA (1975) FEBS Lett 55:38; b) Billington D, Kean E. A, Osmundsen H, Sherratt HSA (1974) IRCS (Research on Biochemistry) Pharmacology 2:1712; c) Billington D, Osmundsen H, Sherratt HSA (1979) Biochem Pharmacol 27:2879
57. Wenz A, Thorpe C, Ghisla SJ (1981) Biol Chem 256:9809
58. a) Tanaka K, Isselbacher KJ, Shoh V (1972) Science 175:69; b) Tanaka K, Kean EA, Johnson B (1976) New Engl J Med 295:461
59. Tanaka KJ (1972) Biol Chem 247:7465
60. Wenz A, Thorpe C, Ghisla SJ (1981) Biol Chem 256:9809; Al-Bassam SS, Sherratt HSA (1981) Biochem Pharmacol 30:2817
61. (a) Stolle A, Salaün J, de Meijere A (1991) Synlett: 327; (b) Stolle A, Ollivier J, Piras PP, Salaün J, de Meijere A (1992) J Am Chem Soc 114:4051
62. Voigt K, Stolle A, Salaün J, de Meijere A (1995) Synlett: 226
63. a) Stolle A (1992) Dissertation, Universität Hamburg; b) Sherratt HSA, Personal communication
64. a) Stammer CH (1990) Tetrahedron 46:2231; b) Williams RM (1989) Synthesis of optically active α-amino acids. In: Baldwin JE, Magnus PD (ed) Organic Chemistry Series.

Pergamon Press, Oxford, pp 30, 59, 113; c) Alami A, Calmes M, Daunis J, Jacquier R (1993) Bull Soc Chim Fr 130:5

65. a) Arenal I, Bernabe M, Fernandez-Alvarez E, Gibello AM (1983) Anales de Quim 79c:65; b) Suzuki M, Kumar SD, Stammer CH (1983) J Org Chem 48:4769
66. Pages RA, Burger A (1967) J Med Chem 10:435
67. Pages RA, Burger A (1966) J Med Chem 9:767
68. Yoshimoto T, Orlowski R, Walter R (1977) Biochemistry 16:2942
69. Koida M, Walter R (1976) J Biol Chem 251:7593
70. Yoshimoto T, Ogita K, Walter R, Koida, M, Tsuru D (1979) Biochem Biophys Acta 569:184
71. Kovacs GL, Bohus B, Versteeg DHG, De Kloet R, De Wied D (1975) Brain Res 175:303; Fukunari A, Kado K, Sakai Y, Yoshimoto T, Ishiura S, Suzuki K, Nakajima T (1994) Neurosci Lett 176:201
72. Toda M, Ochuchida S, Ohno H (1989) European Patent EP 0345 428 A1
73. Yoshimoto T, Kado K, Matsubara F, Koriyama N, Kaneto H, Tsuru D (1987) J Pharmacobio-Dyn 10:730
74. Rennex D, Hemmings BA, Hofsteenge J, Stone SR (1991) Biochemistry 30:2195; Higuchi N (1994) Des Enzyme Inhib Drugs 2:766
75. Saito M, Hashimoto M, Kawaguchi N, Shibta H, Fukami H, Tanaka T, Higuchi N (1991) J Enzyme Inhib 5:51
76. Yoshimoto T, Tsuru D, Yamamoto N, Ikezawa R, Furukawa S (1991) Agric Biol Chem 1:37
77. Portevin B, Benoist A, Rémond G, Hervé Y, Vincent M, Lepagnol J, de Nanteuil G (1996) J Med Chem 39:2379
78. Ueda K, Xia JZ, Doke N, Nakatsuka S (1992) Tetrahedron Lett 33:5377
79. Burroughs LF (1957) Nature (London) 179:360; Vahatalo ML, Virtanen AI (1957) Acta Chem Scand 11:741
80. Kunishi AT, Lieberman M, Mapson LW, Wardale DA (1966) Plant Physiol 41:376; Boller T, Herner R, Kende H (1979) Planta 145:293
81. Lieberman M (1979) Ann Rev Plant Physiol 30:533
82. Wiesendanger R, Martinoni B, Boller T, Arigoni D (1986) J Chem Soc Chem Commun: 238
83. Luerssen K, Naumann K, Schroeder R (1979) Z Pflanzenphysiol Bd 22:285; Kouse JR, Kende M (1979) Planta 146:293
84. Adams DO, Yang SF (1979) Proc Natl Acad Sci USA 76:170; Adlington RM, Aplin RT, Baldwin JE, Rawlings BJ, Osborne D (1982) J Chem Soc Chem Commun: 1086; Prochazka Z, Budesinky M, Smolikova J, Troska P, Jost K (1982) Collect Czech Chem Commun 47:2291; O'Donnell MJ, Bruder WA, Eckrich TM, Shullenberger DF, Staten GS (1984) Synthesis: 127
85. Walsh CT, Pascal RA, Johnston M, Raines R, Dikshit D, Krantz A, Honma M (1981) Biochemistry 20:7509
86. Liu H, Auchus R, Walsh CT (1984) J Am Chem Soc 106:5335
87. Mitchell RE (1985) Phytochemistry 24:1485; Mitchell MC, Pirrung M, McGeehan GM (1987) Phytochemistry 26:2595
88. Rance MJ, Ruddock JC, Pacey MS, Cullen WP, Huang LH, Jefferson MT, Whipple Maeda H, Tone J (1989) J Antibiot 42:206
89. Fox KR (1990) J Antibiot 43:1307
90. Ichihara A, Shiraishi K, Sato H, Sakamura S, Nishiyama K, Sakai R, Furnsaki A, Matsumoto T (1977) J Am Chem Soc 99:636; Mitchell RE (1982) Physiol Plant Pathol 20:83; Mitchell RE, Young H (1978) Phytochemistry 17:2028
91. Sakai R (1978) Proc. 11th Pesticide Sci. Symp. Sapporo, Japan, p 21; Shiraishi K, Konama K, Sato H, Ichihara A, Sakamura S, Nishiyama K, Sakai R (1979) Agric Biol Chem 43:1753
92. Boland W, Hopke J, Donath J, Nüske J, Bublitz F (1995) Angew Chem 107:1715; Angew Chem Int Ed Engl 34:1600
93. Parry RJ, Majoti R (1986) J Am Chem Soc 108:4681; Parry RJ, Lin MT, Walker AE, Mhaskar S (1991) J Am Chem Soc 113:1849
94. Guy M, Kende H (1984) Planta 160:281; Hoffmann NE, Yang S, Ichihara A, Sakamura S (1982) Plant Physiol 70:195

95. a) Wakamiya T, Nakamoto H, Shiha T (1984) Tetrahedron Lett 25:4411; b) Wakamiya T, Oda Y, Fujita H, Shiba T (1986) Tetrahedron Lett 27:2143
96. Salaün J, Marguerite J, Karkour B (1990) J Org Chem 55:4276
97. Fadel A (1990) Synlett: 503
98. Gaucher A, Ollivier J, Marguerite J, Paugam R, Salaün J (1994) Can J Chem 72:1312
99. Gaucher A, Dorizon P, Ollivier J, Salaün J (1995) Tetrahedron Lett 36:2979; Dorizon P, Ollivier J, Salaün J (1996) Synlett 1071; Dorizon P, Su G, Ludvig G, Nikitina L, Ollivier J, Salaün J (1998) Synlett: 483; Darizon P, Su G, Ludvig N, Paugam R, Ollivier J, Salaün J (1999) J Org Chem 64:4712
100. Atlan V, Racouchot S, Rubin M, Bremer C, Ollivier J, de Meijere A, Salaün J (1998) Tetrahedron: Asymmetry 9:1131
101. Aitken DJ, Guillaumne D, Husson HP (1993) Tetrahedron 49:6375
102. Burgess K, Lim D, Ho KK, Ke CY (1994) J Org Chem 59:2179
103. Es-Sayed M, Gratkowski C, Krass N, Meyers AI, de Meijere A (1992) Synlett: 962
104. Es-Sayed M, Devine P, Burgess LE, de Meijere A, Meyers AI (1995) J Chem Soc Chem Commun: 141
105. a) Baldwin JE, Adlington RM, Lajoie GA, Rawlings BJ (1985) J Chem Soc Chem Commun: 1496; b) Hill RK, Prakash SR (1984) J Am Chem Soc 106:795
106. Mc Keabn TA, Hoffman NE, Yang SF (1982) Planta 155:437
107. a) Amrhein N, Schneebeck D, Skorupka H, Tophof S, Stockigt J (1981) Naturwissenschaften 68:619; b) Chkanikov DL, Mikitynk OD, Khokhlov PS, Makoveichuck A Yu (1988) Fiziol Rast (Moscow) 35:122
108. a) Amrhein N, Skompka H, Tophof S (1982) Ger. Offen, DE 3, 122, 240; (1983) Chem Abstr 98:88868w; b) Chen M, Chen J, Zhang L (1982) Huaxue Shijie 23:279
109. Chkanikov DI, Makoveichuck A Yu, Mikitynk OD, Makeev AM (1983) Fiziol Rast (Moscow) 30:406
110. Schroeder R, Luerssen K (1981, 1983) Ger. Offen. DE 3,133,917; (1983) Chem Abstr 99:22004a
111. a) Pirrung MC, McGeehan GM (1986) J Org Chem 51:2103; b) Pirrung MC, McGeehan GM (1985) Angew Chem Int Ed Engl 24:1044; c) Pirrung MC, McGeehan GM (1986) Patent U.S. US 4,851,035; d) (1990) Chem Abstr 112:32151x
112. Pirrung MC, Trinks UP (1989) J Chem Soc Chem Commun: 857
113. a) Elrod LF III, Holt EM, Mapelli C, Stammer CH (1988) J Chem Soc Chem Commun: 252; b) Mapelli C, Elrod LF III, Holt EM, Stammer CH (1989) Tetrahedron 45:4377
114. Switzer FL, Van Halbeek H, Holt EM, Stammer CH (1989) Tetrahedron 45:6091
115. Pirrung MC, Dunlop SE, Trinks UP (1989) Helv Chim Acta 72:1301
116. Bleecker AB, Kenyon WE, Somerville SC, Kende H (1986) Proc Natl Acad Sci USA 83:7755; Nakajima N, Nakagawa N, Imaseki H (1988) Plant Cell Physiol 29:989; Sato T, Theologis A (1989) Proc Natl Acad Sci USA 86:6621; Yip WK, Dong JG, Yang SF (1991) Plant Physiol 95:251
117. Yip W-K, Moore T, Yang SF (1992) Proc Natl Acad Sci USA 89:2475
118. Nakagawa N, Hitoshi M, Kenichi Y, Hidernasa I (1991) Plant Cell Physiol 32:1153
119. Zarembinski TI, Theologis A (1993) Mol Biol Cell 4:363
120. Kornakov M, Tsatsakis A, Shtilman M, Zalukaeva T, Vlahakis J, Assithianakis P, Kurushina N (1993) Proc Int Symp Controlled Release Bioact Mater 20th, 400
121. Babiker AGT, Butler LG, Ejeta G, Woodson WR (1993) Physiol Plant 89:21
122. a) Grombie L, Eliott M (1961) Progr Chem Org Nat Prod 19:121; b) Eliott M, Janes NF (1973) In: Casida JE (ed) Academic Press. New York, p 56
123. a) (1977) Synthetic pyrethroids In: Eliott M (ed) ACS Symposium Series, No. 42. Washington DC; b) Eliott M, James NF (1978) Chem Soc Rev 7:473; c) (1978) Imperial Chemical Industries Ltd. Brit. 1, 537, 499; (1979) Chem Abstr 91:51102w
124. Catteral WA (1980) Ann Rev Pharmacol Toxicol 20:15; Witkop B (1982) Heterocycles 17:431
125. Regel E, Boeckmann K, Buechel KH, Plempel M (1986) Ger Offen. DE 3, 440, 114; (1986) Chem Abstr 105:172468m

126. Regel E, Boeckmann K, Buechel KH, Luerssen K, Brandes W, Konze J, Reinecke P (1986) Ger Offen DE 3, 440, 117; (1986) Chem Abstr 105:133893b
127. Regel E, Luerssen K, Brandes W, Konze J (1985) Ger Offen DE 3, 422, 244; (1986) Chem Abstr 104:186424k
128. Szczepanski H (1984) Eur Pat Appl EP 108,708; (1984) Chem Abstr 101:151888a
129. Boeckmann K, Regel E, Buechel KH, Luerssen K, Konze J, Brandes W (1986) Ger Offen DE 3, 440, 116; (1986) Chem Abstr 105:92922q
130. Kanz E, Buechel KH, Kraatz V, Regel E, Brandes W, Frahberger PE, Luerssen K (1983) Ger Offen DE 3, 145, 864; (1983) Chem Abstr 99:53767f
131. Holmwood G, Kraatz V, Brandes W, Dutzmann S, Haenssler G, Luerssen K (1990) Ger Offen DE 3, 905, 317; (1991) Chem Abstr 114:81839c; Seele R, Kober R, Ammermann E, Lorenz G, Johann J, Rademacher W (1990) Ger Offen DE 3,909,862; (1991) Chem Abstr 114:102007x
132. Yoshida M, Ezaki M, Hashimoto M, Yamashita M, Shigematsu N, Okuhara M, Kohsaka M, Horikoshi K (1990) J Antiobiot 43:748
133. Barrett AGM, Kasdorf K, Tustin GJ, Williams D (1995) J Chem Soc Chem Commun: 1143; Barrett AGM, Hamprecht D, White AJP, Williams DJ (1996) J Am Chem Soc 118:7863
134. Barrett AGM, Doubleday WW, Hamprecht D, Kasdorf K, Tustin GJ, White AJP, Williams DJ (1997) J Chem Soc Chem Commun: 1693; Barrett AGM, Hamprecht D, White AJP, Williams DJ (1997) J Am Chem Soc 119:8608; Barrett AGM, Kasdorf K (1996) J Am Chem Soc 118:11030; Barrett AGM, Doubleday WW, Hamprecht D, Kasdorf K, Tustin GJ (1997) Pure and Appl Chem 69:383
135. Charette AB, Juteau H, Lebel H, Deschenes D (1996) Tetrahedron Lett 37:7925
136. Horikoshi K, Yoshida M (1989) Eur Pat Appl EP 88-302948; Kamuro Y, Kakiuchi T, Yoshida M, Otsuka N, Horoshi K (1990) Jpn Kokai Tokkyo Koho JP 88-253992
137. Dappen MS, Pellicciari R, Natalini B, Monahan JB, Chiarri C, Cordi AA (1991) J Med Chem 34:161
138. Shoji J, Sakazaki R, Kato T, Tori K, Yoshimura Y, Matsuura S (1981) J Antibiot 34:370
139. Umezawa H, Muraoka Y, Fujii A, Naganawa H, Takita T (1980) J Antibiot 33:1079
140. van Tilborg WJM, Schaafsma SE, Steinberg H, De Boer Th (1967) Rec Trav Chim Pays-Bas 86:419; Salaün J, Bennani F, Compain JC, Fadel A, Ollivier J (1980) J Org Chem 45:4129
141. Stork G, Depezay JC, d'Angelo J (1975) Tetrahedron Lett: 389
142. Hanka LJ, Dietz A, Gerpheide SA, Kuentzel SL, Martin DG (1978) J Antibiot 31:1211; Martin DG, Chidester CG, Duchamp DJ, Mizsak SA (1980) J Antibiot 33:902; Martin DG, Biles C, Gerpheide SA, Hanka LJ, Krueger WC, McGovren JP, Mizsak SA, Neil L, Stewart JC, Visser J (1981) J Antibiot 34:1119; Chidester CG, Krueger WC, Mizsak SA, Duchamp DJ, Martin DG (1981) J Am Chem Soc 103:7629
143. (a) Reynolds VL, Molineux IJ, Kaptan DJ, Swenson DH, Hurley LH (1985) Biochemistry 24:6228; (b) Baker BF, Dervan PR (1989) J Am Chem Soc 111:2700
144. Warpehoski MA, Harper DE, Mitchell MA, Monroe TJ (1992) Biochemistry 31:2502
145. Li LH, Kelly RC, Warpehoski MA, McGovren JP, Gebbard I, DeKoning F (1991) Invest New Drugs 9:137
146. Li LH, Swenson DH, Schpok SLF, Kuentzel SL, Dayton CD, Krueger WC (1982) Cancer Res 42:999; Hurley LH, Needham-VanDevanter DR, Lee C-S (1987) J Proc Natl Acad Sci USA 84:6412; Weiland K, Dooley TP (1991) Biochemistry 30:7559; Zsido TJ, Woynarowski JM, Baker RM, Gawron LS, Beerman TA (1991) Biochemistry 30:3733; Sun D, Hurley LH (1992) Biochemistry 31:2822
147. Pratt WB, Ruddon RW (1979) The Anticancer Drugs. Oxford University Press, New York, p 64
148. Warpehoski MA, Harper DE (1994) J Am Chem Soc 116:7573
149. Anchel M, Hervey A, Robbins W (1950) J Proc Natl Acad Sci 36:300; McMorris TC, Anchel M (1963) J Am Chem Soc 85:831; Nakanishi K, Ohashi M, Tada M, Yamada Y (1965) 21:1231; Matsumoto T, Shirahama H, Ichihara A, Tukuoka Y, Kahashi Y, Mori Y, Watanabe M (1965) Tetrahedron 21:2671
150. Kelner MJ, McMorris TC, Taetle R (1990) J Natl Cancer Inst 82:1562

151. Kelner MJ, McMorris TC, Estes L, Starr RJ, Rutherford M, Montoya M, Samson KM, Taetle R (1995) Cancer Res 55:4936
152. Kelner MJ, McMorris TC, Taetle R (1995) Anticancer Res 15:873
153. Kelner MJ, McMorris TC, Estes L, Wang W, Samson KM, Taetle R (1996) Investigational New Drugs 14:161
154. McMorris TC, Hu Y, Yu J, Kelner MJ (1997) J Chem Soc Chem Commun: 315
155. Kinder FR, Kenneth WB (1994) J Org Chem 59:6965; Padwa A, Curtis EA, Sandanayaka VP (1997) J Org Chem 62:1317
156. Harttig U, Anke T, Scherer A, Steglich W (1990) Phytochemistry 29:3942
157. Porter CW, Sufrin JR (1986) Anticancer Res 6:525; Borchardt RT, Creveling CR, Ueland PM (1986) (eds) Biological Methylation and Drug Design. Humana Press, New Jersey
158. Pegg AE, Borchardt RT, Coward JK (1981) Biochem 194:79; Kamatani N, Willis EH, Carson DA (1982) Biochem Biophys Res Commun 104:1335
159. Sufrin JR, Spiess AJ, Karny JF, Kramer DL, Hughes RG, Bernacki RJ, Porter CW (1989) Nucleosides and Nucleotides 8:505
160. Mattia CA, Mazzarella L, Pulti R, Sica D, Zolle F (1978) Tetrahedron Lett 3953; Cho J-H, Djerassi C (1987) J Chem Soc Perkin Trans 1:1307
161. Iguchi K, Fujita M, Nagaoka H, Mitome H, Yamada Y (1993) Tetrahedron Lett 34:6277
162. Buzzetti F, Di Salle E, Lombardi P (1987) Ger Offen De 3,719,913; (1988) Chem Abst 109:110743w
163. Hellmann K, Salsbury AJ, Burrage KS, Leserve AW, James SE (1973) Chemotherapy of Cancer Dissemination and Mutathesis. In: Garattini S, Franchi G (eds) Raven Press, New York, p. 355; Helmann K, James SE, Salsburry AS (1974) Br J Cancer 30:179 and references cited therein
164. Witiak DT, Lee HJ, Goldman HD, Zwilling BS (1978) J Med Chem 21:1194; Lee HJ (1978) Diss Abstr Int B 38:3705
165. Salsbury AJ, Burrage K, Hellmann K (1970) Br Med J 4:344; James SE, Salsbury (1974) Cancer Res 34:839
166. Sunkara PS, Baylin SB, Luk GD (1987) Inhibition of Polyamine Metabolism; Biological Significance and Basis for New Therapies.(eds) McCann PP, Peg AE, Sjoerdsma A. Academic Press, New York, p 121
167. Metcalf BW, Bey P, Danzin C, Jung MJ, Casara P, Vevert JP (1978) J Am Chem Soc 100:2551
168. Casero RA Jr, Celano P, Ervin SJ, Porter CW, Bergeron RJ, Libby PR (1989) Cancer Res 49:3829
169. Porter CW, McManis J, Casero RA Jr, Bergeron RJ (1987) Cancer Res 47:2821
170. Della Ragione F, Pegg AE (1983) Biochem J 213:701
171. Saab NH, West EE, Bieszk NC, Preuss CV, Mank AR, Casero RA, Woster PM (1993) J Med Chem 36:2998
172. Nicholson RI, Borthwick NM, Daniel CP, Syne JS, Davies P (1982) Nonsteroidal Antiestrogens:Molecular Pharmacology and Antitumor Activity. In: Sutherland RL, Jordan VC (ed) Academic Press, New York, p 281
173. Sutherland RL, Watts CK, Hall RE, Ruenitz PC (1987) J Steroid Biochem 27:891
174. Fentiman IS, Powles TJ (1987) Lancet Nov 7:1072
175. Killackey MA, Hakes TB, Pierce VK (1985) Cancer Treat Rep 69:237
176. Jordan V (1982) Hormone Antagonists. In: Agarwal MK (ed) de Gruyter, Berlin, New York, p 109
177. Leclerq G, Devleeschouwer N, Heuson JC (1983) J Steroid Biochem 19:75
178. Pento JT, Magarian RA, King MM (1982) Cancer Lett 15:261
179. Day BW, Magarian RA, Jain PT, Pento T, Mousissian GK, Meyer KL (1991) J Med Chem 34:842; Jain PT, Pento JT, Magarian R (1993) Breast Cancer Res Treat 25:225; Jain PT, Pento JT, Magarian RA (1992) Anticancer Res 12:585; Griffin MT, Magarian RA, Jain PT, Tento JT, Moussissian GK, Graves DC (1992) Anti-Cancer Drug Des 7:49; Jain PT, Pento JT, Magarian RA, Graves DC (1991) Anti-Cancer Drugs 2:487
180. Lerner LJ (1981) Non-Steroidal Antiestrogens. In: Sutherland RC, Jordan VC (eds) Academic Press, Sydney, p 1

181. Jain PT (1992) Diss Abstr Int B 52:4679
182. a) Arnold A, McAuliff JP, Powers LG, Phillillips DK, Beyler AL (1979) Atherosclererosis 32:155; b) Brown DF, Beyler A, Daudiss K (1979) Am J Cardiol 43:409; c) Arnold A, McAuliff JP, Powers LG, Phillillips DK, Beyler AL (1979) Fed Proc Fed Am Exp Biol 38:446
183. Kitatsume T, Takeda M (1990) Jpn Kokai Tokkyo Koto JP 02 49,742 [90 49,742]; (1990) Chem Abstr 113:97086v
184. Zikan V, Semonsky M, Rezabek K (1982) Czech CS 197,616; (1983) Chem Abstr 98:106858Z
185. Ranieri RL, Calton GJ (1978) Tetrahedron Lett: 499
186. Kakiuchi K, Tadaki T, Tobe Y, Odaira Y (1985) Chemistry Lett: 1565
187. Calton GJ, Ranieri RL, Espensade MA (1978) J Antibiot 31:38
188. a) Wang JC (1985) Annu Rev Biochem 54:665; b) Maxwell A, Gellert M (1986) Adv Protein Chem 38:69
189. Potmesil M, Kohn KW (1991) DNA Topoisomerases in Cancer. Oxford University Press, New York
190. Liu LF (1989) Annu Rev Biochem 58:351
191. a) Drlica K, Franco PJ (1988) Biochemistry 27:2253; b) Hooper DC, Wolfson JSN (1991) N Engl J Med 324:384
192. Crobett AH, Guerry P, Pflieger P, Osheroff N (1993) Antimicrob Agents Chemother 37:2599
193. a) Lesher GY, Singh B, Reuman M (1991) United States Patent 5,075,319; b) Wentland MP, Lesher GY, Reuman M, Gruett MD, Singh B, Aldous SC, Dorff PH, Rake JB, Coughlin SA (1992) Proceedings of the Eighty-third Annual Meeting of the American Association for Cancer Research. San Diego CA. May 20–23. Abstract 2577
194. Wentland MP, Lesher, GY, Reuman M, Gruett MD, Singh B, Aldous SC, Dorff PH, Rake JB, Coughlin SA (1993) J Med Chem 36:2801
195. Wentland MP, Aldous SC, Gruett MD, Perni RB, Powles RG, Danz DW, Klingbeil KM, Robinson RG, Rake JB, Coughlin SA (1993) Proceedings of the Eighty-third Annual Meeting of the American Association for Cancer Research. Orlando FL. May 19–22. Abstract 2510
196. Kuo G-H, Eissenstat MA, Wentland MP, Robinson RG, Klingbeil KM, Danz DW, Coughlin SA (1995) Bioorg Med Chem Lett 5:399
197. Hofmann K, Lucas RA (1950) J Am Chem Soc 72:4328; Hofmann K, Jucker O, Miller WR, Young Jr AC, Tansig F (1954) J Am chem Soc 76:1799; Hofmann K, Marco GJ, Jeffrey GA (1958) J Am Chem Soc 80:5717; Tocanne JF (1972) Tetrahedron 28:363; Tocanne JF, Bergmann RG (1972) Tetrahedron 28:373
198. Thiele OW, Lacave C, Asselineau C (1969) Eur J Biochem 7:393
199. Minnikin DE, Polgar N (1967) J Chem Soc Chem Comm: 916; c) Asselineau C, Tocanne G, Tocanne JF (1970) Bull Soc Chim Fr: 1455
200. Domagala JM (1994) J Antimicrob Chemother 33:685
201. Wise R, Andrews JM, Edwards L (1983) J Antimicrob Agents Chemother 23:359
202. Hayakawa I, Hiramitsu T, Tanaka Y (1984) Chem Pharm Bull 32:4907
203. Miyamoto H, Matsumoto J, Chiba K, Egawa H, Shibamori K, Minamida A, Nishimura Y, Okada H, Kataoka M, Fuzita M, Hirose T, Nakano J (1990) J Med Chem 33:1645; Mizuki Y, Fujiwara I, Yamaguchi T (1996) Antimicrob Agents Chemother 40:1875
204. Brighty KE (1991) Patent 91/02526; EU Patent 413455, 1991; (1991) Chem Abstr 115:232216; US patent 5.164.402, 1992; (1993) Chem Abstr 119:117227; Brighty KE, Castaldi MJ (1996) Synlett: 1097; Vilsmaier E, Goerz T (1998) Synthesis: 739
205. a) Gootz TD, McGuirck PR, Moynihan MS, Haskell SL (1994) Antimicrob Agents Chemother 38:130; b) Miyamoto H, Yamashita H, Veda H, Tamaoka H, Ohmori K, Nakagawa K (1995) Bioorg Med Chem 3:1699; c) Klopman G, Fercu D, Renau TE, Jacobs MR (1996) Antimicrob Agents Chemother 40:2637
206. Yoshida T, Yamamoto Y, Orita H, Kakiuchi M, Takahashi Y, Itakura M, Kado N, Mitani K, Yasuda S, Kato H, Itoh (1996) Chem Pharm Bull 44:1074
207. Yoshida T, Yamamoto Y, Orita H, Kakiuchi M, Takahashi Y, Itakura M, Kado N, Yasuda S, Kato H, Itoh Y (1996) Chem Pharm Bull 44:1376

208. Llorente B, Leclerc F, Cedergen R (1996) Bioorganic Med Chem 4:61
209. Fujita M, Egawa H, Miyamoto T, Nakano J, Matsumoto J-I (1996) Chem Pharm Bull 44:987
210. Todo Y, Nitta J, Miyajima M, Fukuoka Y, Ikeda Y, Yamashiro Y, Saikawa I, Narita H (1994) Chem Pharm Bull 42:2055
211. Todo Y, Nitta J, Miyajima M, Fukuoka Y, Yamashiro Y, Nishida N, Saikawa I, Narita H (1994) Chem Pharm Bull 42:2063
212. Renau TE, Sanchez JP, Gage JW, Dever JA, Shapiro MA, Gracheck SJ, Domalaga (1996) J Med Chem 39:729; Renau TE, Gage JW, Dever JA, Roland GE, Joannides ET, Shapiro MA, Sanchez JP, Gracheck SJ, Domalaga JM, Jacobs MR, Reynolds (1996) Antimicrob Agents Chemother 40:2363; Cecchetti V, Clementi S, Cruciani G, Fravolini A, Pagella PG, Savino A, Tabarrini (1995) J Med Chem 38:973; Frigala J, Torrens A, Castrillo JA, Mas J, Vano D, Berrocal JM, Calvet C, Salgado L, Redondo J, Garcia-Granda S, Valanti E, Quintana JR (1994) J Med Chem 37:4195; Ito T, Kohima K, Koizumi K, Nagano H, Nishino T (1994) Biol Pharm Bull 17:927
213. Dürckheimer W, Blumbach J, Lattrell R, Scheunemann KH (1985) Angew Chem Int Ed Engl 24:180
214. Hamashima Y, Kubota T, Minami K, Ishikura K, Konoike T, Yoshioka M, Yoshida T, Nakashimizu H, Motokawa K (1987) J Antibiotics 40:1468
215. Ishikura K, Kubota T, Minami K, Hamashima Y, Nakashimizu H, Motokawa K, Yoshida T (1994) J Antibiotics 47:453
216. Jones MF (1988) Chem Brit: 1122
217. Huryn DM, Okabe M (1992) Chem Rev 92:1745
218. Martin JC (1989) Nucleotide Analogues as Antiviral Agents, ACS Symp. Ser. 401, Washington
219. Harnden MR, Jarvest RL, Bacon TH, Boyd MR (1987) J Med Chem 30:1636; Boyd MR, Bacon TH, Sutton D, Cole M (1987) Antimicrob Agents Chemother 31:1238; Boyd MR, Bacon TH, Sutton D (1988) Antimicrob Agents Chemother 32:358
220. Jacobs GA, Tino JA, Zahler R (1989) Tetrahedron Lett 50:6955
221. Ashton WT, Meurer LC, Cantone CL, Field AK, Hannah J, Karkas JD, Liou R, Patel GF, Perry HC, Wagner AF, Walton E, Tolman RL (1988) J Med Chem 31:2304; Boumchita H, Legraverend M, Huel C, Bisagni E (1990) J Heterocycl Chem 27:1815; Katagiri N, Sato H, Kaneko C (1990) Chem Pharm Bull 38:3184
222. Green GR, Harnden MR, Parratt MJ (1991) Biol and Mec Chem Lett 1:347
223. Isawa T, Nishiyama S, Yamamura S, Kato K, Takita T (1992) J Chem Soc Perkin Trans I: 2519
224. Qui YL, Ksebati MB, Ptak RG, Fan By, Breitenbach JM, Lin JS, Cheng YC, Kern ER, Drach JC, Zemlicka J (1998) J Med Chem 41:10
225. Cheng C, Shimo T, Somekawa K, Baba M (1998) Tetrahedron 54:2031
226. Qui YL, Zemlicka J (1998) Synthesis: 1447
227. Katagiri N, Sato H, Kaneko C (1992) Nucleosides Nucleotides 11:707
228. Hatsuya S, Sekiyama T, Tsuji T, Iwayama S, Okumishi M (1992) Eur Pat Appl EP 502, 690; (1993) Chem Abstr 118:60048 m
229. Norbeck DW, Rosen TJ, Sham HL (1991) US 4,9 88,703; (1991) Chem Abstr 115:50218r
230. Kaneko S, Katagiri S (1992) Jpn Kokai Tokyo Koho JP 0474, 169 [92 74, 169]; (1992) Chem Abstr 117:70271x; Kato K, Takita T (1993) Jpn Kokai Tokyo Koho JP 04,283,571 [92,283, 571]; (1993) Chem Abstr 118:255264 m
231. Ashton W,Walton E, Tolman E (1985) Eur Pat Appl EP 159,264; (1986) Chem Abstr 104:207060
232. Csuk R, von Scholz Y (1994) Tetrahedron 50:10431
233. Chang SH, Bergmeier SC, Frick JA, Bathe A, Rapoport H (1994) J Org Chem 59:5336
234. Tandon M, Singh S, Wiebe LI, Knaus EE, Gati WP, Tempest ML (1991) Drug Des Discovery 7:295; Tandon M, Singh S, Xu L, Kumar P, Wiebe LI, Knaus EE, Gati WP, Tempest ML (1992) Drug Des Discovery 9:79
235. Mitsuya H, Yarchoan R, Broder S (1990) Science 249:1533
236. Lambert JS, Seidlin M, Reichman RC, Plank CS, Laverty M, Morse GD, Knupp C, McLaren C, Pettinelli C, Valentine FF, Dobin E (1990) Engl J Med 322:1333

237. Richman DD, Fischl MA, Grieco MH, Gattlieb MS, Wolderbing PA, Laskin OL, Leedom JM, Groopman JE, Mildvan D, Hirsch MS, Jackson GG, Durack DT, Nusinoff-Lehrman S (1987) Engl J Med 317:192
238. Terrett NK, Bojanic D, Merson JR, Stephenson PT (1992) Bioorg Med Chem Lett 2:1745 and references cited therein
239. Tucker TJ, Lyle TA, Wiscount CM, Britcha SF, Young SD, Sanders WM, Lumma WC, Goldman ME, O'Brien JA, Ball RG, Hommick CF, Schleif WA, Emini EA, Ruff JR, Anderson PS (1994) J Med Chem 37:24; Houpis IN, Molina A, Douglas AW, Xavier L, Lynch J, Volante RP, Reider PJ (1994) Tetrahedron Lett 35:6811
240. Thaisrivongs S, Watenpaugh KD, Howe WJ, Tomich PK, Dolak LA, Chong K-T, Tomich C-S, Tomasselli AG, Turner SR, Strohbach JW, Mulichak AM, Janakiraman MN, Moon JB, Lynn JC, Horng MM, Hinshaw RR, Curry KA, Rothrock DJ (1995) J Med Chem 38:3624
241. Romines KR, Watenpaugh KD, Howe WJ, Tomich PK, Lovasz KD, Morris JK, Janakiraman MN, Lynn JC, Horng M-M, Chong K-T, Hinshaw RR, Dolak LA (1995) J Med Chem 38:4463
242. Mitsuya H, Weinhold KJ, Furman PA, St Clair MH, Nusinoff-Lehrman S, Gallo RC, Bolognesi D, Barry DW, Broder S (1985) Proc Natl Acad Sci USA 82:7096; Yarchoan R, Mitsuya H, Thomas RV, Pluda JM, Hartman NR, Perno C-F, Marczyk KS, Allain J-P, Johns DG, Broder S (1989) Science 245:412
243. Mitsuya H, Broder S (1986) Proc Natl Acad Sci USA 83:1911
244. a) Balzarini J, Kang G-J, Dalal M, Herdewijn P, De Clercq E, Broder S, Johns DG (1987) Mol Pharmacol 32:162; b) Mansuri MM, Starrett JE, Ghazzouli I, Hitchcock MJM, Sterzycki RZ, Brankovan V, Lin T-S, August EM, Prusoff WH, Sommadossi JP, Martin JC (1989) J Med Chem 32:461
245. Merluzzi V, Hargrave KD, Labadia M, Grozinger K, Skoog M, Wu JC, Shih C-K, Eckner K, Hattox S, Adams J, Rosenthal AS, Faanes R, Eckner RJ, Koup RA, Sullivan JL (1990) Science 250:1411
246. a) Miyasaka T, Tanaka H, Baba M, Hayakawa H, Walker RT, Balzarini J, De Clercq EA (1989) J Med Chem 32:2507; b) Tanaka H, Takashima H, Ubasawa M, Sekiya K, Nitta I, Baba M, Shigeta S, Walker RT, De Clercq E, Miyasaka T (1992) J Med Chem 35:337
247. a) Pauwels R, Andries K, Desmyter J, Schols D, Kukla MJ, Breslin HJ, Raeymaeckers A, Van Gelder J, Wiestenborghs R, Heykants J, Schellekens K, Janssen MAC, De Clercq E, Janssen PAJ (1990) Nature (London) 343:470; b) Kukla MJ, Breslin HJ, Pauwels R, Fedde CL, Miranda M, Scott MK, Sherill RG, Raeymaekers A, Van Gelder J, Andries K, Janssen MAC, De Clercq E, Janssen PAJ (1991) J Med Chem 34:746
248. Goldman ME, Nunberg JH, O'Brien JA, Quintero JC, Schleif WA, Freund KF, Gaul SL, Saari WS, Wai JS, Hoffman JM, Anderson PS, Hupe DJ, Emini EA, Stern AM (1991) Proc Natl Acad Sci USA 88:6863
249. a) Romero DL, Busso M, Tan C-K, Reusser F, Palmer JR, Poppe SM, Aristoff PA, Downey KM, So AG, Resnick L, Tarpley WG (1991) Proc Natl Acad Sci USA 88:8806; b) Romero DL, Morge RA, Genin MJ, Biles C, Busso M, Resnick L, Althaus IW, Reusser F, Thomas RC, Tarpley WG (1993) J Med Chem 36:1505; c) Romero DL, Morge RA, Biles C, Berrios-Pena N, May PD, Palmer JR, Johnson PD, Smith HW, Busso M, Tan C-K, Voorman RJ, Reusser F, Althaus IW, Downey KM, So AG, Resnick L, Tarpley WG, Aristoff PA (1994) J Med Chem 37:999
250. Perez-Perez MJ, San-Felix A, Balzarini J, De Clercq E, Camarasa MJ (1992) J Med Chem 35:2988
251. a) Wu JC, Warren TC, Adams J, Proudfoot J, Skiles J, Raghavan P, Perry C, Potoki I, Farina PF, Grob PM (1991) Biochemistry 30:2022; b) Cohen KA, Hopkins J, Ingraham RH, Pargellis C, Wu JC, Palladino DEH, Kinkade P, Warren TC, Rogers S, Adams J, Farina PF, Grob PM (1991) J Biol Chem 266:14670
252. a) Kohlstaedt LA, Wang J, Friedman JM, Rice PA, Steitz TA (1992) Science 256:1783; b) Smerdon SJ, Jager J, Wang J, Kohlstaedt LA, Chirino AJ, Friedman JM, Rice PA, Steitz TA (1994) Proc Natl Acad Sci USA 91:3911

253. Ren J, Esnouf R, Arman E, Somers D, Ross C, Kirby I, Keeling J. Darby G, Jones Y, Stuart D, Stammers D (1995) Nature, Struct Biol 2:293
254. Ding J, Das K, Moereels H, Koymans L, Andries K, Janssen PAJ, Hughes SH, Arnold E (1995) Nature, Struct Biol 2:407
255. Esnouf R, Ren J, Ross C, Jones Y, Stammers D, Stuart D (1995) Nature, Struct Biol 2:303
256. Spence RA, Kati WM, Anderson KS, Johnson KA (1995) Science 267:988
257. a) Larder BA, Darby G, Richman DD (1989) Science 243:1731; b) Gu Z, Gao Q, Li X, Parniak MA, Wainberg MA (1992) J Virol 66:7128; c) Reichman RC, Tejani N, Lambert JL, Strussenberg J, Bonnez W, Bumberg B, Epstein L, Dolin R (1993) Antiviral Res 20:267
258. a) Richman D, Shih C-K, Lowy I, Rose J, Prodanovich P, Goff S, Griffin J (1991) Proc Natl Acad Sci USA 88:11241; b) Richman DD, Havlir D, Corbeil J, Looney D, Inacio C, Spector SA, Sullivan J, Cheeseman S, Barringer K, Pauletti D, Shih C-K, Myers M, Griffin J (1994) J Virol 68:1660
259. a) Nunberg JH, Schleif WA, Boots EJ, O'Brien JA, Quintero JC, Hoffman JM, Emini EA, Goldman ME (1991) J Virol 65:4887; b) Dueweke TJ, Pushkarskava T, Poppe SM, Swaney SM, Zhao JQ, Chen ISY, Stevenson M, Tarpley MG (1993) Proc Natl Acad Sci USA 90:4713
260. Hargrave KD, Proudfoot JR, Grozinger KG, Cullen E, Kapadia SR, Patel UR, Fuchs VU, Mauldin SC, Vitous J, Behnke ML, Klunder JM, Pal K, Skiles JW, McNeil DW, Rose JM, Chow GC, Skoog MT, Wu JC, Schmidt G, Engel WW, Eberlein WG, Saboe TD, Campbell SJ (1991) J Med Chem 34:2231
261. Proudfoot JR, Hargrave KD, Kapadia SR, Patel UR, Grozinger KG, McNeil DW, Cullen E, Cardozo M, Tong L, Kelly TA, Rose J, David E, Mauldin SC, Fuchs VU, Vitous J, Hoermann M, Klunder JM, Raghavan P, Siles JW, Mui P, Richman DD, Sullivan JL, Shih C-K, Grob PM, Adams J (1995) J Med Chem 38:4830
262. Hartmann H, Hunsmann G, Eckstein F (1987) Lancet 1:40
263. Freeman GA, Shaver SR, Rideout JL, Short SA (1995) Bioorg Med Chem 3:447
264. Fogg GE, Foster AC (1983) Neurosci 9:701
265. a) Shinazaki H, Ishida M, Shimamoto K, Ohfurne Y (1989) Brain Res 480:355; b) Pellicciari R, Natalini B, Marinozz M, Selvi L, Chiorri C, Monahan JB, Lanthorn TH, Snyder JP (1988) Neurochem Int 12:26
266. For reviews, see: a) Foster AC, Fogg GE (1987) Nature 329:395; b) Wood PL, Rao TS, Iyengar S, Lanthorn T, Monahan J, Cordi A, Sun E, Vasquez M, Gray N, Contreras P (1990) Neurochem Res 15:217 and references cited therein
267. Monahan JB, Corpus VM, Hood WF, Thomas JW, Compton RPJ (1989) Neurochem 53:370
268. Bonhans DW, Burge BC, McNamara JO (1987) Eur J Pharmacol 142:489
269. a) Swell LD, Johnson KM (1988) Eur J Pharmacol 151:165; b) Hood WF, Sun ET, Compton RP, Monahan JB (1989) Eur J Pharmacol 161:281; c) Kessler M, Terramon T, Lynch G, Bandry M (1989) J Neurochem 52:1319; d) Foster AC, Kemp JA (1989) J Neuroscience 9:2191
270. Hood WF, Compton RP, Monahan JB (1989) Neurosci Lett 98:91
271. Marvizon JCG, Lewin AH, Skolnick P (1989) J Neurochem: 992
272. Dappen MS, Markos CS, Gresk CJ, Cordin AA (1990) J Labell Comp and Radiopharmaceuticals 28:841
273. a) Medrum B (1985) Clin Sci 68:113; b) Schwartz R, Foster AC, Freanch ED, Whetsell WO, Kohler C (1984) Life Sci 35:19
274. Youdim MBH, Kerem D, Duvdevani Y (1988) Eur J Pharmacol 150:38
275. Monahan JB, Handelmann GE, Hood WF, Cordi AA (1989) Pharm Biochem Behav 34:648
276. a) Fowden L, Smith A, Millington DS, Sheppard RC (1969) Phytochemistry 8:437; b) Fujimoto Y, Irreverre F, Karle JM, Karle IL, Witkop B (1971) J Am Chem Soc 93:3471
277. Kurokawa N, Ohfune Y (1985) Tetrahedron Lett 26:83; Yamanoi K, Ohfune Y (1988) Tetrahedron Lett 29:1181
278. Nakajima T, Nomoto K, Ohfune Y, Shiratori Y, Takemoto T, Takenchi H, Watanabe K (1985) Br J Pharmac 86:645
279. Watkins JC, Olverman HJ (1987) Trends Neurosci 10:265; Lehmann J, Schneider JA, Williams M (1987) Annu Rep Med Chem 22:31

280. Pellicciari R, Natalini B, Marizzoni M, Selvi L, Chiarri C, Monahan JB, Lanthorn TH, Snyder JF (1988) In: Cavalheiro EA, Lehman J, Turski L (eds) Frontiers in Excitatory Amino Acids Res AR Liss, Inc New York, p 67
281. Dappen MS, Pellicciari R, Natalini B, Monaham JB, Chiarri C, Cordi AA (1991) J Med Chem 34:161
282. a) Malin DH, Payza K, Lake JR, Corriere LS, Benson TM, Smith DA, Baugher RK, Hon K-K, Burgess K (1993) Peptides 14:47; b) Malin DH, Lake JR, Ho K-K, Corriere LS, Garber TM, Waller M, Benson T, Smith DA, Luu T-A, Burgess K (1993) Peptides 14:731
283. Campbell MM, Horwell DC, Mahon MF, Pritchard MC, Walford SP (1993) Bioorg Med Chem Lett 3:667
284. For recent reviews see: a) Gardner CRA (1992) Prog Neuro-Psychopharmacol Biol Psychiatry 16:755; b) Gammill RB, Carter DB (1993) Annual Reports Medicinal Chemistry. In: Bristol JA, McCall JM (eds) Academic Press, New York, p 19
285. Sieghart W (1992) TiPS 13:446
286. a) Squires R (1988) (ed) CRC Press: Boca Raton FL 1:II; Haefely W (1990) Benzodiazepines. In: Beaumont G, Brandon S, Leonard BE (eds) Wiley, New York, p 1; c) Haefely W (1988) Partial Agonists of the Benzodiazepine Receptor: From Animal Data to Results in Patients. In: Biggio G, Costa E Chloride (eds) Channels and Their Modulation by Neurotransmitters and Drugs. Raven Press, New York, p 275
287. a) Gardner CR (1988) Prog Neurobiol 31:425; b) Braestrup C, Nielsen M (1986) Receptor Biochemistry and Methodology. In: Olsen RW, Venter JC (eds) Liss, New York, p 167; c) Bare TM, Resch JF, Patel JB (1987) Anxiolytics and Sedative-Hypnotics. In: Bailey DM (ed) Annual Reports Medicinal Chemistry. Academic Press. New York, p 11; d) Haefely W, Kyburz E, Gerecke M, Mohler H (1985) Recent Advances in the Molecular Pharmacology of Benzodiazepine Receptors and in the Structure-Activity Relationships of Their Agonists and Antagonists. In: Testa B (ed) Advances in Drug Research. Academic Press. New York, p 165; e) Braestrup C, Honore T, Nielsen M, Petersen EN, Jensen LH (1984) Biochem Pharmacol 33:859
288. a) Petersen EN, Jensen LH, Diejer J, Honore T (1986) Pharmacopsychiatry 19:4; b) Haefely W (1991) Eur Neuropsychopharmacol 1:89
289. a) Tang AH, Franklin SR, Himes CS, Ho PM (1991) J Pharmacol Exp Ther 259:248; b) Petke JD, Im KH, Im WB, Blakeman DP, Pregenzer JF, Jacobsen EJ, Hamilton BJ, Carter DB (1992) Mol Pharmacol 42:294
290. Jacobsen EJ, TenBrink RE, Stelzer LS, Belonga KL, Carter DB, Im HK, Im WB, Sethy VH, Tang AH, Von Voigtlander PF, Petke JD (1996) J Med Chem 39:158
291. TenBrink RE, Im WB, Sethy VH, Tang AH, Carter DB (1994) J Med Chem 37:758
292. Im WB, Im HK, Pregenzer JF, Hamilton BJ, Carter DB, Jacobsen EJ, TenBrink RE, Von Voigtlander PF (1993) Br J Pharmacol 110:677
293. Mickelson JW, Jacobsen EJ, Carter DB, Im HK, Im WB, Schreur JKD, Sethy VH, Tang AH, McGee JE, Petke JD (1996) J Med Chem 39:4654
294. Jacobsen EJ, TenBrink RE, Stelzer LS, Belonga KL, Carter DB, Im HK, Im WB, Sethy VH, Tang AH, Von Voigtlander PF (1996) J Med Chem 39:158
295. a) Crow T (1980) Br Med J 280:66; b) Tarsy D (1983) Clin Neuropharmacol 6:S9
296. a) Creese I, Burt DR, Snyder SH (1976) Science 192:481; b) Seeman P, Lee T, Chau-Wong M, Wong K (1976) Nature 261:717; c) Nordström A, Frade L, Wiesel F-A, Forslund K, Pauli S, Halldin C, Uppfeldt G (1991) Biol Psychiatry 33:227
297. a) Sokoloff P, Giros B, Martres M-P, Bouthenet M-L, Schwartz J-C (1990) Nature 347:146; b) Van Tol HHM, Bunzow JR, Guan H-C, Sunahara RK, Seeman P, Niznik HB, Civalli O (1991) Nature 350:610
298. Sunahara RK, Guan H-C, O'Dowd BF, Seeman P, Laurier LG, Ng G, George SR, Torchia J, Van Tol HHM, Niznik HB (1991) Nature 350:614
299. a) Bouthenet M-L, Souil E, Martres M-P, Sokoloff P, Giros B, Schwartz J-C (1991) Brain Res 564:203; b) Seeman P (1992) Neuropsychopharmacology 7:261
300. a) Ebstein R, Novick O, Umansky R, Priel B, Osher Y, Blaine D, Bennett ER, Nemanov L, Katz M, Belmaker RH (1996) Nature:Genet 12:78; b) Seeman P, Guan H-C, Van Tol HHM

(1993) Nature 365:441; c) Lahti RA, Evans DL, Stratman NC, Gigur LM (1993) Eur J Pharmacol 236:483
301. Ohmori J, Maeno K, Hidaka K, Nakato K, Matsumoto M, Tada S, Hattori H, Sakamoto S, Tsukamoto S, Usuda S, Mase T (1996) J Med Chem 39:2764
302. Tsang JW, Schmeid M, Nyfeler R, Goodman M (1984) J Med Chem 27:1663
303. Mapelli C, Newton MG, Ringold CE, Stammer CH (1987) Int J Pept Protein Res 30:498
304. Mapelli C, Stammer CH, Lok S, Mierke DF, Goodman M (1988) Int J Pept Protein Res 32:484
305. Zhu YF, Yamazaki T, Tsang JW, Lok S, Goodman M (1992) J Org Chem 57:10744. For a review see: Yamazaki T, Benedetti E, Kent D, Goodman M (1994) Angew Chem Int Ed Engl 33:1437
306. a) Petrillo EW, Ondetti MA (1982) Med Res Rev 2:1; b) Greenlee WJ (1990) J Med Res Rev 10:174
307. Mantlo NB, Chakravarty PK, Ondeyka DL, Siegl PKS, Chang RS, Lotti VJ, Faust KA, Chen T-B, Shorn TW, Sweet CS, Emmert SE, Patchett AA, Greenlee WJ (1991) J Med Chem 34:2919; Weinstock J, Keenan RM, Samanen J, Hempel J, Finkelstein JA, Franz RG, Gaitanopoulos DE, Girard GR, Gleason JG, Hill DT, Morgan TM, Peishoff CE, Aiyar N, Brooks DP, Fredrickson TA, Ohlstein EH, Ruffolo RR, Stack EJ, Sulpizio AC, Weidley EF, Edwards RM (1991) J Med Chem 34:1514; Middlemiss D, Drew GM, Ross BC, Robertson MJ, Scopes IC, Dowle MD, Akers J, Cardwell K, Clark KL, Coote S, Eldred CD, Hamblett J, Hilditch A, Hirst GC, Jack T, Montana J, Panchal TA, Paton JMS, Shah P, Stuart G, Travers A (1991) Biorg Med Chem Lett 1:711
308. Llyod J, Ryono DE, Bird JE, Buote J, Delaney CL, Dejneka T, Dickinson KEJ, Moreland S, Normandin DE, Skwish S, Spitzmiller ER, Waldron TL (1994) Bioorg Med Chem Lett 4:195
309. Ryono DE, Llyod J, Poss MA, Bird JE, Buote J, Chong S, Dejneka T, Dickinson KEJ, Gu Z, Mathers P, Moreland S, Morrison RA, Petrillo EW, Powell JR, Schaeffer T, Spitzmiller ER, White RE (1994) Bioorg Med Chem Lett 4:201
310. Evans RM (1988) Science 240:889
311. Brand N, Petkovich M, Krust A, Chambon P, de The H, Marchio A, Dejean A (1988) Nature 332:850; Krust A, Kastner P, Petkoch M, Zelent A, Chambon P (1989) Proc Natl Acad USA 86:5310
312. Mangelsdorf DJ, Ong ES, Dyck JA, Evans RM (1990) Nature 345:224; Mangelsdorf DJ, Borgmeyer U, Heyman RA, Zhou JY, Ong ES, Oro AE, Kakizuka A, Evans R (1992) Gen Dev 6:329
313. Sporn MB, Roberts AB, Goodman DS (1984) The Retiniolds (eds) Academic Press, Orlando FL, Vols 1–2
314. Hong WK, Itri LM (1994) Retinoids and Human Cancer. In: Sporn MB, Roberts AB, Goodman DS (eds). The Retinoids:Biology, Chemistry, and Medicine, 2nd ed Raven Press, New York, p 597
315. a) de The H, Chomienne C, Lanotte M, Degos L, Dejean A (1990) Nature 347:448; b) Gebert JF, Moghal N, Frangioni JV, Sugarbaker DJ, Neel BG (1991) Oncogene 6:1859
316. Shalita AR, Fritsch PO (eds) (1992) Retinoids: Present and Future. Proceedings of a Symposium Held at the 18th World Congress of Dermatology Supplement to J Am Acad Dermatol 27:part 2
317. a) Smith MA, Parkinson DR, Cheson BD, Friedman MA (1992) J Clin Oncol 10:839; b) Lippman SM, Benner SE, Hong WK (1994) J Clin Oncol 12:851
318. Verstraeten T, Hartzer M, Wilcox DK, Cheng M (1992) Invest Ophthalomol Vis Sci 33:2830
319. Rottman JN, Widom RL, Nadal-Ginard B, Mahdavi V, Karathanansis SK (1991) Mol Cell Biol 11:3814
320. a) Heyman RA, Mangelsdorf DJ, Dyck JA, Stein RB, Eichele G, Evans RM, Thaller C (1992) Cell 68:397; b) Levin AA, Sturzenbecker LJ, Kazmer S, Bosakowski T, Huselton C, Allenby G, Speck J, Kratzeisen Cl, Rosenberger M, Lovey A, Grippo (1992) Nature 355:359
321. Vuligonda V, Lin Y, Chandraratna AS (1996) Bioorg Med Chem Lett 6:213

322. Chapleo CB, Doxey JC, Meyers PL, Roach AG (1981) Br J Pharmacol 74:842P
323. Kilpatrick AT, Brown CC, MacKinnon AC (1992) Biochem 20:113
324. Ruffolo RR, Waddell JE (1982) J Pharmacolo Exp Ther 222:29
325. Brasili L, Pigini M, Marucci G, Quaglia W, Malmusi L, Lanier SM, Lanier B (1995) Bioorg Med Chem 3:1503
326. Musacchio JM, Klein M, Paturzo JJ (1989) Mol Pharmacol 35:1; Hudkins RL, DeHaven-Hudkins DL (1991) Life Sci 49:1229
327. Tortella FC, Musacchio JM (1986) Brain Res 383:314
328. Eddy NB, Friebel H, Hahn K-J, Halbach H (1969) Bull WHO 40:721
329. Wellens D (1967) Arzneim Forsch 7:495
330. Calderon SN, Izenwasser S, Heller B, Gutkind JS, Mattson MV, Su T-P, Newman AH
331. Vicini P, Amoretti L, Ballaberi V, Barocelli E, Chiavarini M (1995) Eur J Med Chem 30:809
332. Gerwick WH, Proteau PJ, Nagle DG, Hamel E, Bloklin A, Slate DL (1994) J Org Chem 59:1243
333. Nagle DG, Geralds RS, Yoo HD, Gerwick WH, Kim TS, Nambu M, White JD (1995) Tetrahedron Lett 36:1189
334. Onoda T, Shirai R, Koiso Y, Iwasaki S (1995) Tetrahedron Lett 36:5765
335. Kuo MS, Zielinski TJ, Cialdella JI, Marschke CK, Dupuis MJ, Li GP, Kloosterman DA, Spilman CH, Marshall VP (1995) J Am Chem Soc 117:10629
336. Yoshida M, Ezaki M, Hashimoto M, Yamashita M, Shigematsu N, Okuhara M, Kohsaka M, Horikoshi K (1990) J Antibiot 18:748
337. For a recent review see: Yalpani M (1996) Chem Ind: 85
338. Barrett AGM, Hamprecht D, White AJP, Williams DJ (1996) J Am Chem Soc 118:7863; Charette AB, Lebel H (1996) J Am Chem Soc 118:10327

Transition Metal Promoted Ring Expansion of Alkynyl- and Propadienylcyclopropanes

Nobuharu Iwasawa[1] · Koichi Narasaka[2]

[1] Department of Chemistry, Tokyo, Institute of Technology, 2-12-1, O-okayama, Meguro-ku, Tokyo, 152-8551, Japan. *E-mail: niwasawa@chem.titech.ac.jp*

[2] Department of Chemistry, School of Science, The University of Tokyo, 7-3-1, Hongo, Bunkyo-ku, Tokyo 113-0033, Japan. *E-Mail: narasaka@chem.s.u-tokyo.ac.jp*

Transition metal promoted ring expansion of alkynyl- and propadienylcyclopropanes is surveyed with emphasis on the $Co_2(CO)_8$-mediated reactions developed by the author's group. A novel rearrangement of 1-(1-alkynyl)cyclopropanols to 2-cyclopenten-1-ones proceeds on complexation of their alkynyl part with $Co_2(CO)_8$. In the case of the reactions of 1-alkynylcyclopropanols with an alkyl substituent on the cyclopropane, either 4-substituted or 5-substituted 2-cyclopenten-1-ones can be selectively obtained by appropriate choice of stereochemistry and protective group of the substrates. This rearrangement is successfully applied to cyclopentenone annelation reactions onto cycloalkenes. The rearrangement proceeds catalytically on addition of triaryl phosphite as ligand. The same type of rearrangement proceeds when 1-[*o*-(1-alkynyl)phenyl]cyclopropanols are employed as substrates; these are converted to 2,3-dihydro-1-naphthalenone derivatives on heating their hexacarbonyldicobalt complexes in 2-propanol. Furthermore, a new type of isomerization-cyclization reaction proceeds to give 3a,4-dihydro-3*H*-cyclopenta[*a*]inden-2-one derivatives when the same reaction is carried out in the presence of 1,4-diazabicyclo[2.2.2]octane (DABCO). A novel transformation of 1-propadienylcyclopropanols into substituted 1,4-hydroquinones is developed utilizing the interaction of the 1,2-propadienyl group and $Co_2(CO)_8$. This reaction is applied to the synthesis of vitamin E and K analogs.

Keywords: Alkynylcyclopropanes, Propadienylcyclopropanes, Octacarbonyldicobalt, Cyclopentenones, Hydroquinones

1 Introduction

Since the pioneering work by Sarel and co-workers on the iron carbonyl promoted transformation of vinylcyclopropanes and related compounds [1], a variety of transition metal complexes have been examined to achieve effective activation of the vinylcyclopropane-cyclopentene rearrangement which usually requires pyrolytic conditions. These reactions have been applied to natural product synthesis in some cases and have already been reviewed in several excellent articles [2–4].

Contrary to the well-established chemistry of the vinylcyclopropanes, the corresponding reactions of alkynyl- and propadienylcyclopropanes have not, until recently, received much attention. We present here a summary of the recent efforts towards the development of transition metal promoted transformation of these molecules with a brief survey of the corresponding thermal reactions.

2 Rearrangement of Alkynylcyclopropanols

Unlike the well-known chemistry of the vinylcyclopropane-cyclopentene rearrangement, there is no general method for the rearrangement of alkynylcyclopropane to cyclopentene derivatives. One specific example is the pyrolysis of 1-ethynyl-2-methylcyclopropane to methylenecyclopentene and other compounds [5]. At 530°C, 1-ethynyl-2-methylcyclopropane (**1**) undergoes a [1,5]-hydrogen shift to give hexa-1,2,5-triene (**2**), which further isomerizes to methylenecyclopentenes **3** and **4** in 38 and 29% yield, respectively (Scheme 1).

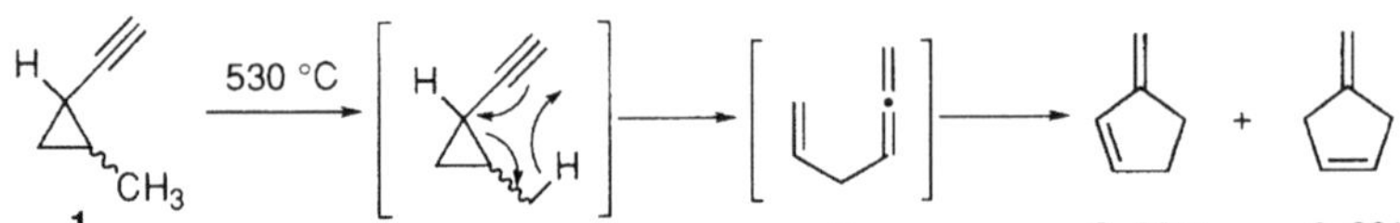

1 **2** **3**, 38% **4**, 29%

Scheme 1

Sarel and co-workers have examined some reactions of alkynylcyclopropanes with iron carbonyl compounds [1]. Treatment of cyclopropylacetylene (**5**) with iron pentacarbonyl under photolytic conditions gives, after cerium(IV) oxidation, isomeric quinones **6** and **7**, derived from two molecules of **5** and two carbonyls with both cyclopropane rings intact [6]. Furthermore, the photoreaction of dicyclopropylacetylene (**8**) with iron carbonyl gives some ten different products depending on the reagents and the reaction conditions, and some of them have the cyclopentenone skeleton formed by the opening of cyclopropane ring coupled with carbonyl insertion [7] (Scheme 2).

It is obvious from these trials that it is rather difficult to realize smooth transformation of alkynylcyclopropanes to cyclopentene derivatives. This difficulty presumably arises from the longer distance between the alkyne terminus and the cyclopropane ring compared with that between the latter and an alkene ter-

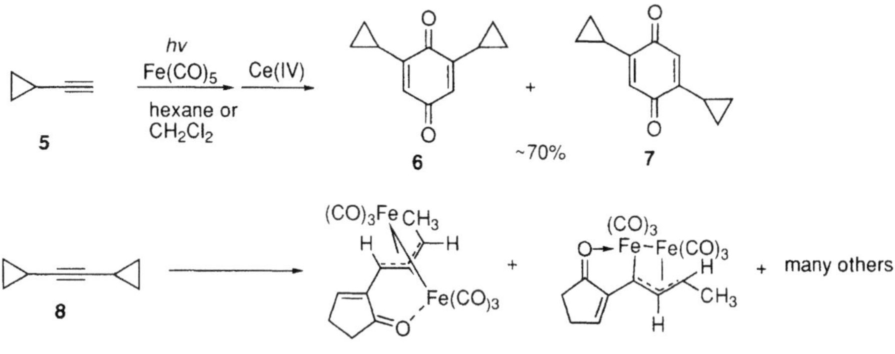

Scheme 2

minus. Thus it is expected that by appropriate complexation of the alkynyl part with a transition metal compound, alkynylcyclopropanes could also be made to rearrange, provided that the complexed transition metal moiety can activate the neighboring cyclopropane ring effectively (Scheme 3).

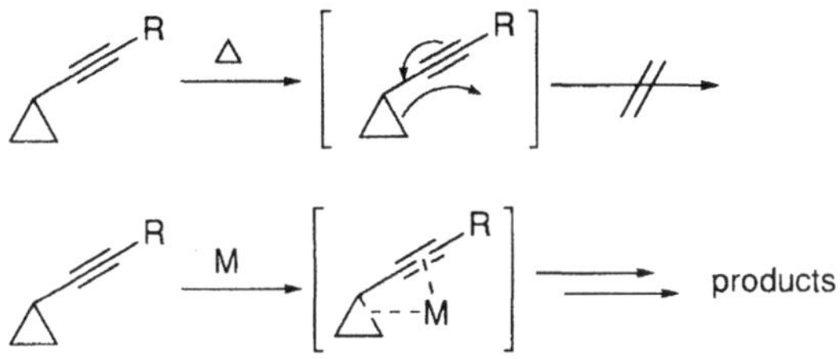

Scheme 3

Alkyne-hexacarbonyldicobalt complexes (alkyne-$Co_2(CO)_6$) can easily be synthesized from alkynes and octacarbonyldicobalt ($Co_2(CO)_8$) simply by mixing, and these complexes are usually chromatographically isolable, stable compounds [8]. These alkyne-$Co_2(CO)_6$ complexes have been employed in organic synthesis mostly for three types of reaction, that is, (1) as a protective group for alkynyl functionality [9]; (2) as a means to stabilize propargylic cations for nucleophilic attack (Nicholas-type reaction) [10]; and (3) as a component in the Pauson–Khand reaction to give 2-cyclopenten-1-ones by reaction with alkenes [11]. Although the unique properties of alkyne-$Co_2(CO)_6$ complexes are obvious from these reactions, it is expected that other useful reactions using these complexes still remain to be explored.

2.1 Stoichiometric Reaction

Treatment of 1-(1-octynyl)cyclopropanol (**11a**) ($R = n\text{-}C_6H_{13}$) with 1.1 molar amounts of $Co_2(CO)_8$ in tetrahydrofuran (THF) at room temperature gives the corresponding $Co_2(CO)_6$ complex **12a**. This complex is stable at room temperature but when the solution is heated to reflux for 8 h under argon, 3-hexyl-2-

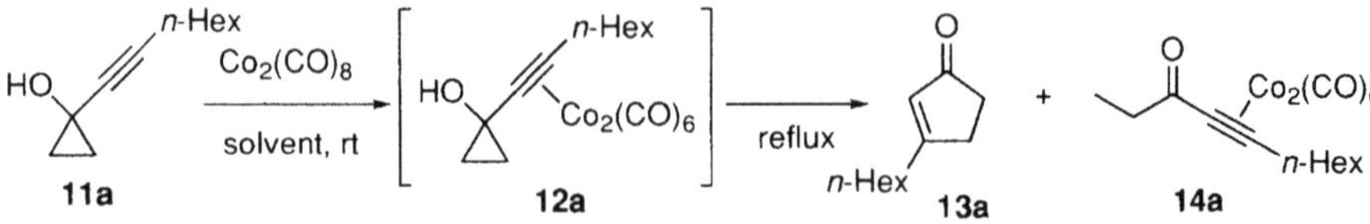

Scheme 4

cyclopenten-1-one (**13a**) is produced in 62% yield (Scheme 4). In general, 1-(1-alkynyl)cyclopropanols themselves are thermally more stable than 1-alkyl- or 1-alkenylcyclopropanols and, in practice, **11a** is recovered unchanged when it is heated in refluxing benzene for 2.5 h. Thus, the complexation of $Co_2(CO)_6$ to the alkynyl part of **11a** is essential for the rearrangement.

Examination of the effect of solvents on this reaction revealed that use of ethereal solvents such as THF and dimethoxyethane (DME) generally favors the formation of the cyclopentenone **13a**, while use of hydrocarbon solvents favors another reaction pathway, that is, rearrangement of **12a** to an ethyl 1-octynyl ketone-$Co_2(CO)_6$ complex **14a**. For example, by carrying out the reaction in refluxing hexane, the complex **14a** is obtained in approximately 70% yield, along with the cyclopentenone **13a** in 13% yield [12, 13].

The results of the reactions of 1-alkynylcyclopropanols with various substituents on the alkyne moiety using a stoichiometric amount of $Co_2(CO)_8$ are summarized in Scheme 5. Not only alkyl-substituted alkyne derivatives, but also aryl- or silyl-substituted alkyne derivatives give the corresponding 3-substituted 2-cyclopenten-1-ones in good to high yields. Interestingly, bulkiness of the substituent on the alkyne has a large effect on the reaction rate. In the case of the reaction of the trimethylsilyl derivative **11c**, it takes about 1.5 h for the reaction in refluxing DME to go to completion. On the other hand, the reaction of the triphenylsilyl derivative **11d** needs only 0.5 h and, furthermore, the reaction goes to completion within 10 min when the triisopropylsilyl (TIPS) derivative **11e** is employed as substrate. Thus, the bulkier the substituent on the alkyne, the faster the reaction proceeds. Various alkynyl substituents containing a functional group can also be employed in this reaction, as shown in Scheme 5.

In the case of transformation of alkynylcyclopropanols with a substituent on the 2-position of the cyclopropane ring, two regioisomers, the 4-substituted and/or 5-substituted 2-cyclopentenones, can be formed depending on which bond of the cyclopropane is cleaved [14].

In the case of the reaction of *cis*-(1*R**, 2*S**)-2-methyl-1-phenylethynylcyclopropanol (*cis*-**15a**), almost equal amounts of 4-methyl-3-phenyl-2-cyclopenten-

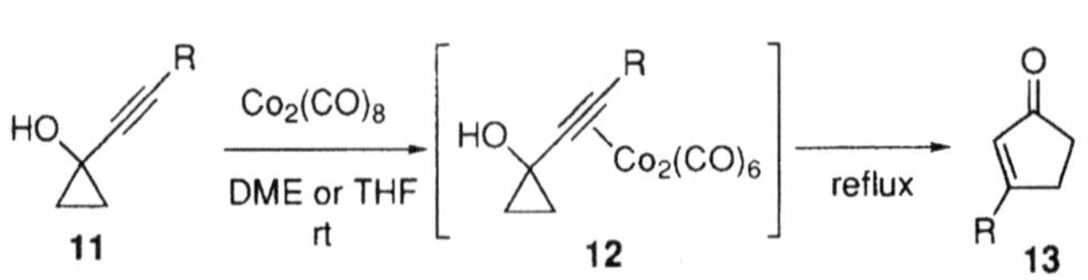

Scheme 5

b: R=Ph, 91%; **c**: R=$SiMe_3$, 77%; **d**: R=$SiPh_3$, 73%; **e**: R=Si*i*-Pr_3, 62%
f: R=CH_2St-Bu, 68%, **g**: R=CH_2CH_2OH, 68%; **h**: R=COOEt, 33%

1-one (**16a**) and 5-methyl-3-phenyl-2-cyclopenten-1-one (**17a**) are obtained in a combined yield of 75%. Examination of the reaction of various silyl ethers of *cis*-**15a** revealed that the reactions proceed in the same manner when siloxycyclopropanes are employed as substrates and that the use of bulky silyl ether groups greatly enhances the regioselectivity of the rearrangement. By using *tert*-butyldimethylsilyl or triisopropylsilyl ethers, and by carrying out the reaction in refluxing THF, the regioselectivity was improved to 94:6 without lowering the yields of the products. In the reaction of *trans*-(1*R**, 2*R**)-2-methyl-1-phenylethynylcyclopropanol (*trans*-**15a**), the 5-substituted isomer **17a** is obtained as the sole isomer, whereas when silyl ethers of *trans*-**15a** are employed, the same 5-substituted isomer **17a** is also obtained with high regioselectivity (Scheme 6). The reactions of 2,2-dimethyl-1-phenylethynylcyclopropanol (**18a**) and its TBS ether revealed that the substituent *cis* to the alkynyl group plays a dominant role in determining the regioselectivity of the reaction. When **18a** itself is employed as substrate, 5,5-dimethyl-3-phenyl-2-cyclopenten-1-one (**20a**) is obtained as the sole product. The reaction of the corresponding TBS ether gives the same regioisomer as the major product with a small amount of the other regioisomer **19a**.

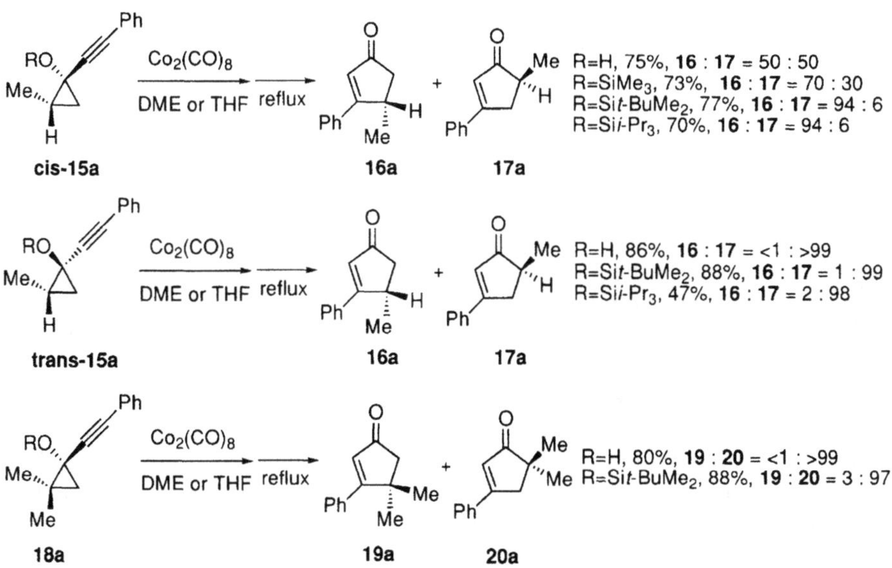

Scheme 6

It is assumed that the reaction proceeds by the oxidative addition of the carbon–carbon bond of the cyclopropane to cobalt species to form a metallacyclic intermediate. The regioselectivity of this reaction is controlled by the relative ease of insertion of the cobalt species into the C1–C2 and the C1–C3 bonds of the cyclopropane (Scheme 7). In the case of the reaction of the (1*R**, 2*R**)-isomer *trans*-**15a**, 5-methyl-3-phenyl-2-cyclopenten-1-one (**17a**) is produced via **TS2** without any serious steric repulsion, while the 4-methyl isomer **16a** must be pro-

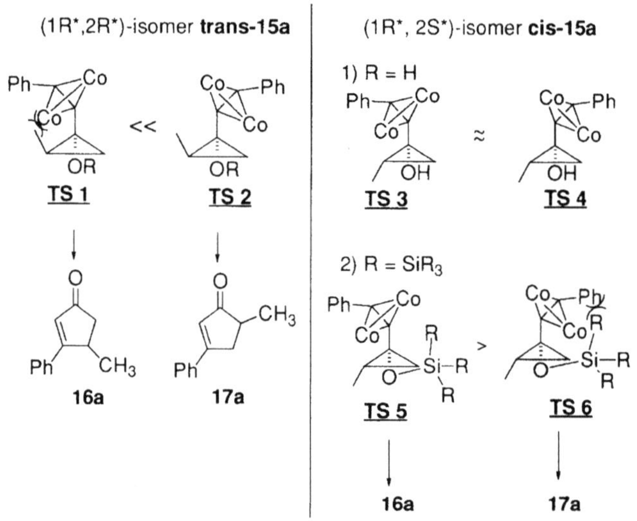

Scheme 7 Carbonyl ligands are omitted from the figure for the sake of simplicity.

duced via **TS1** in which severe steric repulsion between the methyl group and the cobalt moiety occurs. In the case of the reaction of the (1*R**, 2*S**)-isomer *cis*-**15a**, the unprotected derivative *cis*-**15a** gives equal amounts of **16a** and **17a** because the reaction can proceed either via **TS3** or **TS4** with equal efficiency. When the silyl derivatives of *cis*-**15a** are employed, the silyl groups take a position to minimize the repulsion with the methyl group as shown in **TS5** and **TS6**. Of these two conformations, **TS5** is thought to be favored as, in this case, there is no steric repulsion between the phenyl group on the alkyne and the silyl group. Thus, the bulkier the silyl group becomes, the higher the selectivity becomes in favor of the formation of **16a**. As can clearly be seen in Scheme 7, the methyl group *cis* to the alkynyl group plays a dominant role in determining the regioselectivity.

The generality of these observations is supported by the data in Scheme 8. In the case of the reactions of monosubstituted 1-(1-alkynyl)cyclopropanols **15**, 5-substituted cyclopentenones **17** are obtained with complete selectivity when *trans*-**15** are used as substrates. On the other hand, 4-substituted cyclopentenones **16** are obtained with good to high regioselectivity by employing *tert*-butyldimethylsilyl ethers of *cis*-**15** as substrates. Furthermore, 5,5-disubstituted cyclopentenones **20** including a spirocyclic derivative are obtained selectively starting from 2,2-disubstituted 1-(1-alkynyl)cyclopropanols **18**.

The reaction of three diastereomers of 2,3-dimethyl-1-(phenylethynyl)cyclopropanol, i.e. (2*S**, 3*S**)-**21**, (1*S**, 2*R**, 3*S**)-**22** and (1*R**, 2*R**, 3*S**)-**23**, revealed that the rearrangement proceeds with retention of configuration at the migrating carbon [13]. The reaction of the 2,3-*trans*-isomer **21** gives *trans*-4,5-dimethyl-3-phenylcyclopentenone (**24**) almost exclusively in good yield. When the reaction of the 2,3-*cis*-isomer **22** is carried out in refluxing THF for 15 min, the 4,5-*cis*-isomer **25** is obtained as the major product with a substantial amount of recovered **22**, while prolonged reaction time results in the formation of an inse-

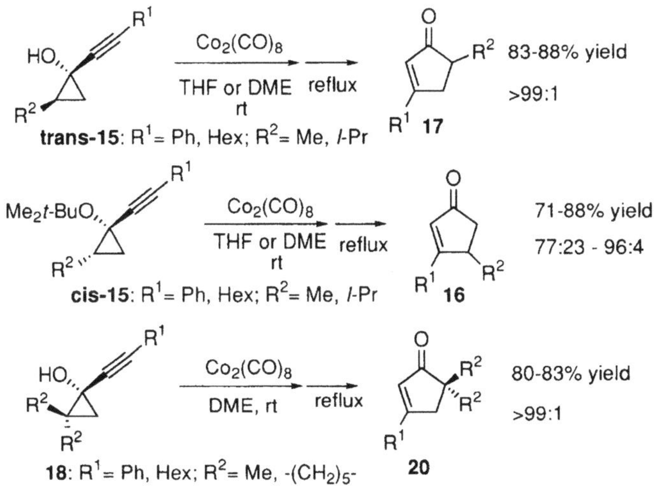

Scheme 8

parable mixture of 4,5-*trans*-isomer **24**, *cis*-isomer **25** and 2,3-dimethyl-4-phenyl-2-cyclopenten-1-one (**26**) in 79% yield in a ratio of 15:33:52. Rearrangement of the other 2,3-*cis*-isomer **23** does not proceed smoothly and decomplexation occurs presumably due to the severe steric hindrance imposed on the alkyne-$Co_2(CO)_6$ moiety by the two methyl groups on the same side of the cyclopropane ring (Scheme 9).

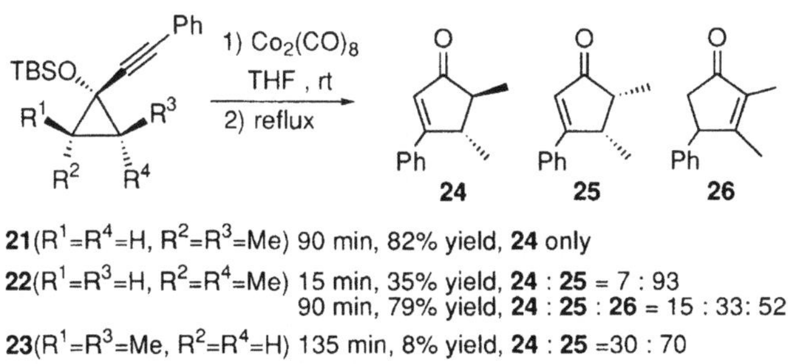

Scheme 9

2.2 Cyclopentenone Annelation Reaction

This reaction has been successfully applied to cyclopentenone annelation onto cycloalkenes. By modifying Danheiser's protocol [15], alkynyl-substituted bicyclo[n.1.0]alkanol derivatives are prepared by the reaction of *gem*-lithiobromocyclopropanes **27** with alkynylborons, as shown in Scheme 10.

As shown in Scheme 11, the alkynylcyclopropanol to cyclopentenone rearrangement of these compounds proceeds smoothly in most cases. Interestingly, the reactions of the silyl ethers of the *exo*-isomers **29c** and **29d** proceed much faster than those of the corresponding *endo*-isomers **28c, d,** and the reaction goes to completion within several hours in refluxing THF. This clear differ-

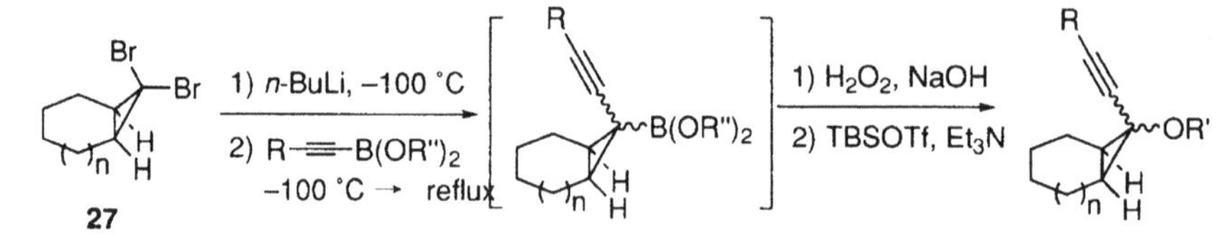

Scheme 10

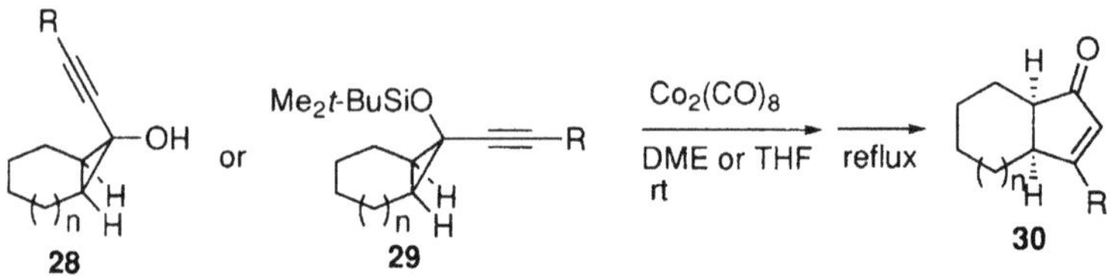

28a (n=1, R = Ph)	DME, refl. 2 h	86 %
28b (n=1, R=*n*-Bu)	DME, refl. 4.5 h	19 %
28c (n=2, R = Ph)	DME, refl. 4 h	75 %
28d (n=2, R=*n*-Bu)	DME, refl.12 h	58 %
29c (n=2, R = Ph)	THF, refl. 2 h	81 %
29d (n=2, R=*n*-Bu)	THF, refl. 3 h	49 %

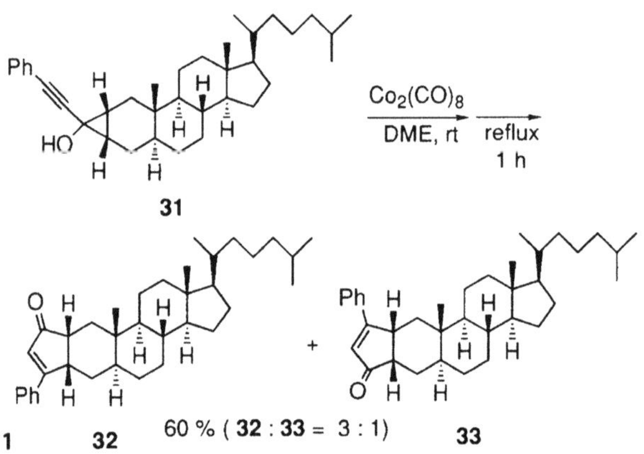

Scheme 11

ence in reactivity of the *endo*- and *exo*-isomers is probably due to easy access of the cobalt moiety of the complexed *exo*-isomers **29c, d** to the carbon–carbon bonds of the cyclopropane. The cobalt moiety is severely hindered from approaching these in the *endo*-isomers **28c, d.** This annelation reaction is further applied to the steroidal skeleton. The rearrangement of **31** gives two isomeric cyclopentenones **32** and **33** in a ratio of about 3:1 in 60% yield.

As dibromocyclopropanes can easily be synthesized by reacting a cycloalkene with bromoform in the presence of a base [16], this method affords an alternative procedure for cyclopentenone annelation onto cyclic alkenes. It should be noted that in the Pauson–Khand reaction, which is probably the most direct cyclopentenone annelation reaction, the reaction using cyclohexene gives the product only in very low yield [11, 17]. Also, the position of the original alkynyl substituent on the product double bond is opposite to that in the present reaction. Thus the two reactions are complementary.

2.3
Catalytic Reaction

All the reactions described so far are carried out using stoichiometric amounts of $Co_2(CO)_8$. However, a coordinatively unsaturated $Co_2(CO)_6$ species would be liberated when the rearrangement proceeds, and if this cobalt species could form an alkyne-$Co_2(CO)_6$ complex with another molecule of 1-(1-alkynyl) cyclopropanol, the reaction should proceed with only a catalytic amount of $Co_2(CO)_8$ [13, 18]. In practice, when 1-phenylethynylcyclopropanol (**11b**) is treated with a 10 or 20 mol% amount of $Co_2(CO)_8$, the rearranged product **13b** is obtained in 43 and 59% yield, respectively. Thus, the reaction in fact proceeds with a catalytic amount of $Co_2(CO)_8$; however, the efficiency is low and a complex mixture of by-products is also obtained. As this low efficiency could be ascribed to instability of the unsaturated cobalt species, the catalytic reaction could be made more efficient by the addition of a stabilizing additive.

When typical phosphines or phosphites are added to 1-(phenylethynyl) cyclopropanol **11b**-$Co_2(CO)_6$ complex **12b**, complete formation of new complexes are observed within a few minutes at reflux temperature, and the reactions proceed from these new complexes to give the cyclopentenone **13b**. In the reaction using triphenyl phosphite as an additive, the newly formed complex is shown on the basis of integration of its NMR spectrum to be the ligand-exchanged complex **34b** (R = OPh) where two carbonyl ligands of the complex **12b** have been exchanged for two molecules of triphenyl phosphite [8]. These results indicate that the reaction also proceeds from the ligand-exchanged complexes **34b** with comparable yields to the reaction of the hexacarbonyldicobalt complex **12b** (Scheme 12).

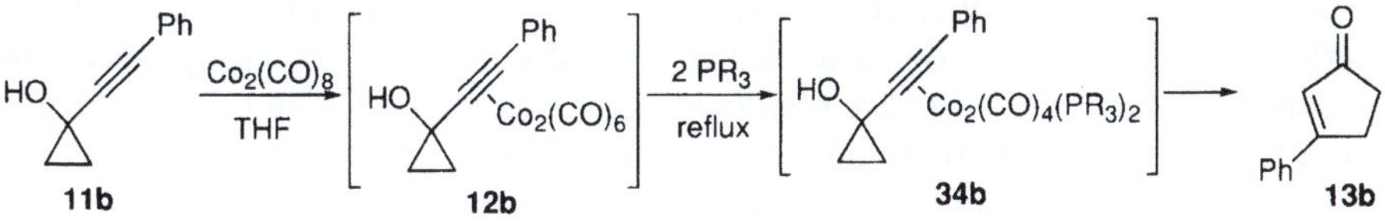

Scheme 12

Examination of the reactions using a 10 mol% amount of $Co_2(CO)_8$ and 20 mol% amount of phosphines or phosphites revealed that with triphenyl phosphite as an additive, **13b** is obtained in 91% yield, just as in the stoichiometric reaction. In these reactions, the presence of ligand-exchanged complex **34b** is observed during the reaction, and $Co_2(CO)_4(P(OPh)_3)_2$ is thought to act as the real catalyst. Use of a bulkier aryl phosphite promotes the reaction more efficiently, and when tri(*o*-tolyl) phosphite is used as the additive, the reaction goes to completion within 2 h even with only a 5 mol% amount of $Co_2(CO)_8$ and 10 mol% amount of phosphite, giving the product in 84% yield. Furthermore, when a 10 mol% amount of tri(*o*-isopropylphenyl) phosphite is added, the reaction is dramatically accelerated and the reaction is complete after 15 min, the cyclopentenone **13b** being obtained in 95% yield (Scheme 13).

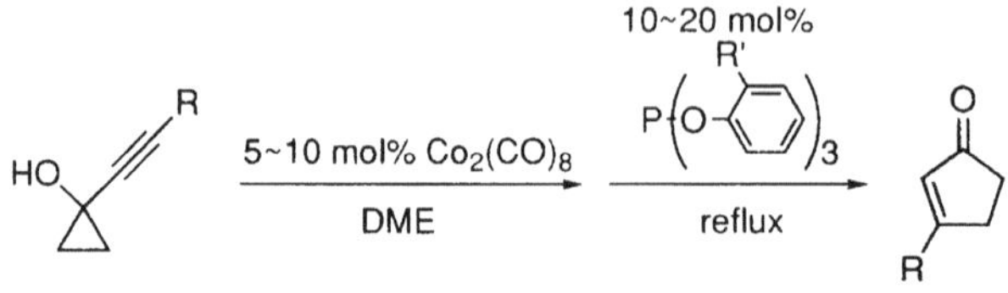

R=Ph, R'=H, reflux 420 min, 76%
R'=Me, reflux 120 min, 84%
R'=*i*-Pr, reflux 15 min, 95%

R=*n*-Hex, R'=*i*-Pr, reflux 30 min, 82%
R=CH_2CH_2OSi*t*-$BuMe_2$, R'=*i*-Pr, reflux 30 min, 85%
R=$SiMe_3$, R'=*i*-Pr, reflux 30 min, 63%

Scheme 13

The reactions of various 1-(1-alkynyl)cyclopropanols were examined using 5–10 mol% of $Co_2(CO)_8$ and 10 mol% of tri(*o*-isopropylphenyl) phosphite. 1-Phenylethynylcyclopropanol (**11b**), as well as the alkyl- or silyl-substituted derivatives, gave the corresponding 3-substituted 2-cyclopentenones in good to high yields (Scheme 13).

2.4 Mechanism of the Reaction

Several reports have appeared on the effect of additives on the Pauson–Khand reaction employing an alkyne-$Co_2(CO)_6$ complex. For example, addition of phosphine oxide improves the yields of cyclopentenones [19], while addition of dimethyl sulfoxide accelerates the reaction considerably [20]. Furthermore, it has been reported that the Pauson–Khand reaction proceeds even at room temperature when a tertiary amine *N*-oxide, such as trimethylamine *N*-oxide or *N*-methylmorpholine *N*-oxide, is added to the alkyne–$Co_2(CO)_6$ complex in the presence of alkenes [21]. These results suggest that in the Pauson–Khand reaction generation of coordinatively unsaturated cobalt species by the attack of *oxides* on the carbonyl ligand of the alkyne-$Co_2(CO)_6$ complex [22] is the key step. With this knowledge in mind, we examined further the effect of various other additives on the reaction to obtain information on the mechanism of this rearrangement.

The reaction using **11a** as a substrate in the presence of several oxides as additives revealed that addition of tributylphosphine oxide, hexamethylphosphoric triamide, and dimethyl sulfoxide all accelerate the reaction considerably. Furthermore, when about 10 molar amounts of *N*-methylmorpholine *N*-oxide (NMO) is added to the alkyne–cobalt complex **12b** in THF, the reaction proceeds even at room temperature and cyclopentenone **13b** is obtained in 37% yield accompanied by another rearranged product, the methylenecyclobutanone **35**, obtained in 23% yield as a mixture of (*E*)-and (*Z*)-isomers (Scheme 14). These facts indicate that dissociation of the carbonyl ligand of the alkyne–cobalt complex **12** is the rate-determining step in this rearrangement. This is also supported by the fact that under a CO atmosphere in refluxing THF the reaction is completely suppressed.

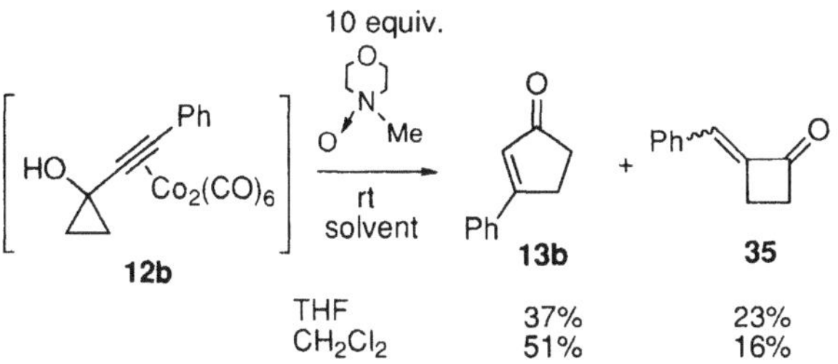

Scheme 14

These results suggest that the rearrangement of 1-(1-alkynyl)cyclopropanols to 2-cyclopentenones is closely related to the Pauson–Khand reaction in its reaction mechanism [11]. Although none of the intermediate complexes has been isolated or detected, we currently suppose the following mechanism for this reaction (Scheme 15). The first step of the reaction is the dissociation of a carbonyl ligand from the alkyne-$Co_2(CO)_6$ complex **12** to generate a coordinatively unsaturated alkyne-$Co_2(CO)_5$ complex **A**, which is formed thermally or by the addition of a tertiary amine *N*-oxide. The coordinatively unsaturated cobalt species thus generated inserts into the carbon–carbon bond of the cyclopropanol to give a four-membered metallacyclic intermediate **B**, which rearranges into a metallacyclohexanone intermediate **C** by C–Co bond cleavage with proton transfer as indicated by path a. Reductive elimination between C1 and C2 gives a cyclopentenone-cobalt carbonyl complex **D**, which in turn gives the free cyclopentenone **13** with liberation of $Co_2(CO)_5L$ (L = CO or a solvent). The fact that the reaction using tertiary amine *N*-oxide at room temperature gives rise to the formation of the methylenecyclobutanone **35** indicates that the reaction can also proceed via

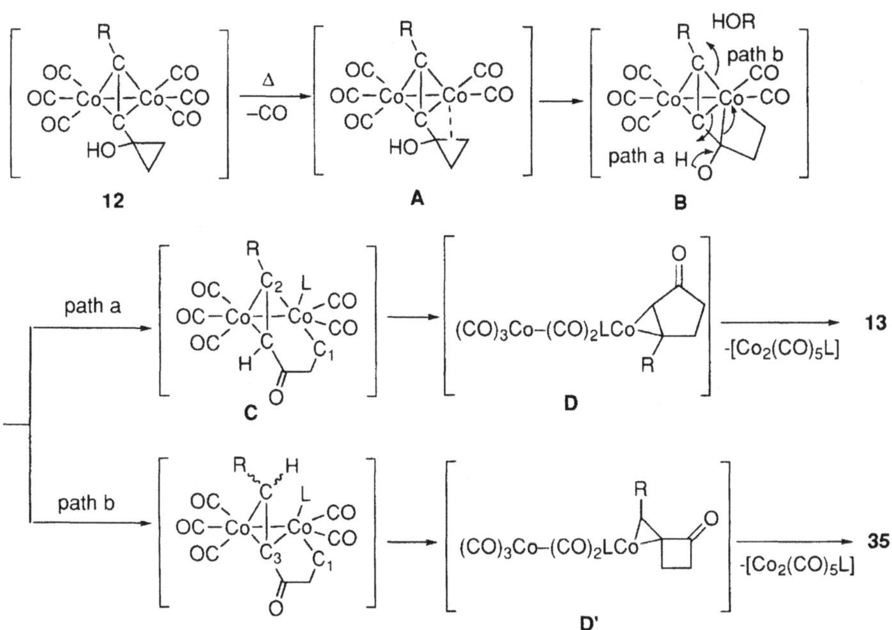

Scheme 15

path b to give **C′** and that, in this case, reductive elimination between C1 and C3 occurs at room temperature to give a methylenecyclobutanone-cobalt carbonyl complex **D′**. The fact that both bulky substituents on the alkyne and bulky ligands accelerate the reaction can be ascribed to facilitation of ligand dissociation due to steric congestion.

Oxidative addition of the carbon–carbon bond of cyclopropanes to zero-valent cobalt species is not in general a facile process. It is assumed that in this reaction the alkynyl part of the molecule works as an anchor for the cobalt carbonyl, which enables an efficient insertion of the cobalt moiety into the proximal carbon–carbon bond of the cyclopropane to proceed. It therefore became a matter of interest to see whether direct connection of the alkynyl part with the cyclopropanol is essential or not for this type of reaction.

3 Reaction of $Co_2(CO)_6$-Complexed 1-[*o*-(1-Alkynyl)phenyl]cyclopropanols

1-[*o*-(1-Alkynyl)phenyl]cyclopropanols **36**, in which the alkyne and the cyclopropane ring are connected through a benzene ring, were chosen as substrates. If the same type of reaction proceeds, **36** should give 2,3-dihydro-1-naphthalenone derivatives **37**, as shown in Scheme 16.

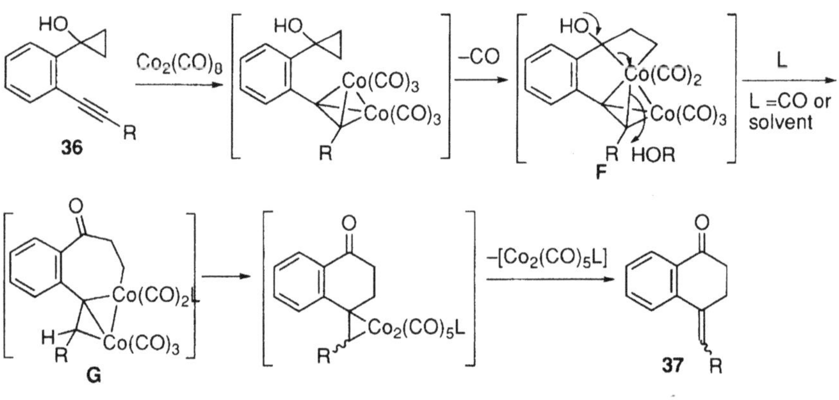

Scheme 16

In fact, the reaction proceeded as expected [23]. Thus, by heating the 1-[*o*-(1-alkynyl)phenyl]cyclopropanol complexes **36**-$Co_2(CO)_6$ in refluxing 2-propanol, 2,3-dihydro-1-naphthalenone derivatives **37** are obtained as a mixture of (*E*)- and (*Z*)-isomers in moderate yield accompanied by a substantial amount of an ethyl ketone derivative formed by ring opening of the cyclopropanol moiety. Furthermore, when an analogous naphthyl derivative **38** was employed, the reaction proceeded cleanly and the 2,3-dihydrophenanthren-1-one derivative **39** was obtained in 83% yield (Scheme 17). The obvious difference in reactivity between phenyl and naphthyl derivatives is probably due to the presence of hydrogen at the *peri* position of the latter. To avoid steric repulsion between the alkyne-$Co_2(CO)_6$ moiety and this hydrogen, the molecule adopts a conforma-

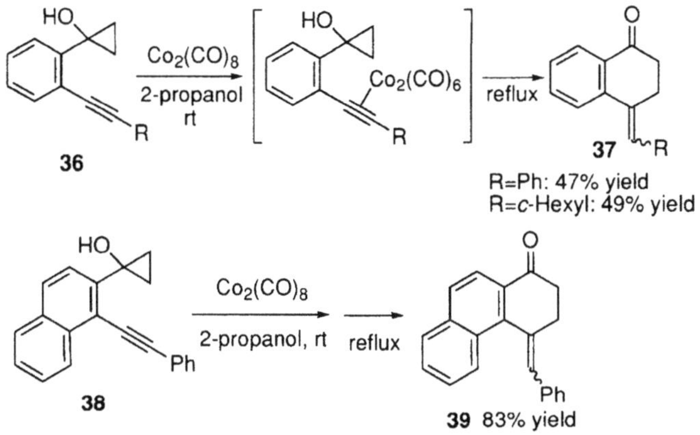

Scheme 17

tion more suitable for insertion of the cobalt carbonyl into the carbon–carbon bond of the cyclopropane.

When 1-[*o*-(phenylethynyl)phenyl]cyclopropanol-$Co_2(CO)_6$ complex (**36**) is heated at 50 °C in 2-propanol under argon in the presence of DABCO, a completely different product, 3a,4-dihydro-3*H*-cyclopenta[*a*]inden-2-one derivative **40**, is produced as a 95:5 diastereomeric mixture in 72% yield. As shown in Scheme 18, not only aryl-substituted alkynyl derivatives, but also alkyl-substituted alkynyl derivatives, give the corresponding cyclopenta[*a*]inden-2-one derivatives **40** in moderate to good yields.

When the reaction of a triethylsilyl-substituted derivative **42** is examined, an allylic alcohol **43** is isolated albeit in low yield. This result suggests that in the presence of a tertiary amine isomerization of the cyclopropanol to an allylic al-

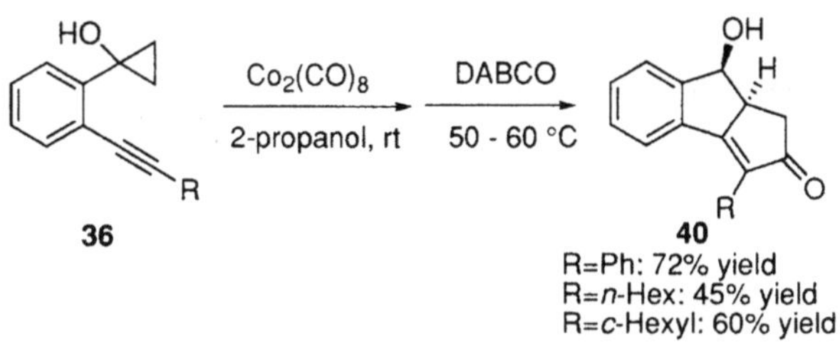

Scheme 18

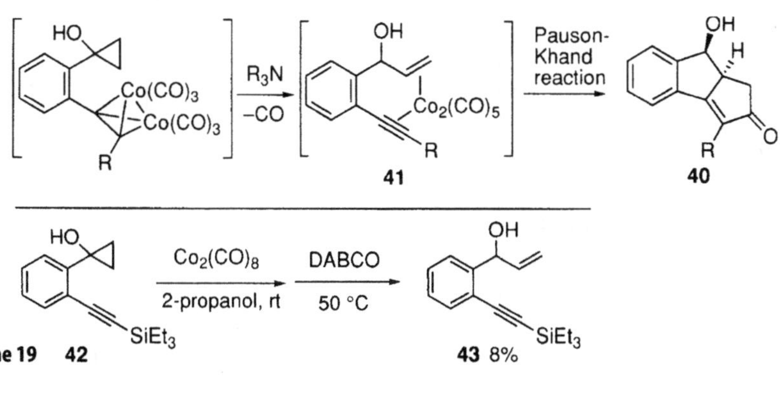

Scheme 19

cohol occurs, which is promoted by the neighboring alkyne-$Co_2(CO)_6$ moiety. The enyne-cobalt complex **41** produced then undergoes a Pauson–Khand reaction to give the product **40** (Scheme 19). There are a few examples of transition metal catalyzed isomerizations of silyl ethers of cyclopropanols to silyl ethers of allylic alcohols using Rh or Pt complexes [24]. This reaction is probably the first example of the transition metal catalyzed rearrangement of unprotected cyclopropanols to allylic alcohols. Although the role of the tertiary amine is not obvious, it is noteworthy that two different types of products, that is 2,3-dihydro-1-naphthalenones and 3a,4-dihydro-3*H*-cyclopenta[*a*]inden-2-ones, are obtained depending on the presence or absence of the tertiary amine.

4 Transformation of Propadienylcyclopropanes

4.1 Rearrangement of Propadienylcyclopropanes to Methylenecyclopentenes

Thermal rearrangement of propadienylcyclopropanes to methylenecyclopentenes has been examined in several cases; however, selective transformation to the product has not necessarily been easy due to the harsh reaction conditions required for the rearrangement. The first example of this type of reaction was reported by Dewar, Fonken, and co-workers in a paper on the kinetics of the thermal reaction of 3-cyclopropyl-1,2-butadiene (**44**), and the reaction was found to proceed much faster (activation energy difference 8.2 kcal) than that of the corresponding vinylcyclopropane [25]. Several examples have appeared since this initial work, most of which have dealt with the mechanistic aspect of the reaction, but none of them has reached a synthetically useful level [26]. For example, thermal reaction of 3-(2-methylcyclopropyl)-1,2-butadiene (**45**) gives a mixture of five products, as shown in Scheme 20 [27].

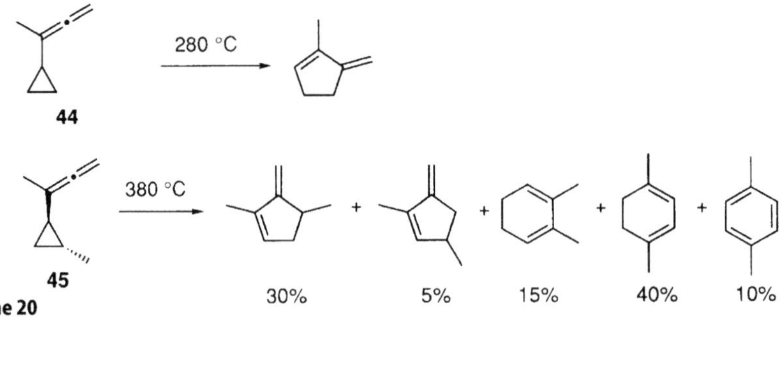

Scheme 20

Recently, Hayashi and Saigo reported the first example of transition metal catalyzed rearrangement of allenylcyclopropanes to methylenecyclopentenes [28]. Heating allenylcyclopropane (**46a**) (R^1 = H, R^2 = $COOCH_2Ph$) in refluxing benzene for 1.5 h in the presence of $RhCl(PPh_3)_3$ gave the corresponding methylenecyclopentene in 88% yield (Scheme 21). When the cyclopropane ring

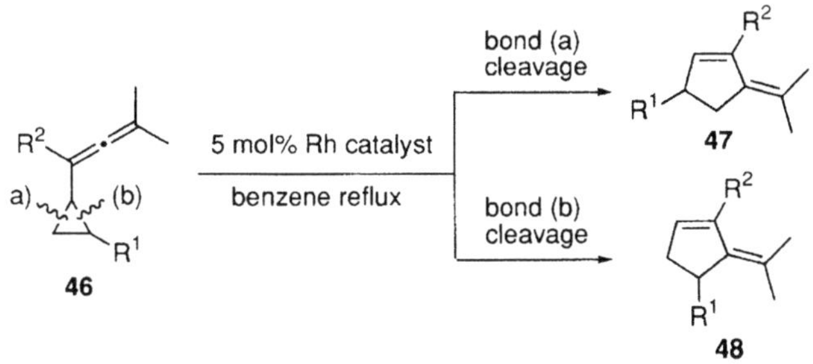

R^1=H, R^2=COOBn, $RhCl(PPh_3)_3$, 88%
R^1=Ph, R^2=COOMe, $[Rh(PPh_3)_3]^+BF_4^-$, 98%, **47** : **48**=92 : 8
R^1=Ph, R^2=COOMe, $[Rh(cod)_2]^+BF_4^-$, 95%, **47** : **48**=5 : 95

Scheme 21

bears an alkyl or aryl substituent, the regioselectivity of the cyclopropane ring cleavage is dependent on the type of the substituents R^1 and R^2 and the nature of the catalyst. For example, the reaction of a substrate with a phenyl group on the cyclopropane ring (R^1) and a methoxycarbonyl group on the allene moiety (R^2) gives **47** as the major product when $[Rh(PPh_3)_3]^+BF_4^-$ is employed, while the same reaction using $[Rh(cod)_2]^+BF_4^-$ gives the other regioisomer **48** with high selectivity.

4.2
Carbonylative Ring Expansion of Propadienylcyclopropanes

Murakami and Ito reported a novel iridium-catalyzed carbonylative ring expansion of allenylcyclopropanes [29]. When a mixture of substituted allenylcyclopropane **49a** ($R^1 = R^2 =$ Et, $R^3 = R^4 =$ H) and 5 mol% of $IrCl(CO)(PPh_3)_2$ in xylene is heated at 130°C under 5 atm pressure of CO for 35 h, cyclohexenone **53a** is obtained in 81% yield. The reaction is proposed to proceed through in-

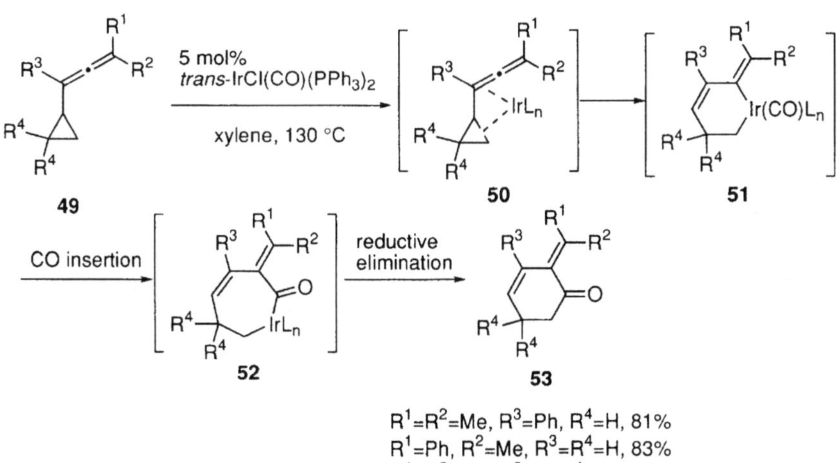

R^1=R^2=Me, R^3=Ph, R^4=H, 81%
R^1=Ph, R^2=Me, R^3=R^4=H, 83%
R^1=R^2=Me, R^3=H, R^4=Ph, 63%

Scheme 22

itial coordination of the allenyl group to the metal, which enables effective activation of the cyclopropane ring. Ring opening by the metal occurs to form a six-membered metallacycle **51**, and then insertion of CO into the Ir–C bond, followed by reductive elimination, affords the product **53** (Scheme 22). Allenylcyclopropanes with a disubstituted allenic terminus afford the products in good yields.

4.3 Transformation of 1-Propadienylcyclopropanols to Hydroquinones

Use of $Co_2(CO)_8$ in reactions involving 1,2-propadienes remains for the most part unexplored. It has been reported that terminal 1,2-propadienes react with $Co_2(CO)_8$ to form unidentified complexes, and that excess 1,2-propadiene is polymerized concurrently [30]. It has also been reported by Nakamura that a novel dimeric complex **54**, in which a carbonyl ligand is connected to the central carbon of 1,2-propadiene, is produced by the reaction of 1,2-propadiene itself with $Co_2(CO)_8$ (Scheme 23) [31]. However, unlike the well-known chemistry of alkyne-$Co_2(CO)_6$ complexes, these 1,2-propadiene-cobalt carbonyl complexes have rarely been applied in synthetic reactions, probably due to their high activity in catalyzing the polymerization of 1,2-propadienes [32].

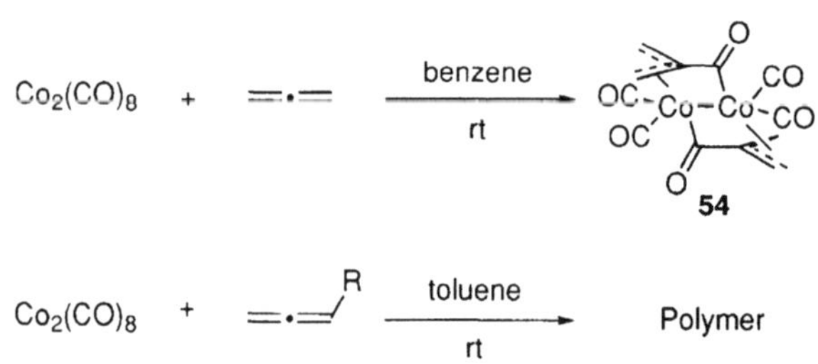

Scheme 23

It was surmised that a similar proximity effect to that observed in the 1-(1-alkynyl)cyclopropanol to cyclopentenone rearrangement might well induce novel ring transformation reactions when cyclopropanols containing an allenic, instead of an acetylenic, moiety are exposed to $Co_2(CO)_8$. When 1-(1-phenyl-1,2-propadienyl)cyclopropanol (**55a**) was treated with a 1.1 molar amount of $Co_2(CO)_8$ in THF at 0 °C and subsequently warmed to room temperature under an argon atmosphere, all of the **55a** disappeared within a few hours to give 3-methyl-2-phenyl-1,4-hydroquinone (**56a**) and 3-methyl-2-phenyl-1,4-benzoquinone (**57a**) in 35 and 20% yield, respectively. A much more successful work-up procedure that completely suppressed formation of quinone **57a** involved acetylation of the intermediate hydroquinone **56a**. Thus, addition of acetic anhydride and triethylamine to the reaction mixture after complete disappearance of **55a** furnished the acetylated hydroquinone **58a** in 53% yield (Scheme 24) [33, 34]. A plausible mechanism for the reaction is as follows. Combination of **55a** with $Co_2(CO)_8$ may lead to the carbonylated intermediate **59a**, similar to the complex proposed by Nakamura [31]. This complex, however, is not detectable by TLC during the course of the reaction. The cyclopropane ring in **59a**, now

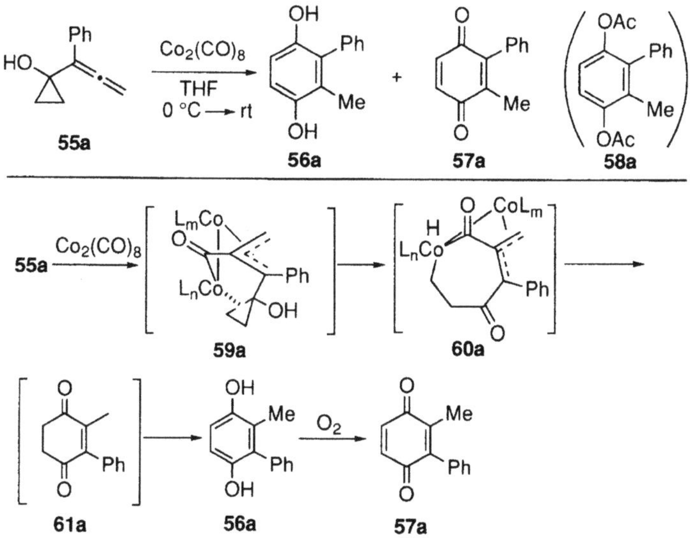

Scheme 24

efficiently activated by the cobalt moiety, undergoes ring expansion to give a metallacyclic intermediate **60a**, which is further transformed to the cyclohexenedione **61a** by reductive elimination. Aromatization of **61a** gives hydroquinone **56a**, which is subsequently oxidized by air to the benzoquinone **57a**. In ethyl acetate and ethereal solvents such as THF good yields of the acetylated hydroquinone **58a** are obtained, and the formation of oligomers can be suppressed by carrying out the reaction under dilute conditions. When the reaction is conducted in 0.01 M THF solution at between 0 °C and room temperature, the acetylated hydroquinone **58a** is obtained in 78% yield.

Reactions of various 1-(1,2-propadienyl)cyclopropanols having a substituent at the 1- or 3-position of the propadienyl moiety **55** proceed smoothly at between 0 °C and room temperature with 1.1 mole amounts of $Co_2(CO)_8$ in either THF or ethyl acetate, and various 2-monosubstituted or 2,3-disubstituted 1,4-hydroquinone derivatives **58** are obtained in good yields (Scheme 25). In particular, 1-(1,2-propadienyl)cyclopropanol having the *tert*-butyldimethylsilyl group at the 1-position of the 1,2-propadienyl moiety gives a high yield of the silylated hydroquinone.

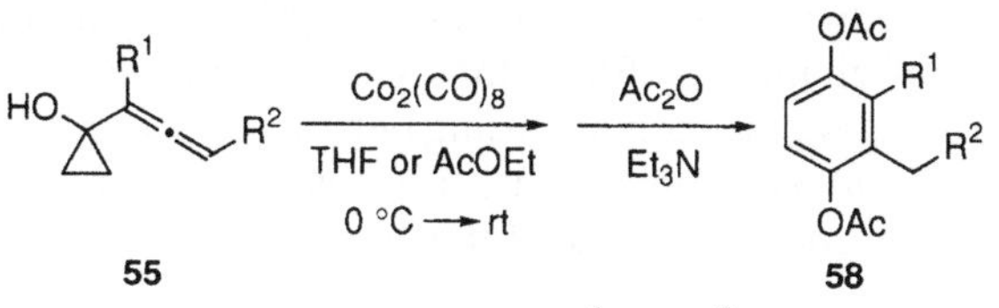

R^1=Ph, R^2=H, 78%
R^1=*n*-Hex, R^2=H, 60%
R^1=*t*-BuMe$_2$Si, R^2=H, 87%
R^1=H, R^2=Ph, 56%

Scheme 25

This reaction was applied to the synthesis of quinonoid natural products [34]. Propargyl bromides **62a** and **62b**, which were prepared from prenyl bromide and phytyl bromide by a standard procedure, were converted to the corresponding aluminum reagents by reaction with powdered aluminum and a catalytic amount of mercuric chloride [35]. Then the iodomagnesium salt of cyclopropanol hemiacetal was treated with these reagents [36] affording the prenyl derivative **63a** and the phytyl derivative **63b** in 49 and 50% yield, respectively (Scheme 26).

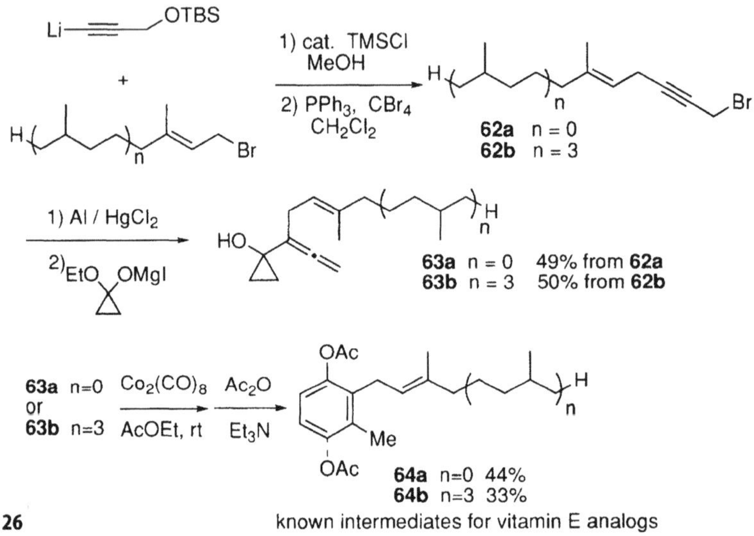

Scheme 26

When the prenyl derivative **63a** was treated with $Co_2(CO)_8$ in ethyl acetate at room temperature for 2 h and then acetic anhydride and triethylamine were added to the reaction mixture, the desired acetylated hydroquinone **64a**, a known vitamin E analog intermediate [37], was obtained in 44% yield. The same reaction employing 1-(1-phytyl-1,2-propadienyl)cyclopropanol (**63b**) gave a known vitamin E analog intermediate **64b** [37] in 33% yield. Benzene analogs of vitamin K were also prepared by carrying out the quenching under oxidative conditions. After the treatment of the prenyl derivative **63a** with $Co_2(CO)_8$ in ethyl acetate, triethylamine was added to the mixture followed by the addition of iron(III) chloride solution to give 2-methyl-3-(3-methylbut-2-enyl)-*p*-benzoquinone (**65a**), a benzoquinone analog of vitamin K_2(5), in 45% yield. The same reaction using the phytyl derivative **63b** gave the corresponding benzoquinone **65b**, a benzoquinone analog of vitamin K_1, in 34% yield. Thus, either vitamin E or vitamin K analogs can be obtained by appropriate choice of the quenching conditions.

References

1. Sarel S (1978) Acc Chem Res 11:204
2. Hudlicky T, Reed JW (1991) Rearrangements of vinylcyclopropane and related systems. In: Trost BM (ed) Comprehensive organic synthesis, vol 5. Pergamon Press, Oxford, p 899; Hudlicky T, Becker DA, Fan RL, Kozhushkov S (1997) Carbocyclic three-and four-membered ring compounds. In: de Meijere A (ed) Houben–Weyl methods of organic chemistry, vol El7c. Thieme, Stuttgart, p 2538
3. Goldschmidt Z, Crammer B (1988) Chem Soc Rev 17:229
4. Hudlicky T, Kutchan TM, Naqvi SM (1985) Org React 33:247
5. Dalacker V, Hopf H (1974) Tetrahedron Lett 15
6. Victor R, Ben-Shoshan R, Sarel S (1973) Tetrahedron Lett 4211
7. Victor R, Usieli V, Sarel S (1977) J Organomet Chem 129:387
8. Kemmitt RDW, Russell DR. (1982) In: Wilkinson G (ed) Comprehensive organometallic chemistry, vol 5. Pergamon Press, Oxford, p 1
9. Seyferth D, Wehman AT (1970) J Am Chem Soc 92:5520; Nicholas KM, Pettit R (1971) Tetrahedron Lett 3475; Melikyan GG, Vostrowsky O, Bauer W, Bestmann HJ, Khan M, Nicholas KM (1994) J Org Chem 59:222; Milgrom LR, Rees RD, Yahioglu G (1997) Tetrahedron Lett 38:4905
10. Varghese V, Saha M, Nicholas KM (1988) Org Synth 67:141; Nicholas KM (1987) Acc Chem Res 20:207; Caffyn AJM, Nicholas KM (1995) Transition metal alkyne complexes: transition metal stabilized propargyl system. In: Abel EW, Stone FGA, Wilkinson G (eds) Comprehensive organometallic chemistry II, vol 12. Pergamon Press, Oxford, p 685; Schreiber SL, Klimas MT, Sammakia T (1987) J Am Chem Soc 109:5749; Nakamura T, Matsui T, Tanino K, Kuwajima I (1997) J Org Chem 62:3032
11. Schore NE (1991) Org React 40:1; Schore NE (1991) The Pauson-Khand reaction. In: Trost BM (ed) Comprehensive organic synthesis, vol 5. Pergamon Press, Oxford, p 1037; Pauson PL (1985) Tetrahedron 41:5855; Schore NE (1995) Transition metal alkyne complexes: Pauson-Khand reaction. In: Abel EW, Stone FGA, Wilkinson G (eds) Comprehensive organometallic chemistry II, vol 12. Pergamon Press, Oxford, p 703
12. Iwasawa N (1992) Chem Lett 473
13. Iwasawa N, Matsuo T, Iwamoto M, Ikeno T (1998) J Am Chem Soc 120:3903
14. Iwasawa N, Iwamoto M (1993) Chem Lett 1257
15. Danheiser RL, Savoca AC (1985) J Org Chem 50:2401
16. Skattebøl L (1963) Acta Chem Scand 17:1683; Last LA, Fretz ER, Coates RM (1982) J Org Chem 47:3211
17. Khand IU, Pauson PL (1978) J Chem Res (M) 168
18. Iwasawa N, Matsuo T (1993) Chem Lett 997
19. Billington DC, Helps IM, Pauson PL, Thomson W, Willison D (1988) J Organomet Chem 354:233
20. Chung YK, Lee BY, Jeong N, Hudecek M, Pauson PL (1993) Organometallics 12:220
21. Shambayati S, Crowe WE, Schreiber SL (1990) Tetrahedron Lett 31:5289; Jeong N, Chung YK, Lee SH, You SE (1991) Synlett 204; Ornum SGV, Cook M (1997) Tetrahedron Lett 3657; Jamison TF, Shambayati S, Crowe WE, Schreiber SL (1997) J Am Chem Soc 119:4353
22. Alper H, Edward JT (1970) Can J Chem 48:1543; Shi YL, Gao YC, Shi QZ, Kershner DL, Basolo F (1987) Organometallics 6:1528
23. Iwasawa N, Matsuo T (1997) Chem Lett 341
24. Ikura. K, Ryu I, Ogawa A, Kambe N, Sonoda N (1989) Tetrahedron Lett 30:6887; Ikura K, Ryu I, Kambe N, Sonoda N (1992) J Am Chem Soc 114:1520
25. Dewar MJS, Fonken GJ, Kirschner S, Minter DE (1975) J Am Chem Soc 97:6750
26. Roth WR, Schmidt T, Humbert H (1975) Chem Ber 108:2171; Minter DE, Fonken GJ (1977) Tetrahedron Lett 4149; Minter DE, Fonken GJ, Cook FT (1979) Tetrahedron Lett 714; Olson LP, Gajewski JJ (1993) Tetrahedron Lett 34:2895
27. Minter DE, Fonken GJ (1977) Tetrahedron Lett 1717
28. Hayashi M, Ohmatsu T, Meng YP, Saigo K (1998) Angew Chem Int Ed Engl 37:837
29. Murakami M, Itami K, Ubukata M, Tsuji I, Ito Y (1998) J Org Chem 63:4

30. Greenfield H, Wender I, Wotiz JH (1956) J Org Chem 21:875; Mushina EA, Makhina TK, Kiriya I, Artamonova SD, Dyshlevskii YN, Krentsel BA, Taber AM (1984) Vysokomol Soedin Ser B 26:654 Chem Abstr (1985) 102:46292z
31. Nakamura A (1966) Bull Chem Soc Jpn 39:543; Caffyn AJM, Mays MJ, Solan GA, Conole G, Tiripicchio A (1993) J Chem Soc Dalton Trans 2345; Sóvágó J, Newton MG, Mushina EA, Ungváry F (1996) J Am Chem Soc 118:9589
32. Kim PJ, Hagihara N (1965) Bull Chem Soc Jpn 38:2022; Trifonov LS, Orahovats AS, Heimgartner H (1987) Helv Chim Acta 70:1070; Ahmar M, Antras F, Cazes B (1995) Tetrahedron Lett 36:4417
33. Iwasawa N, Owada Y, Matsuo T (1995) Chem Lett 115
34. Owada Y, Matsuo T, Iwasawa N (1997) Tetrahedron 53:11069
35. Gandemar M (1956) Ann Chim 13:1161; Daniels RG, Paquette LA (1981) Tetrahedron Lett 22:1579
36. Salaün J, Bennani F, Compain JC, Fadel A, Ollivier J (1980) J Org Chem 45:4129
37. Inoue S, Saito K, Kato K, Nozaki S, Sato K (1974) J Chem Soc Perkin Trans 1 2097

Bicyclopropylidene – A Unique Tetrasubstituted Alkene and a Versatile C_6-Building Block

Armin de Meijere · Sergei I. Kozhushkov · Alexander F. Khlebnikov

Institut für Organische Chemie der Georg-August-Universität Göttingen, Tammannstrasse 2, D-37077 Göttingen, Germany. *E-mail: ameijer1@uni-goettingen.de*

Strain in an organic molecule often correlates with increased reactivity, at least for certain types of reactions. Thus, the elucidation of the chemistry of highly strained alkenes like the unusual tetrasubstituted alkene, bicyclopropylidene (**1**), which is a particularly strained derivative of methylenecyclopropane (**2**), has proved to be fruitful both with respect to synthetic applications as a C_6 building block, as well as understanding certain reaction principles. The different, steadily improved methods which have been developed for the preparation of this unusual alkene in the last thirty years as well as rather recent methods for the synthesis of functionally substituted, and spirocyclopropanated derivatives as well as bis(bicyclopropylidenyls) are presented. The special aspects of molecular geometry in bicyclopropylidenes as well as strain and its influence on chemical properties are discussed.The rich chemistry of bicyclopropylidene beginning with its well-known thermal rearrangement and dimerization, its dihalocarbene additions all the way to its recently developed organometallic chemistry, especially its reactions under the catalysis of palladium and other transition metals, is covered. Finally, some synthetically useful chemical transformations of bicyclopropylidene derivatives, e.g., synthetic approaches to the cyclopropanated analogs of natural products, are presented.

Keywords: Absolute configuration, Amino acids, Bicyclopropylidene, Coupling reactions, Cycloadditions, Cyclopropanation, Cyclopropanes, Organolithium derivatives, Palladium catalysis, Radical reactions, Small ring polycycles, Spiro compounds, Strain energy, Sulfides

1 Introduction

Bicyclopropylidene (**1**) may be regarded as a more highly strained cyclopropanated derivative of methylenecyclopropane (**2**) (Fig. 1). As a consequence of its high-lying HOMO in addition to its high total strain energy, bicyclopropylidene is uniquely reactive towards a wide range of electrophiles and cyclophiles. Starting from the first preparation of unsubstituted bicyclopropylidene (**1**) in 1970 [1], this uniquely tetrasubstituted, strained alkene has attracted a good deal of attention in theoretical and experimental studies. In the beginning exotic, this compound became more and more accessible and as such a versatile substrate in organic chemistry. Particularly remarkable progress in the development of its chemistry has been made within the last decade.

The chemistry of the closest relative of hydrocarbon **1**– methylenecyclopropane (**2**) – has recently been reviewed extensively [2]. The presence of the second three-membered ring in **1** strongly increases the total strain of this molecule, this enhances its chemical reactivity and leads to specific physical properties. The meanwhile large variety of experimental material obtained for compound **1**, some of its derivatives and analogs prompted us to summarize the results in order to help set the stage for future developments.

1 **2**

Fig. 1. Bicyclopropylidene (**1**) and methylenecyclopropane (**2**)

2 The Syntheses of Bicyclopropylidenes

2.1 Methods for the Preparation of Unsubstituted Bicyclopropylidene

The first report on the preparation of bicyclopropylidene (**1**) achieved by a Simmons-Smith-type monocyclopropanation of the terminal double bond in vinylidenecyclopropane (**3**) (Scheme 1) is only of historical interest [1, 3].

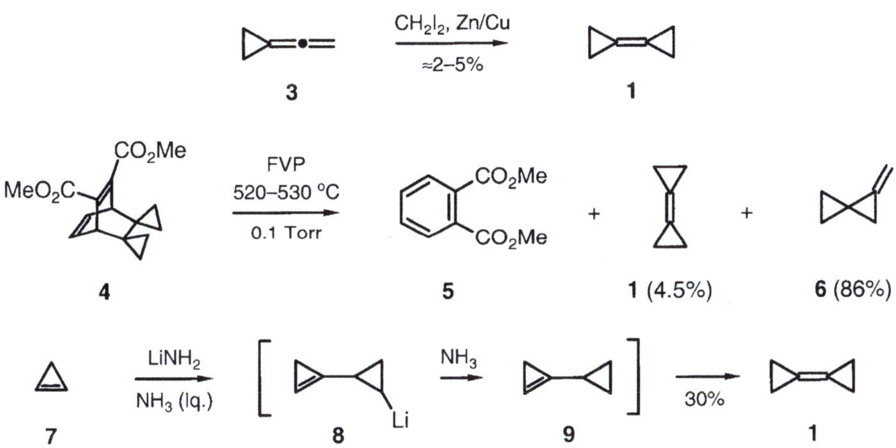

Scheme 1. The first preparations of bicyclopropylidene (**1**) [1, 3–7]

Although the yield for this reaction was not reported, and the isolation of **1** had to be performed by preparative gas chromatography, it was at best 2–5%, since the conversion of **3** was only of the order of 15% as reported in a later communication [3]. Approximately the same range of yield (4.5%) was achieved in the flash vacuum pyrolysis of **4**, the Diels-Alder adduct of dimethyl acetylenedicarboxylate and dispiro[2.0.2.4]deca-7,9-diene, which gave dimethyl phthalate (**5**), bicyclopropylidene (**1**), and methylenespiropentane (**6**), the rearrangement product of **1** (see below) [4]. Subsequent attempts to improve the cyclopropanation of **3** did not lead to an essential progress: application of the modified Simmons-Smith procedure with Zn/Ag couple [5] gave a mixture of **1** (36%), **3** (47%), and dispiro[2.0.2.1]heptane ([3]triangulane, see below) resulting from twofold cyclopropanation. The attempted activation by ultrasound only complicated the situation in that it led to the formation of side-products [6]. Bicyclopropylidene (**1**) was also formed upon treatment of cyclopropene with lithium amide in liquid ammonia, however, the isomerization of 1-cyclopropylcyclopropene (**9**) to **1** proceeds very slowly so that only after one month can **1** be isolated in 30% yield (Scheme 1) [7]. In spite of this time requirement, this method was the first one by which preparatively useful quantities of **1** could be obtained [8].

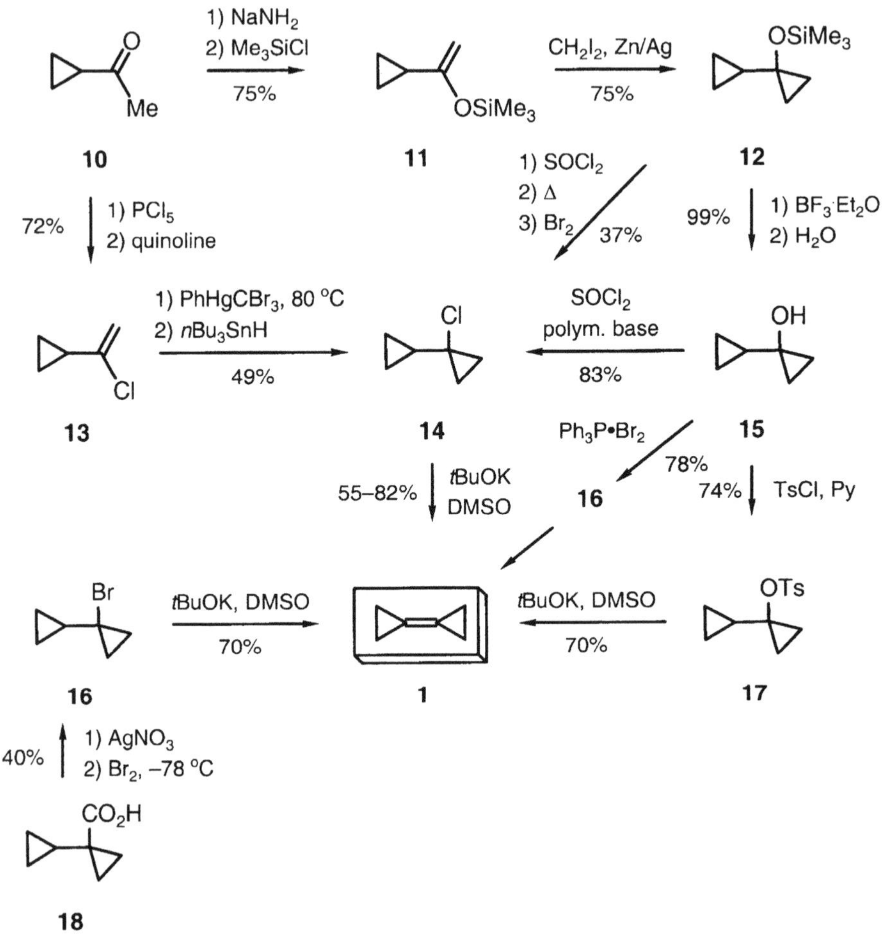

Scheme 2. Preparations of bicyclopropylidene (**1**) starting from acetylcyclopropane (**10**) and 1-cyclopropylcyclopropanecarboxylic acid (**18**) [9–12]

The fourth reported strategy approached **1** in a five-step sequence starting from acetylcyclopropane (**10**) via **11**, **12**, **15**, and **14** [9] (Scheme 2).

1-Chloro-1-cyclopropylcyclopropane (**14**) can also be obtained from 1-cyclopropyl-1-trimethylsilyloxycyclopropane (**12**) directly by treatment with thionyl chloride [9], yet this transformation turned out to be tricky. The conversion of the alcohol **15** to the chloride **14** was not easily reproduced and mostly accompanied by rearrangement. This drawback could be overcome by the use of a polymeric base to scavenge the hydrogen chloride formed in the conversion of **15** to **14** [10] and thus avoid aqueous work-up. An alternative four-step sequence from **10** to the chloride **14** via 1-chloro-1-cyclopropylethene (**13**) was also developed [9], but has major disadvantages (poor reproducibility, unpleasant reagents, poor overall yield) over the five-step approach. The dehydrochlorination of **14** to give **1** was achieved in 40% yield with sodium amide in liquid ammonia [9] and in 55–82% yield with potassium *tert*-butoxide in DMSO [10, 11].

The dehydrotosylation of the tosylate **17**, which is readily obtained from the alcohol **15** in 74% yield, has been reported to give **1** in 70% yield [11], but could not be reproduced in several laboratories. However, the dehydrobromination of the bromide **16** gives reproducibly good results (up to 81% yield), and the bromide **16** is easily prepared from the alcohol **15** by treatment with the triphenylphosphane/bromine reagent [12]. The alternative preparation of the bromide **16** by Hunsdiecker degradation of 1-cyclopropylcyclopropanecarboxylic acid (**18**) is only of basic interest, as **18** is not readily accessible [11]. A significant improvement in the preparation of the trimethylsilyl enol ether **11** from **10** using Et_3N as a base in the presence of $ZnCl_2$ made it possible to prepare **11** on a large scale [10], so that 18 years after the first report on its successful synthesis, bicyclopropylidene (**1**) became available in reasonable quantities to study its chemical properties on a broad scope.

However, the real breakthrough came with the drastically facilitated preparation of 1-cyclopropylcyclopropanol (**15**) from methyl cyclopropanecarboxylate (**19**) applying the transformation of an alkoxycarbonyl group into a cyclopropanol fragment with ethylmagnesium bromide in the presence of $Ti(iPrO)_4$ as developed by Kulinkovich et al. [13]. The optimized conversion of the alcohol **15** to the bromide **16** and its dehydrobromination makes the alkene **1** available in synthetically useful quantities of 40–55 g within one week (Scheme 3) [14]. This sequence is also applicable to prepare substituted, especially spirocyclopropane-annelated, bicyclopropylidenes [14a].

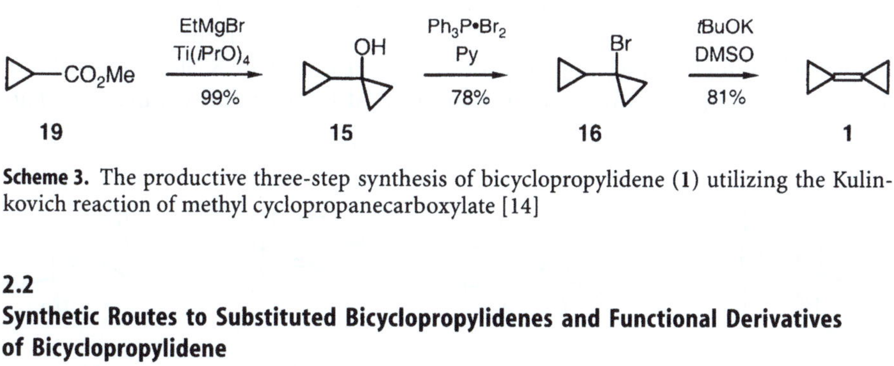

Scheme 3. The productive three-step synthesis of bicyclopropylidene (**1**) utilizing the Kulinkovich reaction of methyl cyclopropanecarboxylate [14]

2.2
Synthetic Routes to Substituted Bicyclopropylidenes and Functional Derivatives of Bicyclopropylidene

Although the unsubstituted bicyclopropylidene (**1**) has been known for only about 30 years, the first compound containing a bicyclopropylidene moiety had been reported as early as 1960 by Moore and Ward [15]. Over the years compounds **22a–m** were obtained by treatment of 1,1-dihalocyclopropanes **20** with alkyllithium reagents to yield cyclopropylidenoids **21**, mostly 1-halo-1-lithiocyclopropanes, which underwent dimerization and β-elimination (Scheme 4) [15–31], preferentially in cases where R^1 and R^2 were part of a cyclic fragment so that ring-opening of an intermediate cyclopropylidene into an allene was prevented [32]. In a few cases, other metals (magnesium/potassium [18], germanium/magnesium [26]) have been employed in the intermediate carbenoids **21**.

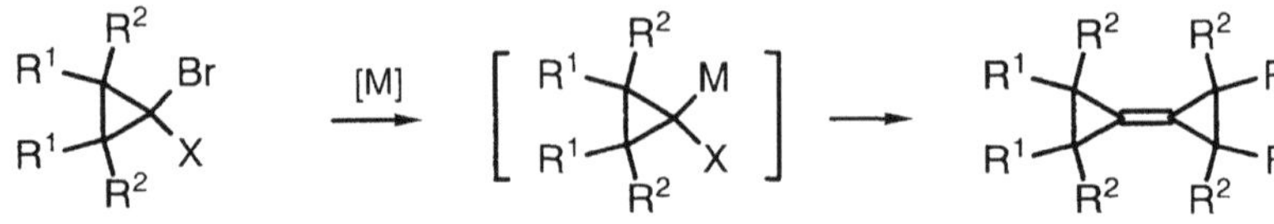

20 21 22

Scheme 4. Synthetic approach to substituted bicyclopropylidenes **22** by dimerization of cyclopropylidenoids **21** [15–31]

A number of ring-annelated bicyclopropylidenes has been obtained by this approach in moderate to good yields (Table 1). The main shortcoming of this method is a certain unpredictability and a variety of side reactions (see, e.g., [33, 34]), especially for the conversion of monocyclic dibromocyclopropanes. Only in a few cases have such monocyclic dibromocyclopropanes been reported to yield bicyclopropylidenes upon treatment with alkyllithium reagents, e. g., the formation of divinylbicyclopropylidenes **23**, mostly as mixtures of isomers (R^1 = H, Me; R^2 = Me, H; R^3 = H, Cl; yields were not reported) [35]. The debrominating dimerization of 1,1-dibromotetramethylcyclopropane upon treatment with methyllithium to give permethylated bicyclopropylidene **24a** went surprisingly well (41 % yield) [36a], and the yield could not be improved (9 %) by running the reaction in the presence of 12-crown-4 [36b]. The (*Z*)- and (*E*)-isomeric tetramethylbicyclopropylidenes **24b** and **24c** can also be easily prepared in almost quantitative yield by the dimerization of carbenoid generated in situ from 1,1-dibromo-2,2-dimethylcyclopropane and MeLi at –78 °C; it is important to add MeLi as quickly as possible [37]. A complex mixture of stereoisomeric bis(phenylthiomethyl)bicyclopropylidenes was formed in 85% yield without any precautions [38] with the (*E*,*E*)-isomer **24d** as the main component. The

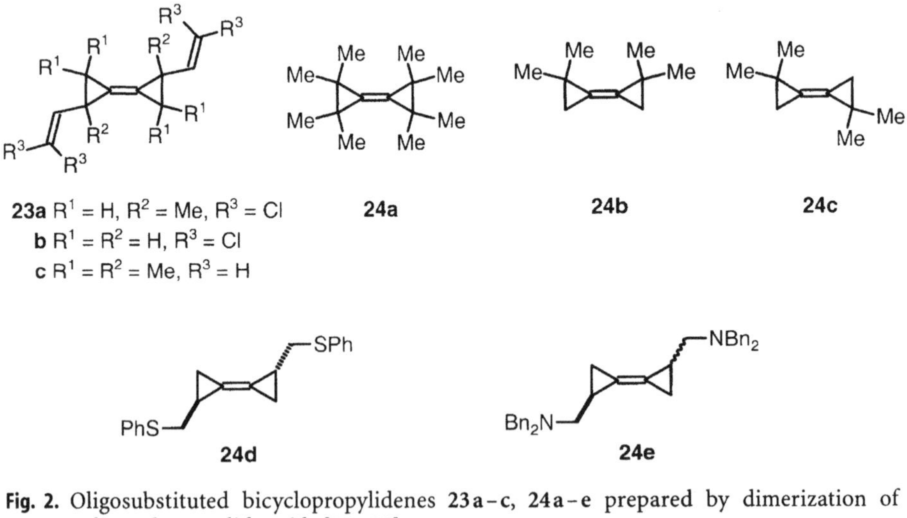

Fig. 2. Oligosubstituted bicyclopropylidenes **23a–c**, **24a–e** prepared by dimerization of monocyclic cyclopropylidenoids [35–39]

Table 1. Bicyclopropylidene derivatives **22a–m** by dimerization of bicyclic cyclopropylidenoids

20	X	R^1	R^2	T [°C]	Solvent	Metallating Reagent	Product **22**	Yield (%)	Ratio *syn/anti*	Ref.
a	Br	$-(CH_2)_4-$	H	-80	Et_2O	MeLi	**a**	30	n. r.	[15]
a	Br	$-(CH_2)_4-$	H	-90	Et_2O	MeLi	**a**	40	10:1	[16]
a	Br	$-(CH_2)_4-$	H	-70	Pentane	MeLi	**a**	53	10:1	[17]
a	Br	$-(CH_2)_4-$	H	-40	THF	$MgCl_2/K$	**a**	1	n. r.	[18]
a	Br	$-(CH_2)_4-$	H	-80	Et_2O	MeLi/LiI	**a**	30	n. r.	[19]
b	Cl	$-(CH_2)_4-$	H	-107	THF/Et_2O	BuLi	**a**	29	1:1	[20]
c	Br	$-(CH_2)_6-$	H	-78	Et_2O	MeLi/LiI	**b**	16	1:1	[21]
c	Br	$-(CH_2)_6-$	H	-50→-25	Et_2O	MeLi	**b**	44	1:1	[22]
d	Cl	$-CH_2OCH_2OCH_2-$	H	-78	Et_2O	MeLi	**c**	60	1:1	[23]
d	Cl	$-CH_2OCH_2OCH_2-$	H	-78	Et_2O	MeLi/LiBr	**c**	5	1:1	[24]
e	Br	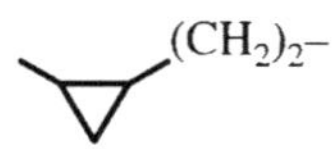	H	0	Pentane	MeLi	**d**	16	n. r.	[17]
f	Br	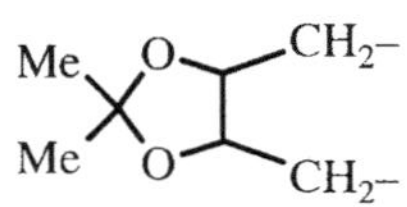	H	-30	Et_2O	MeLi/LiI	**e**	58	n. r.	[24]
f	Br		H	-30	Et_2O	MeLi	**e**	81	1.7:1	[25]
g	$Me_2Ge/2$	$-(CH_2)_4-$	H	20	THF	Mg	**a**	43–73	n. r.	[26]
h	Br	$-(CH_2)_3-$	$-(CH_2)_3-$	20	Et_2O	MeLi	**f**	6	–	[27]
i	Br	$-(CH_2)_4-$	$-(CH_2)_3-$	20	Et_2O	MeLi/12-c-4	**g**	2	n. r.	[27]
j	Br	$-CH_2CH=CHCH_2-$	$-(CH_2)_3-$	20	Et_2O	MeLi/12-c-4	**h**	2	n. r.	[27]

Table 1 (continued)

20	X	R^1	R^2	T [°C]	Solvent	Metallating Reagent	Product **22**	Yield (%)	Ratio *syn*/*anti*	Ref.
k	Cl	=C(Ph)–CH=CH–C(Ph)=		-78	THF	BuLi	**i**	23	n. r.	[28]
l	Cl	=C(Ph)–CH=CH–C= 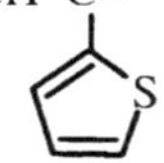		-78	THF	BuLi	**j**	19	1:1	[28]
m	Br	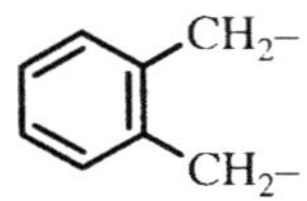		-80	Et_2O	MeLi	**k**	18	1:1.6	[25]
n	Br	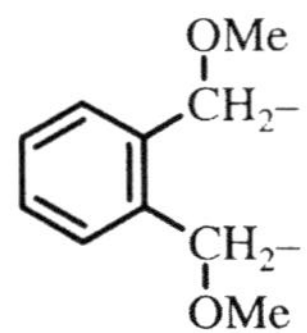		-80	Et_2O	MeLi	**l**	25	1:1.3	[29]
o	Br			-90	THF	BuLi	**m**	trace	-	[30]

analogous (*E*)-bis(dibenzylaminomethyl)bicyclopropylidene **24e** was also prepared by this method, albeit in only 5.4% yield (*syn*/*anti* ratio 2:1) [39].

A significant improvement was eventually reported by Neuenschwander et al. who found that copper(II) salts assist the coupling of bromolithiocyclopropanes **26** generated from dibromides **25** to give a variety of substituted bicyclopropylidenes of type **24** in reproducible and reasonable yields, however, as mixtures of diastereomers only (Scheme 5) [40]. This coupling probably proceeds via coppercarbenoids, and cross-coupling between two different carbenoids has also been achieved, but not with any reasonable selectivity.

Another approach to substituted bicyclopropylidenes is by carbene addition onto butatrienes and alkenylidenecyclopropanes bringing the substituents in with the allene or with the carbene moiety as well as with both fragments. This way, Skattebøl et al. [41] and later Kostikov et al. [42] have prepared tetrahalote-

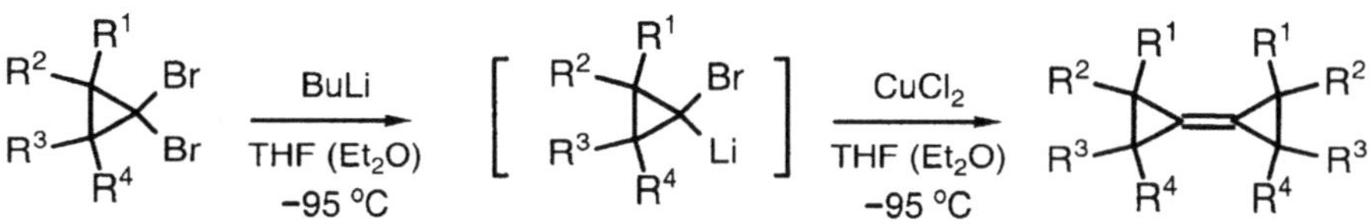

R^1	R^2	R^3	R^4	Product	Yield (%)
Me	Me	Me	Me	**24a**	30
$PhCH_2$	H	H	H	**24f**	87
$-(CH_2)_4-$		H	H	**24g**	87
Me	Et	H	H	**24h**	65
H	$-(CH_2)_4-$		H	**24i**	85
Ph	H	H	H	**24j**	50
Me	H	H	H	**24k**	75
H	Ph	Ph	H	**24l**	60
BuO	H	H	H	**24m**	21
H	$-O-(CH_2)_3-$		H	**24n**	26
PhO	H	H	H	**24o**	56
PhS	H	H	H	**24p**	44
H	9,10-Dihydro-anthracene-9,10-diyl		H	**24q**	25
Ph	H	Me	H	**24r**	75
Ph	Ph	H	H	**24s**	60

Scheme 5. Copper(II)-assisted coupling of bromolithiocyclopropanes **26** generated from dibromocyclopropanes **25** [40]

tramethylbicyclopropylidenes of type **33** via dihalocarbene addition onto 2,5-dimethylhexa-2,3,4-triene (**32**) (Scheme 6). Addition of dichlorocarbene onto di- and tetraphenylsubstituted butatrienes **34** also led to bicyclopropylidenes **35** in poor yields (Scheme 6) [43, 44].

A 1:1 mixture of (*Z*)- and (*E*)-tetramethylbicyclopropylidenes **24b, c** was obtained by dihalocyclopropanation of dimethyl(dimethylethenylidene)cyclopropane **27** [45, 46] followed by reduction of the adducts with sodium in methanol (Scheme 6). Addition of monochlorocarbene onto 2-(trimethylsilyl)-1-ethenylidenecyclopropane (**29**) proceeds with low diastereo- and regioselectivity to give a mixture of bicyclopropylidene and methylenespiropentane derivatives **30**, **31** in poor yield [47]. Upon treatment of 1,1-dibromo-2-methylpropene (**36**) with butyllithium at –110 °C the unique diisopropylidenetetramethylbicyclopropylidene **37** was formed by addition of isobutylidene to in situ generated tetramethylbutatriene (**32**), albeit in very low yield [48] (Scheme 7).

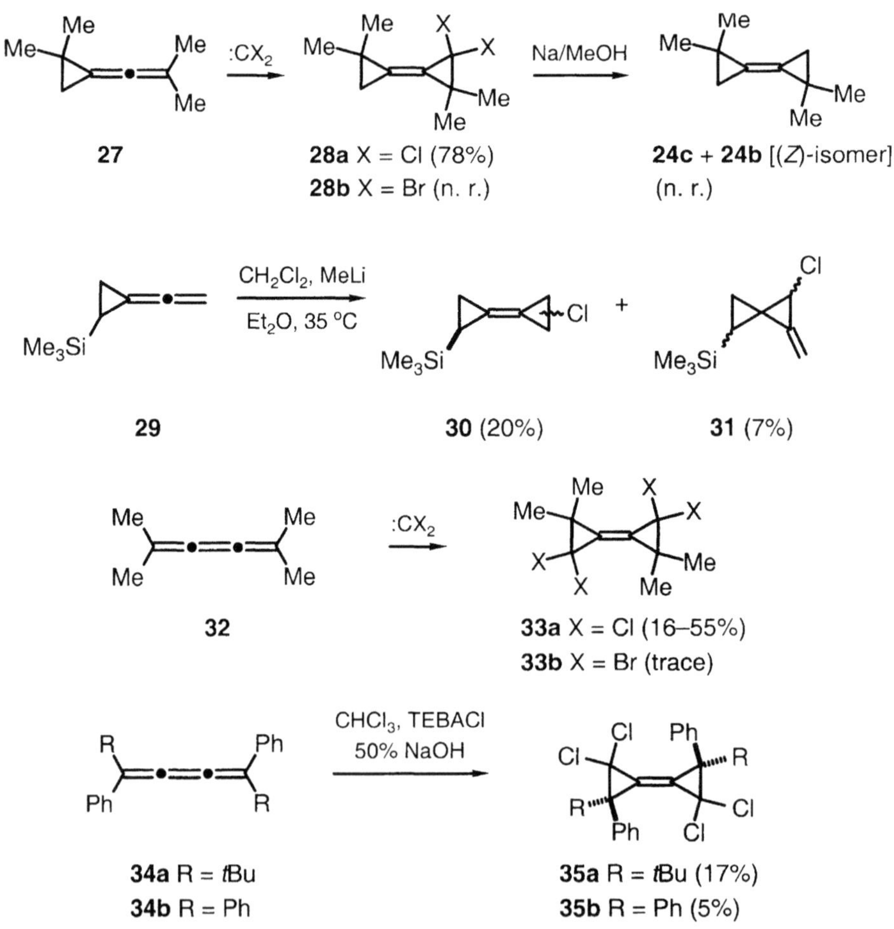

Scheme 6. Carbene additions onto allenes as a synthetic approach to substituted bicyclopropylidenes (n. r. = yield not reported) [41–47]

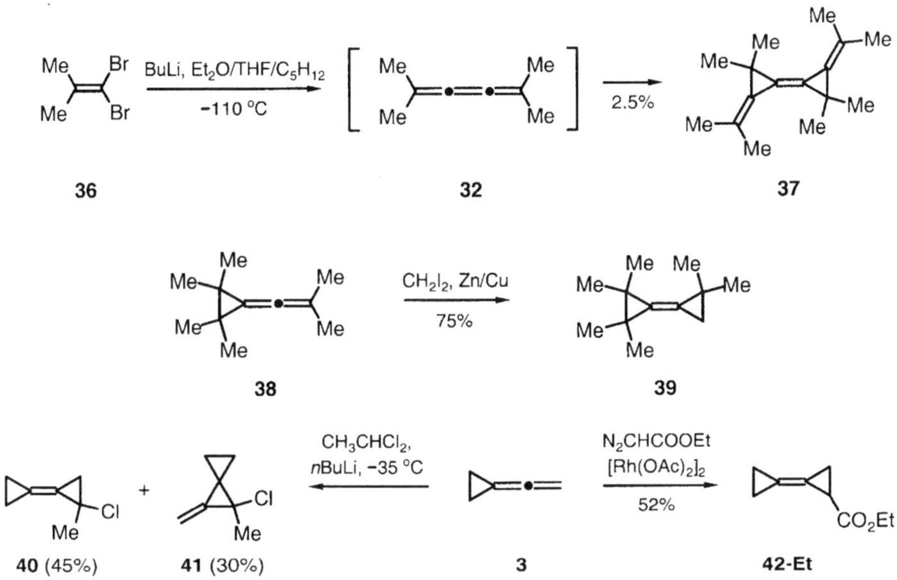

Scheme 7. Cyclopropanation of allenes as a synthetic approach to substituted bicyclopropylidenes [48–50, 52, 53]

Hexamethylbicyclopropylidene **39** has been prepared in 75% yield by applying the Simmons-Smith protocol to isobutenylidenetetramethylcyclopropane **38** (Scheme 7) [49, 50]. Analogously, 1,1-dideuteriobicyclopropylidene was obtained starting from (2,2-dideuterioethenylidene)cyclopropane [D_2]-**3** [51]. In contrast to the Simmons-Smith reagent methylchlorocarbene does not add to **3** in a regioselective way and thus gives a mixture of both possible monocyclopropanation products **40** and **41** in a ratio of 3:2 (Scheme 7) [52].

The rhodium(II)-catalyzed cyclopropanation of **3** with ethyl diazoacetate gave ethyl bicyclopropylidenecarboxylate **42**-Et as a single product, but in moderate yield (Scheme 7) [53].

A large variety of functionally monosubstituted bicyclopropylidenes **43a–v** can be prepared directly from bicyclopropylidene (**1**) in moderate to excellent yields by deprotonation with butyllithium in THF at 0 °C and electrophilic substitution of the lithiobicyclopropylidene with appropriate reagents (Scheme 8) [54–58].

Several of these new derivatives **43** could be further transformed with retention of the bicyclopropylidene moiety (Scheme 9) [53, 55–58].

Whereas the repeated lithiation-trimethylsilylation sequence of trimethylsilylbicyclopropylidene **43a** yielded predominantly the cyclopropene derivative **51** [54], bicyclopropylidenecarboxylates **42**-Me, **42**-*t*Bu after repeated deprotonation and carboxylation retain the bicyclopropylidene moiety and give 2,2-disubstituted products **52**-R only (Scheme 9) [55]. So far, alkylbicyclopropylidenes **43e, f, g** have not been induced to undergo deprotonation and a second substitution [56a]. The urethane **53** with a nitrogen directly attached to the skeleton, more easily than any other bicyclopropylidene derivative, rearranges to

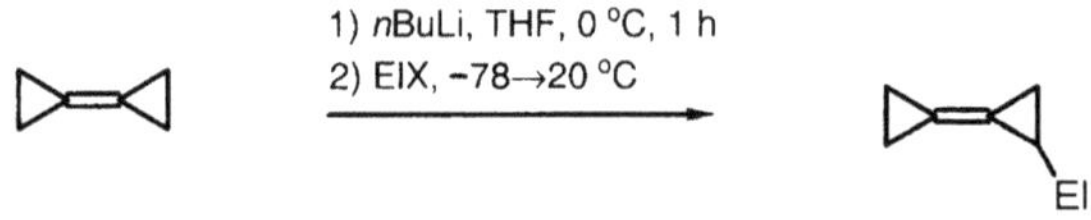

1 42, 43

ElX	Me_3SiCl	PhSSPh	C_2Cl_6	$(CH_2Br)_2$	nC_4H_9I	$nC_5H_{11}I$	$nC_7H_{15}I$	Me_2CO
El	$SiMe_3$	SPh	Cl	Br	nC_4H_9	nC_5H_{11}	nC_7H_{15}	CMe_2OH
Product	**43a**	**43b**	**43c**	**43d**	**43e**	**43f**	**43g**	**43h**
Yield (%)	85	77	73	65	72	80	74	82
ElX	oxirane	DMF	DMA	D_2O	$ICH_2CH{=}CH_2$	O_2, AcCl	O_2	CO_2
El	CH_2CH_2OH	CHO	C(O)Me	D	$CH_2CH{=}CH_2$	OAc	OH	CO_2H
Product	**43i**	**43j**	**43k**	**43l**	**43m**	**43n**	**43o**	**42-H**
Yield (%)	50	73	71	80	50	32	22	95
ElX			$I(CH_2)_nOTHP$, $n = 3–8$					Boc_2O
El	$(CH_2)_3OTHP$	$(CH_2)_4OTHP$	$(CH_2)_5OTHP$	$(CH_2)_6OTHP$	$(CH_2)_7OTHP$	$(CH_2)_8OTHP$		CO_2tBu
Product	**43p**	**43r**	**43s**	**43t**	**43u**	**43v**		**42-tBu**
Yield (%)	63	85	82	88	68	65		69

Scheme 8. Preparation of functionalized bicyclopropylidene derivatives **42, 43** from bicyclopropylidene (**1**) via lithiobicyclopropylidene [54–58]

the corresponding methylenespiropentane derivative **54** (see below). Thus, under conditions of the Curtius degradation of the acid azide of **52**-Me, a 1:1 mixture of carbamates **53** and **54** was obtained in 37% yield [55].

Since the enantiomers of the carboxylic acid **42**-H can easily be separated via its diastereomeric salts (Scheme 10) [59], many of the other bicyclopropylidene derivatives can also be obtained in enantiomerically pure form by transformations of the acids (*R*)- and (*S*)-**42**-H. The absolute configuration of (*R*)-**42**-H was proved by an X-ray crystal structure analysis of its (*R*)-α-phenylethylamide [59].

2.3
Spirocyclopropanated and Ring-Annelated Bicyclopropylidenes as well as Bis(bicyclopropylidenyls)

Commonly known methods for the preparation of bicyclopropylidenes of types **55–62** (Fig. 3) are the cyclopropanations of appropriately cyclopropanated allenes according to the Gaspar-Roth [60] or modified Simmons-Smith protocol [61] or addition of cyclopropylidene generated in situ from *N*-nitroso-*N*-cyclopropylurea [62, 63]). Along these routes, the compounds **55** [63], **58–60** [33, 64, 65], and **62** [33, 64] were obtained in low to moderate yields (Scheme 11).

As well, the functionalized spirocyclopropanated bicyclopropylidene **68** has been isolated as a by-product from the rhodium(II)-catalyzed cyclopropanation of dicyclopropylidenemethane (**63**) with an excess of ethyl diazoacetate [14].

In spite of low yields and possible difficulties in separating the mixtures of final products, this approach was the only one known so far for the compounds **58–60, 62**.

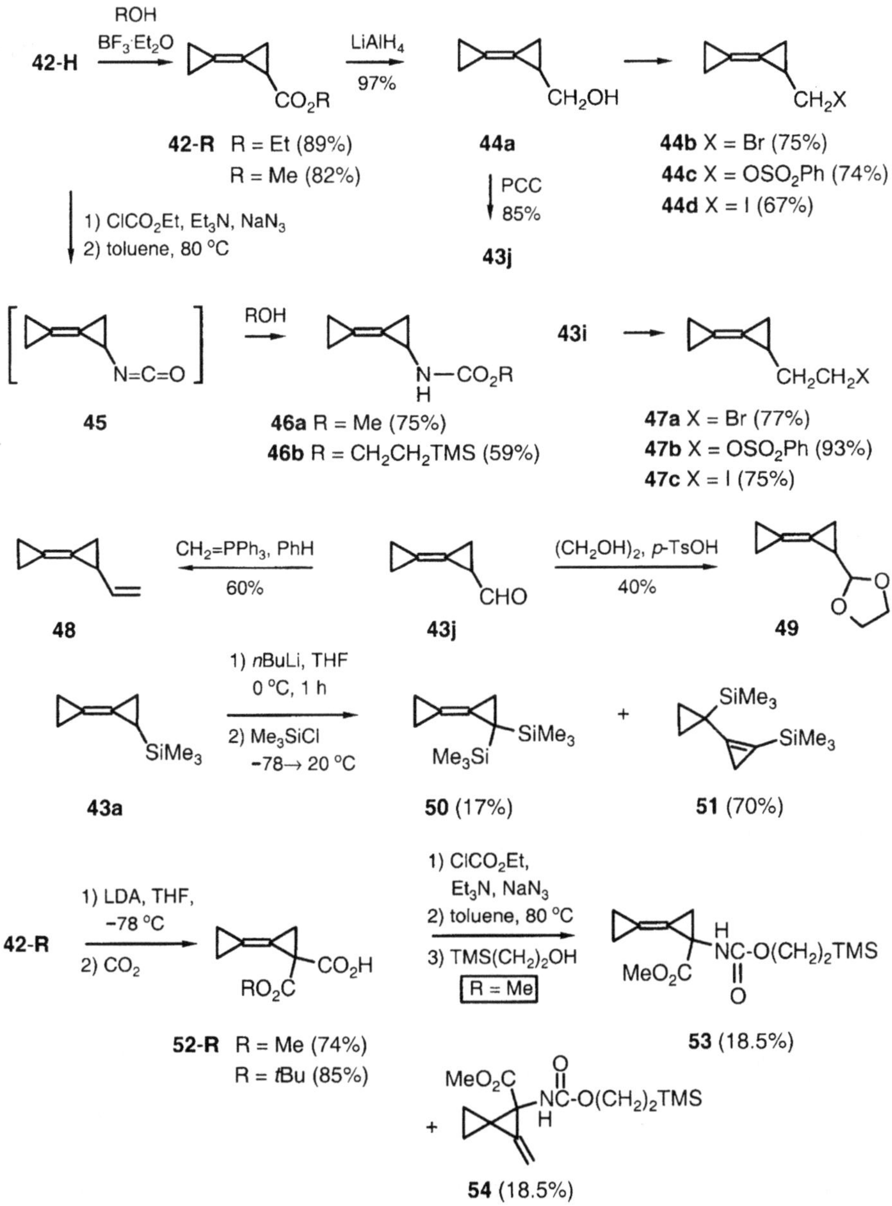

Scheme 9. Some transformations of functionally substituted bicyclopropylidenes [53, 55–58]

7-Cyclopropylidenedispiro[2.0.2.1]heptane (**56**) was initially prepared by reductive cyclization of the dibromide **72** under the action of phenyllithium [66, 67] (in this case **56** was obtained as a mixture with bromobenzene) or magnesium [68] (Scheme 12).

The ultimate methodology developed for the preparation of bicyclopropylidene (**1**) [14] turned out to also be appropriate for the synthesis of the spiro-

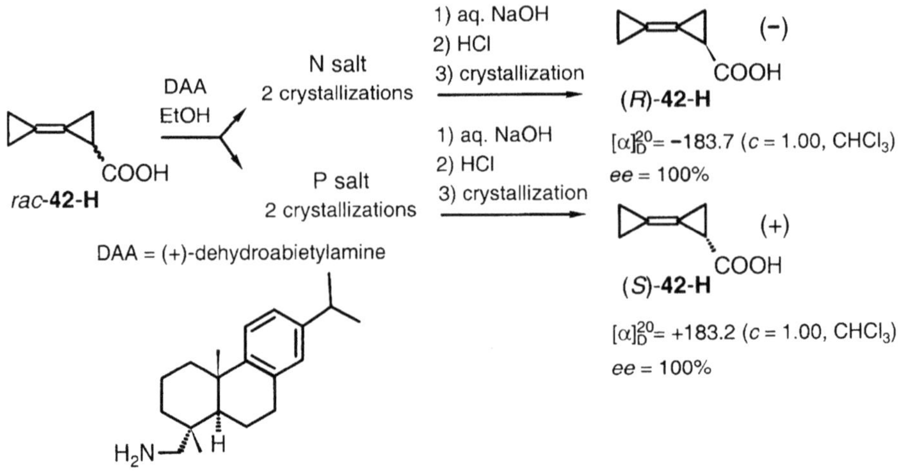

Scheme 10. Optical resolution of bicyclopropylidenecarboxylic acid *rac*-**42**-H with dehydroabietylamine [59]

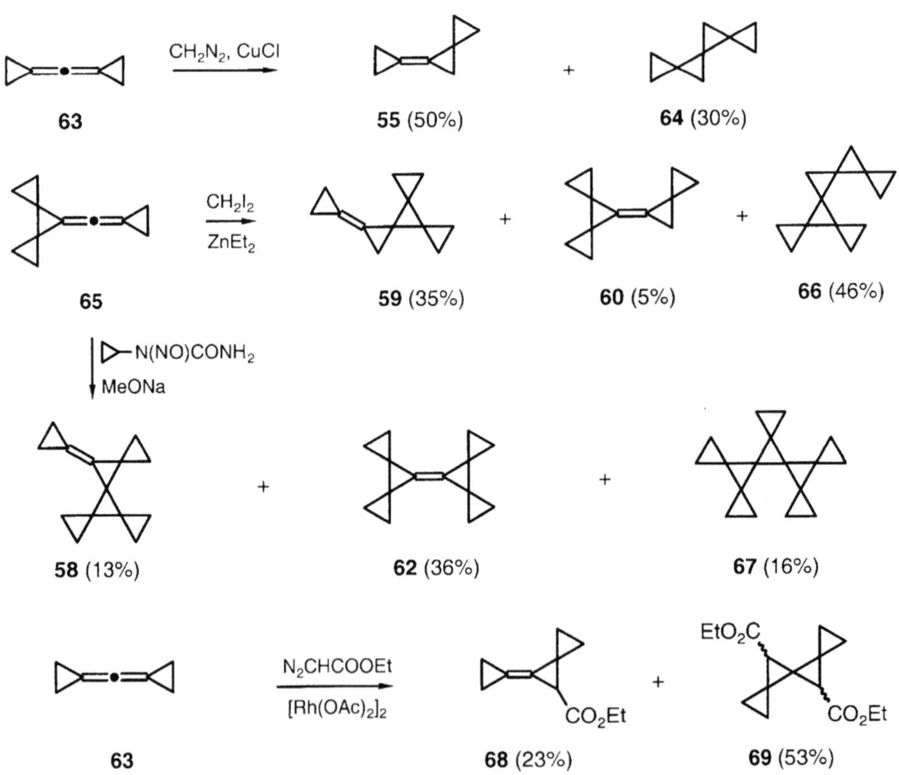

Scheme 11. Cyclopropanation of dicyclopropylidenemethanes **63**, **65** as a synthetic route to various spirocyclopropanated bicyclopropylidenes [14, 60–64]

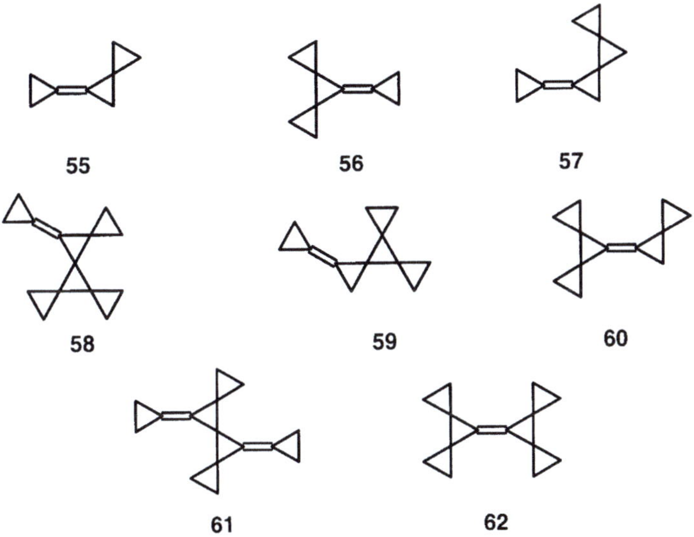

Fig. 3. Some oligospirocyclopropanated bicyclopropylidenes

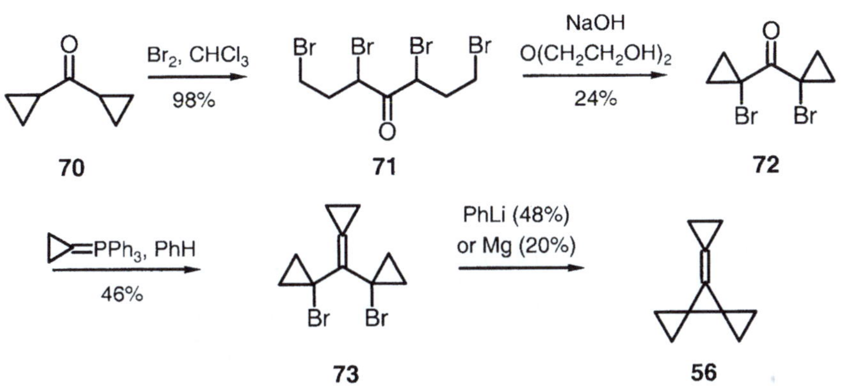

Scheme 12. The first preparation of unsymmetrically bisspirocyclopropanated bicyclopropylidene **56** [66–68]

cyclopropanated bicyclopropylidenes **55**–**57** and **61**, as an alkoxycarbonylcarbene addition onto bicyclopropylidenes creates a new cyclopropanecarboxylate of type **75** (Scheme 13), and the whole sequence can be applied in a repetitive fashion. This allows the preparation of a large variety of compounds with a bicyclopropylidene moiety on a preparative scale [14].

Other interesting spirocyclopropanated bicyclopropylidenes, the ring-annelated compounds **80 a, b** have also been prepared applying the Kulinkovich reaction [69, 70] (Scheme 14). The starting materials **79** were obtained by alkoxycarbonylcarbene addition onto cyclooctene and bicyclo[6.1.0]non-1-ene, respectively.

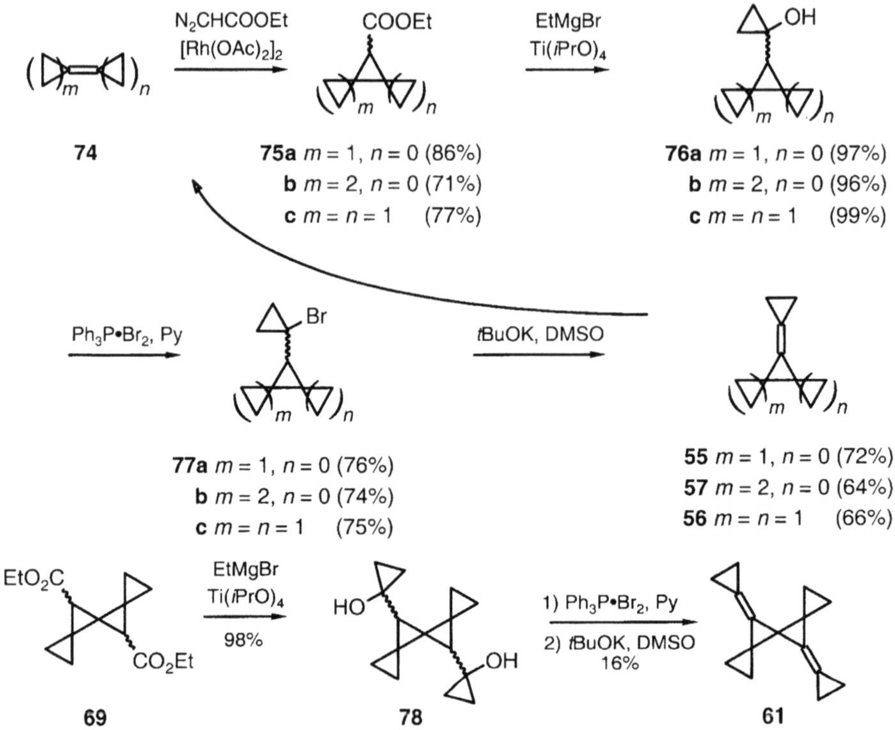

Scheme 13. Convergent synthesis of spirocyclopropanated bicyclopropylidenes **55**–**57** and **61** via repetitive application of the Kulinkovich reaction [14]

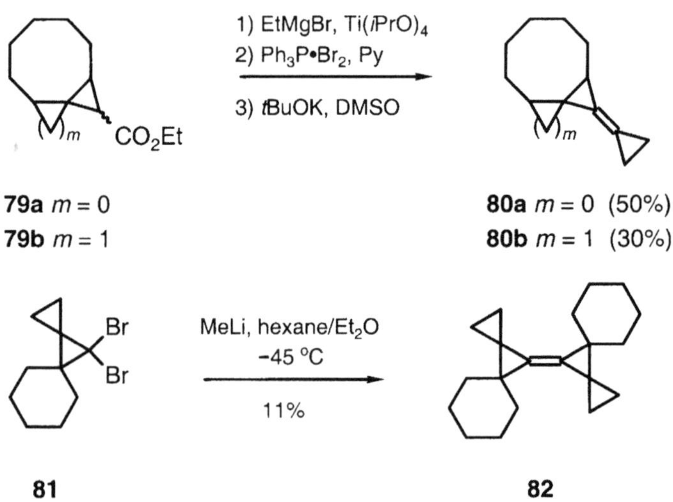

Scheme 14. Preparation of spirocyclopropanated bicyclopropylidenes **80** and **82** [31, 69, 70]

The interesting bis(spirocyclohexane)-annelated bisspirocyclopropanated bicyclopropylidene **82** was prepared by dimerization of the cyclopropylidenoid generated from the dibromide **81** under the conditions discussed above [31].

Oligospirocyclopropanated bicyclopropylidenes can be functionalized in the same manner as the parent compound **1** (see Scheme 8), but in the deprotonations and subsequent electrophilic substitutions on unsymmetrical hydrocarbons like **55** and **57**, very little selectivity at best was observed [54].

An attempted approach to bis(bicyclopropylidenyls) of type **87** utilizing the Kulinkovich reaction turned out to be unfruitful; although the twofold cyclopropanation of the diester **84** prepared from butadiene, worked perfectly well, and the conversion of the biscyclopropanol **85** to the dibromide **86** also gave a good yield (60 %), the twofold dehydrobromination of **86** afforded only 1.7 % of a mixture of *meso*- and *d,l*-**87** (Scheme 15) [56a]. However, the direct oxidative coupling of two molecules of lithiobicyclopropylidene was accomplished under

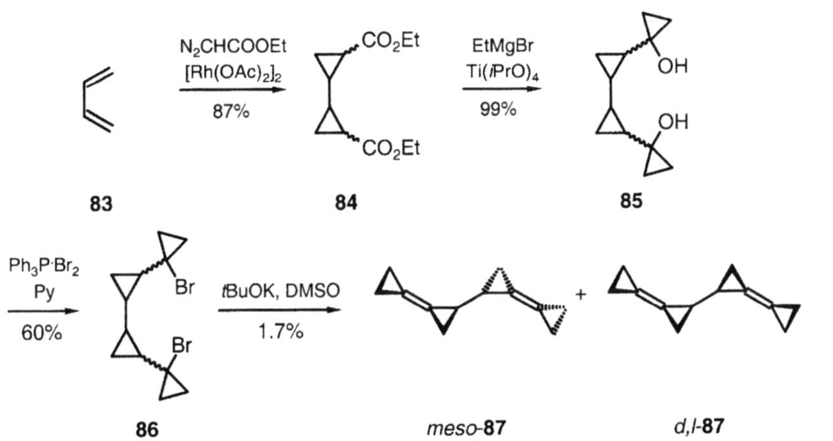

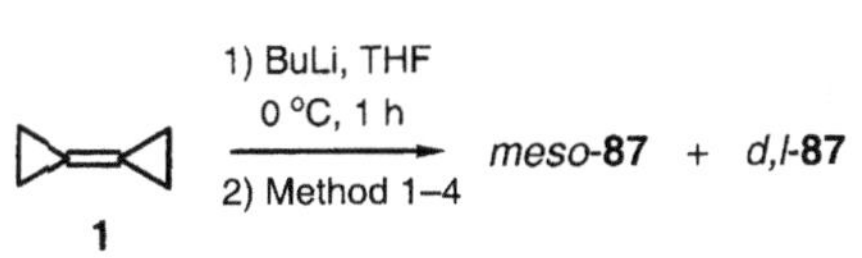

Entry	Method of Coupling	Yield (%)
1	1) $MgBr_2$; 2) **43d**, $NiCl_2(PPh_3)_2$ 20 °C	20
2	1) CuCN; 2) O_2, −78 °C	38
3	1) $MgBr_2$; 2) $CuCl_2$, 20 °C	40–50
4	1) $[CuIBu_3P]_4$; 2) O_2, −78 °C	80

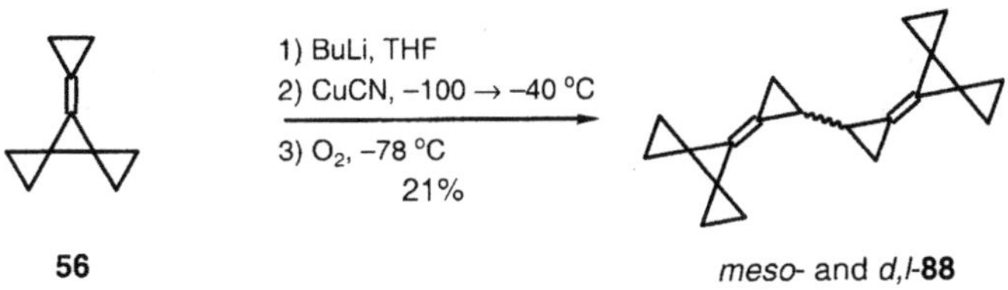

Scheme 15. Syntheses of the bis(bicyclopropylidenyls) **87** and **88** [56]

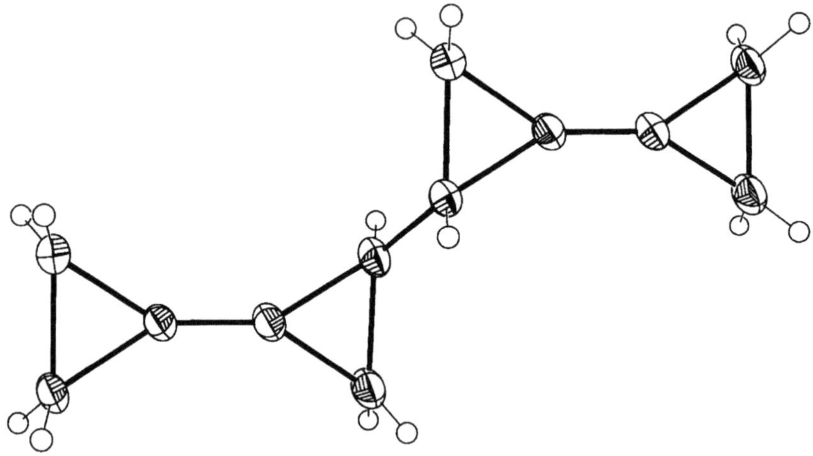

Fig. 4. Structure of *meso*-bis(bicyclopropylidenyl), *meso*-**87**, in the crystal [56]

different conditions (Scheme 15) and in the best case gave an 80% yield of **87** as a 1.6:1 mixture of *meso*- and *d,l*-diastereomers [56]. Since the *meso*-**87** crystallized well, it could easily be separated from the *d,l*-diastereomer and characterized by an X-ray crystal structure analysis (Fig. 4). Under analogous conditions, the bisspirocyclopropanated bicyclopropylidene **56** could also be dimerized to give *meso*- and *d,l*-**88**, but in poorer yield [56].

Bicyclopropylidenes with two larger rings annelated at both cyclopropyl moieties are sufficiently well documented (Table 1). But compounds in which the larger ring bridges the two ends of a bicyclopropylidene are usually more highly strained and, as a consequence, not as easily accessible. The first such compounds **89** and **91** were obtained by Diels-Alder reaction of the bis(cyclopropenyl) derivative **90** with dimethyl acetylenedicarboxylate (DMAD) and te-

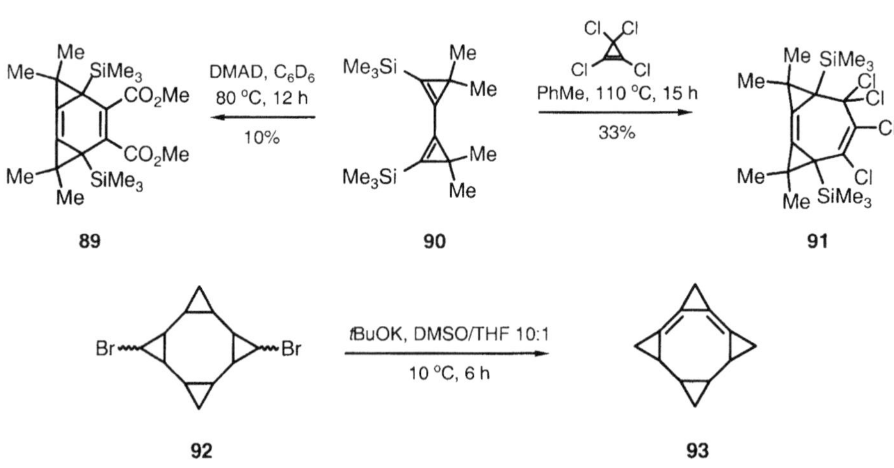

Scheme 16. Synthetic approaches to ring-annelated bicyclopropylidenes [71, 72]

trachlorocyclopropene, respectively (Scheme 16) [71]. It has been reported that the treatment of the 6,12-dibromopentacyclo[9.1.0.0^{2,4}.0^{5,7}.0^{8,10}]dodecane **92** with potassium *tert*-butoxide in DMSO/THF gave the bridged dicyclopropylidenecyclopropane **93**, but no yield and no experimental procedure have been published [72].

3 Physical Properties of Bicyclopropylidenes

3.1 Spectral, Bonding and Structural Properties

The IR [1, 3, 7, 73, 74] and Raman [7, 74] spectra of unsubstituted bicyclopropylidene (**1**) have been published several times and with some discrepancies. Compound **1** reveals peaks with chemical shifts of 1.16 (^{1}H) [1, 3, 7, 74–76] and 2.85, 110.21 ppm (^{13}C) [76] in the NMR spectra in CCl_4 or $CDCl_3$ solutions. For monodeuteriobicyclopropylidene (**431**) the ^{1}H-NMR spectrum exhibits a broad singlet with essentially the same chemical shift, but in the ^{13}C-NMR spectrum the weak influence of the symmetry distortion is more pronounced, and signals at 2.71 (CHD, t, $J = 24.8$ Hz), 2.87 (CH_2), 3.10 (2 CH_2), 110.27 and 110.36 ppm have been observed [56b]. The NMR spectra of functionally monosubstituted bicyclopropylidenes as reported along with their preparations are very diagnostic. The spectra of oligospirocyclopropanated bicyclopropylidenes combine features of the spectra of bicyclopropylidene (**1**) and of triangulanes [77].

Upon γ-irradiation of **1** in a CF_3CCl_3 matrix at 77 K [78], a radical cation was formed, the ESR spectrum of which consisted of nine broad hyperfine components spaced by ca. 0.75 mT (g = 2.0029 ± 0.002), and the corresponding proton ENDOR spectrum exhibited two essentially isotropic signals at 25.83 and 24.58 MHz. The detailed analysis of the ESR and ENDOR spectra disclosed that the initially formed radical cation **1**$^{+\bullet}$ had transformed into the tetramethyleneethane radical cation **94**$^{+\bullet}$ (Scheme 17). In $CFCl_3$ and $CF_2ClCFCl_2$ matrices **1**$^{+\bullet}$ persists up to 100 K [79]. On going from **1** to **1**$^{+\bullet}$, the set of eight equivalent protons splits

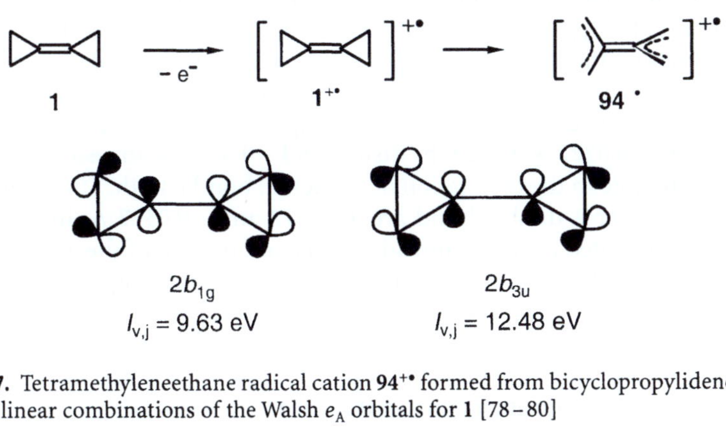

Scheme 17. Tetramethyleneethane radical cation **94**$^{+\bullet}$ formed from bicyclopropylidene (**1**) and bonding linear combinations of the Walsh e_A orbitals for **1** [78–80]

into two sets of four, and the two protons in each CH_2 group have greatly different coupling constants of 2.24 and 0.27 mT due to considerable changes in geometry upon ionization, such as a twist about the central C-C bond. Similar results were obtained also for the dispirocyclopropanated bicyclopropylidene **56** [79].

The photoelectron spectrum of bicyclopropylidene (**1**) reveals a split between the bonding linear combinations of the Walsh e_A orbitals of 2.85 eV ($2b_{1g}-2b_{3u}$) (Scheme 17) and a value for the resonance integral between linked $2p$ atomic orbitals of the adjacent cyclopropane rings in **1** of –2.14 eV [80–82a]. The fact that bicyclopropylidene has such a high-lying HOMO is responsible for its uniquely enhanced reactivity towards a wide range of electrophiles and cycloaddends. The lowest ionization energy (π-IE$_v$) is equal to 8.93 for **1** [80–82a], 8.70 for **55** [82b], and 8.50 eV for **56** [33, 64], i.e., the consecutive spirocyclopropanation of **1** leads to a significant decrease of its π-IE$_v$.

The determination of precise parameters describing the shape and size of the bicyclopropylidene moiety in different environments has been of constant interest ever since the first examination of the substituted bicyclopropylidenes **22f** [27] and **24d** [38]. Structural analyses have been performed also for the unsubstituted bicyclopropylidene (**1**) (several different studies) [83, 84, 85, 86], perspirocyclopropanated bicyclopropylidene **62** [33, 64], and the oligospirocyclopropanated bicyclopropylidenes **55**–**57** [87]. Corresponding theoretical investigations have been sparse [88, 89, 90]. A comparison of the structural parameters within the series exhibits the same effects as known for the triangulanes [77, 87, 91]: a consistent difference between longer distal and shorter proximal bonds in the outer spirocyclopropane units (Table 2). This is due to hybridization changes which cause increased angular strain [77, 91]. The influence of a double bond upon the proximal and distal bond lengths, in accord with theoretical calculations [89, 90], should be even more pronounced, and it was indeed observed for the compounds **55**–**57**. The abundance of experimental data prompted the development of a general increment scheme (GIS) for the quantitative prediction of different C-C bond lengths in bicyclopropylidenes and spirocyclopropanated bicyclopropylidenes [87].

In this scheme (GIS), all bonds are divided into five categories depending on their relative position in a molecule, the number and locations of the spiro-fused three-membered rings or double bonds: distal (A), proximal (B), distal-proximal-proximal (C), proximal-proximal (D) and distal-proximal (E) (Fig. 5).

Assuming that the influences of all cyclopropanes and double bonds are additive and independent of each other, the structures can be described using an initial value of 1.5008 Å for the C-C bond length in a cyclopropane unit and an increment of +0.0227 Å for each "distal" and –0.0234 Å for each "proximal" spirocyclopropane in the estimation of the actual bond length. The corresponding increments for a double bond are +0.0372 and –0.0327 Å. In the case of spirocyclopropanated bicyclopropylidenes, a combination of the above two sets of parameters has to be used. A comparison of the bond lengths estimated according to this scheme and the experimentally determined ones is presented in Table 2. All theoretical calculations predict a lengthening of the distal and a shortening of the proximal bonds except for those by the MINDO/3 method, which is particularly unsatisfactory in this case.

Table 2. Calculated and experimentally determined bond lengths in compounds **1**, **2**, **22f**, **24n**, **55**–**57**, **62**, and *meso*-**87** (all values in Å), mean values (X-ray) according to the given symmetry (GIS = general increment scheme)

Compound	Method	Length of CC Bond				Ref.
	IMOA	**a** 1.522	**b** 1.475	**c** 1.316		[89]
	MNDO	**a** 1.536	**b** 1.492	**c** 1.319		[89]
	MNDO	**a** 1.537	**b** 1.492			[87]
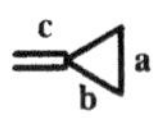	MINDO/3	**a** 1.486	**b** 1.495	**c** 1.318		[89]
	STO-3G	**a** 1.521	**b** 1.474	**c** 1.298		[89]
2	3-21 G	**a** 1.545	**b** 1.472	**c** 1.301		[89]
	4-21 G	**a** 1.546	**b** 1.476	**c** 1.300		[89]
(C_{2v})	6-31 G*	**a** 1.527	**b** 1.462	**c** 1.308		[87]
	GIS	**a** 1.538	**b** 1.468			[87]
	X-ray (125 K)	**a** 1.526(1)	**b** 1.460(1)	**c** 1.316(1)		[91]
	IMOA	**a** 1.522	**b** 1.475	**c** 1.298		[89]
	MNDO	**a** 1.538	**b** 1.496	**c** 1.302		[89]
	MNDO	**a** 1.538	**b** 1.492			[87]
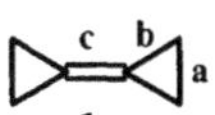	MINDO/3	**a** 1.496	**b** 1.485	**c** 1.322		[89]
	STO-3G	**a** 1.521	**b** 1.476	**c** 1.287		[89]
1	3-21 G	**a** 1.545	**b** 1.475	**c** 1.284		[89]
(D_{2h})	4-21 G	**a** 1.544	**b** 1.478	**c** 1.286		[89]
	6-31 G*	**a** 1.527	**b** 1.463	**c** 1.296		[87]
	GIS	**a** 1.538	**b** 1.468			[87]
	Electron diffr.	**a** 1.554(2)	**b** 1.468(1)	**c** 1.314(1)		[83]
	X-ray (233 K)	**a** 1.504(7)	**b** 1.442(6)	**c** 1.304(8)		[83]
	X-ray (173 K)	**a** 1.539(2)	**b** 1.467(2)	**c** 1.304(2)		[84]
	X-ray (245 K)	**a** 1.534(2)	**b** 1.467(1)	**c** 1.304(2)		[85,86]
	X-ray (140 K)	**a** 1.544(1)	**b** 1.469(1)	**c** 1.314(1)		[85,86]
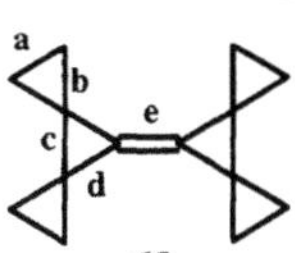	MNDO	**a** 1.529	**b** 1.512	**c** 1.503	**d** 1.476	[87]
	6-31G*	**a** 1.514	**b** 1.480	**c** 1.479	**d** 1.456	[87]
		e 1.301				
	GIS	**a** 1.524	**b** 1.477	**c** 1.491	**d** 1.467	[87]
62	X-ray (210 K)	**a** 1.507(3)	**b** 1.472(4)	**c** 1.475(3)	**d** 1.454(3)	[33,64]
(D_{2h})		**e** 1.305(4)				
	MNDO	**a** 1.538	**b** 1.492	**d** 1.495	**e** 1.521	[87]
		f 1.474	**g** 1.511	**h** 1.528		
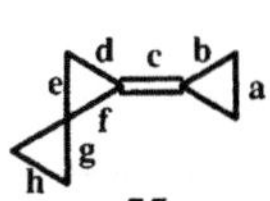	6-31 G*	**a** 1.527	**b** 1.463	**c** 1.297	**d** 1.482	[86,87]
		e 1.501	**f** 1.439	**g** 1.480	**h** 1.512	
55	GIS	**a** 1.538	**b** 1.468	**d** 1.491	**e** 1.515	[87]
(C_s)		**f** 1.445	**g** 1.477	**h** 1.524		
	X-ray (115 K)	**a** 1.543(1)	**b** 1.470(1)	**c** 1.313(1)	**d** 1.493(1)	[87]
		e 1.514(1)	**f** 1.444(1)	**g** 1.490(2)	**h** 1.527(1)	

Table 2 (continued)

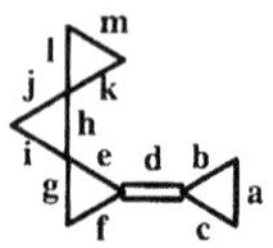 **57** (C_1)	MNDO	**a** 1.538	**b** 1.492	**c** 1.492	**e** 1.475	[87]
		f 1.496	**g** 1.521	**h** 1.514	**i** 1.493	
		j 1.511	**k** 1.510	**l** 1.510	**m** 1.529	
	6-31 G*	**a** 1.527	**b** 1.463	**c** 1.463	**d** 1.298	[86,87]
		e 1.440	**f** 1.482	**g** 1.502	**h** 1.459	
		i 1.498	**j** 1.488	**k** 1.474	**l** 1.475	
		m 1.517				
	GIS	**a** 1.538	**b** 1.468	**c** 1.468	**e** 1.445	[87]
		f 1.491	**g** 1.515	**h** 1.454	**i** 1.500	
		j 1.500	**k** 1.477	**l** 1.477	**m** 1.524	
	X-ray (110 K)	**a** 1.537(1)	**b** 1.469(1)	**c** 1.466(1)	**d** 1.312(1)	[87]
		e 1.442(1)	**f** 1.495(1)	**g** 1.515(1)	**h** 1.468(1)	
		i 1.514(1)	**j** 1.499(1)	**k** 1.480(1)	**l** 1.483(1)	
		m 1.531(1)				
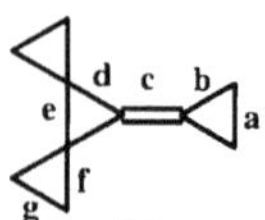 **56** (C_{2v})	MNDO	**a** 1.538	**b** 1.492	**d** 1.477	**e** 1.503	[87]
		f 1.511	**g** 1.529			
	6-31 G*	**a** 1.527	**b** 1.463	**c** 1.298	**d** 1.456	[86,87]
		e 1.480	**f** 1.480	**g** 1.513		
	GIS	**a** 1.538	**b** 1.468	**d** 1.467	**e** 1.491	[87]
		f 1.477	**g** 1.524			
	X-ray (120 K)	**a** 1.538(1)	**b** 1.465(1)	**c** 1.309(1)	**d** 1.462(1)	[87]
		e 1.481(1)	**f** 1.485(1)	**g** 1.521(1)		
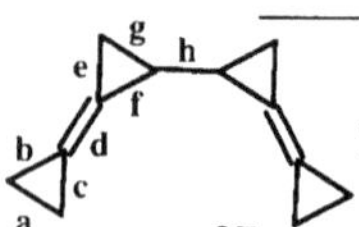 *meso*-**87** (C_s)	X-ray (113 K)	**a** 1.532(1)	**b** 1.466(1)	**c** 1.461(1)	**d** 1.304(1)	[56b]
		e 1.466(1)	**f** 1.471(1)	**g** 1.538(1)	**h** 1.494(1)	
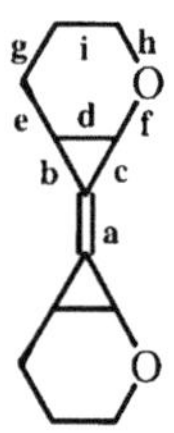 **24n** (C_s)	X-ray (295 K)	**a** 1.287(6)	**b** 1.476(6)	**c** 1.479(6)	**d** 1.535(6)	[40b]
		e 1.518(4)	**f** 1.415(4)	**g** 1.530(5)	**h** 1.439(4)	
		i 1.518(6)				
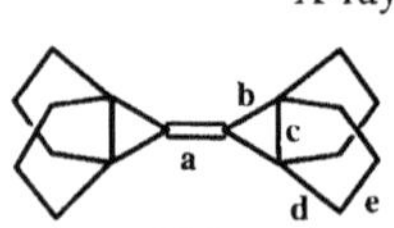 **22f** (D_{2h})	X-ray (295 K)	**a** 1.307(3)	**b** 1.470(3)	**c** 1.549(3)	**d** 1.512(3)	[27]
		e 1.538(4)				

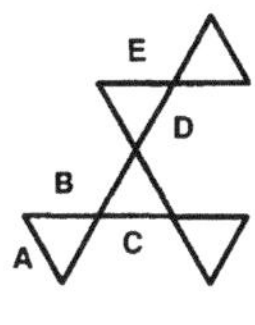

95

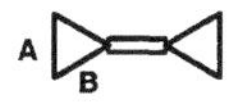

1

Fig. 5. Classification of the C–C bonds in triangulanes and bicyclopropylidenes in the general increment scheme (GIS) for the prediction of C–C bond lengths [87]

Bicyclopropylidene (**1**) with its melting point of –10.4 °C undergoes a solid state phase transition at –40.2 °C with $\Delta H = 0.038$ kcal/mol, and the two polymorphous structures possess different structural parameters (Table 2) [85, 86]. A similar behavior was also observed for *meso*-bis(bicyclopropylidenyl) *meso*-**87** [56]. The structure of *meso*-**87** represents essentially a combination of the two bicyclopropylidene units; the central bond length **h** (1.494 Å, Table 2) which is practically the same as that in bicyclopropyl (1.492 Å) indicates the absence of any strong electronic interaction between the two bicyclopropylidene units in *meso*-**87** which should thus react independently of each other.

3.2 Thermochemical Properties: Strain in Bicyclopropylidene

Although strain cannot be rigorously defined, the concept of strain and strain energies provides a basis that helps to correlate the structures, stabilities, and reactivities of molecules [92, 93]. A quantitative assessment of strain and strain energies (SE) can only be made by taking the difference between the enthalpy of formation $\Delta H_f^\circ(g)$ of the substance under consideration (theoretically calculated or experimentally determined) and that of a hypothetical strain-free model [93], for example, a group increments (GI) model like the one proposed by Schleyer et al. [94, 95]. However, experimental energetic parameters have been reported only for the parent bicyclopropylidene (**1**) [77, 96] (Table 3). The attempted evaluation of $\Delta H_f^\circ(g)$ for **1** using the homodesmotic Eq. (1) led to an overestimate (for a definition of the term "homodesmotic" see [92], p. 7).

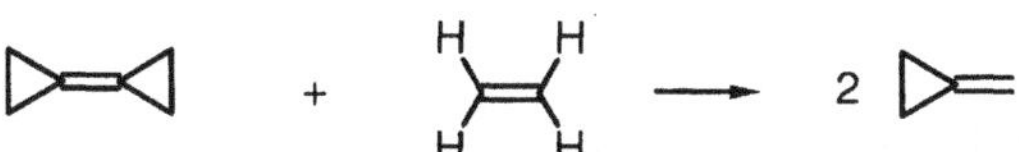

1 2 (1)

The comparison of SEs for alkenes **1**, **2**, and methylenespiropentane **6** with those for cyclopropane and the linear triangulanes **96**, **97** is shown in Table 4.

The additivity of strain energy increments appears to be a general feature for the triangulanes [95] and, most probably, also for their synthetic precursors [89], bicyclopropylidene (**1**) and spirocyclopropanated bicyclopropylidenes. As much as the strain energy of spiropentane (**96**) is more than twice that of cyclopropane, i.e., the spirojunction of two cyclopropane rings leads to an additional

Table 3. Calculated and experimentally determined ΔH_f^o(g) and SE for bicyclopropylidene (1) (all values in kcal/mol)

Method	4-31G	Eq. (1)[a]	Exp.
ΔH_f^o	75.5	83.5	77.5
SE	77.4[b]	83.4	77.4
Ref.	[89]	this work	[96]

[a] Calculated according to Eq. (1) with ΔH_f^o (g =48.0 for **2** [94] and 12.52 kcal/mol for ethylene [97]. A group increments model [94] was used to estimate SE.

[b] A strain-free model proposed by Schulman et al. [93b] was used to estimate SE.

Table 4. Strain energies (SE) in cyclopropane and compounds **1**, **2**, **6**, **96**, **97** (all values in kcal/mol)

Compound			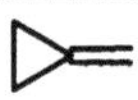 **2**		 **96**		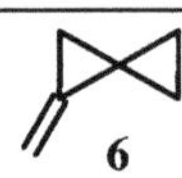 **6**		 **1**		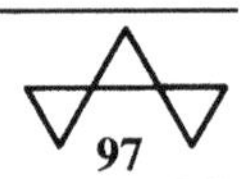 **97**
SE	28.1		41.7		65.1		74.6		77.4		98.5
ΔSE		13.6		23.4		9.5		2.8		21.1	
Ref.	[94]		[94]		[94]		[96]		[96]		[95]

strain energy increment ΔSE of 8.6 kcal/mol [95], the SE of bicyclopropylidene (**1**) exceeds the sum of SEs of methylenecyclopropane (**2**) and cyclopropane by 7.6 kcal/mol (77.4−41.7−28.1), and the SE for methylenespiropentane (**6**) exceeds the sum of those for cyclopropane and methylenecyclopropane (**2**) by 4.8 kcal/mol (74.6−28.1−41.7). Assuming that additivity also holds for oligospirocyclopropanated bicyclopropylidenes, their SEs can be assessed using a number of basic parameters such as 28.1 (SE of cyclopropane), 41.7 (SE of methylenecyclopropane 2), 8.6, 7.6, and 4.8 kcal/mol (excess increments ΔSE for spiropentane, bicyclopropylidene, and methylenespiropentane, respectively), but the lack of experimental data does not allow us to verify this scheme for spiropentane, bicyclopropylidene, and methylenespiropentane linkages.

On the other hand, as much as the increase in SE on going from methylenecyclopropane (**2**) to spiropentane (**96**) is substantial (23.4 kcal/mol), ΔSE is 21.1 kcal/mol on going from bicyclopropylidene (**1**) to [3] triangulane (**97**) (Table 4). As an expression of its increased SE, the double bond in **1** demonstrates a uniquely enhanced reactivity compared to the one in methylenecyclopropane (**2**) (see below).

4
Chemical Behavior

4.1
Thermal Rearrangements and Electron-Transfer Reactions

Crandall et al. first showed that hexamethylbicyclopropylidene **39** rearranges at 400 °C in a flow pyrolysis system to yield a 10:1 mixture of two isomeric hexamethylmethylenespiropentanes **98** and **99** (Scheme 18) [49, 50] formed via trimethylenemethane diradical intermediates. In close analogy to this result, the photolysis of the diazo compound **100** did not only give permethylbicyclopropylidene **24a**, but a 45:55 mixture of **24a** and **101**, and upon flash vacuum pyrolysis at 400 °C only **101** was formed (the yields were not reported) (Scheme 18) [98].

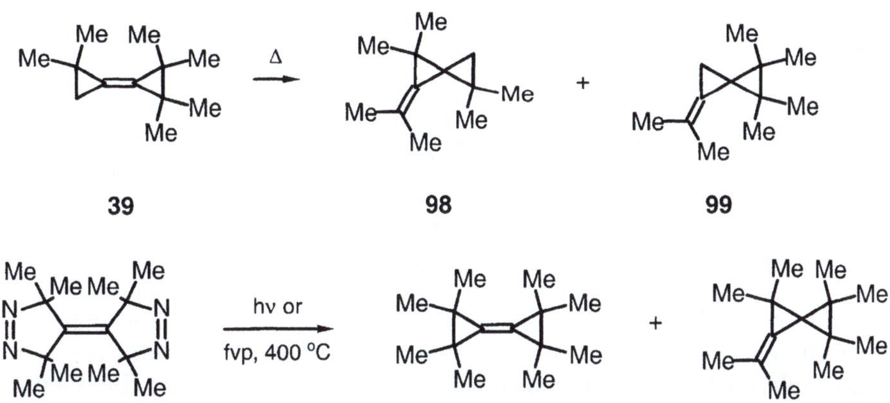

39 **98** **99**

100 **24a** **101**

Scheme 18. Thermal rearrangement of oligomethylated bicyclopropylidenes **39** and **24a** [49, 50, 98]

Analogous transformations have also been reported for the tetramethylbicyclopropylidenes **24b,c** [99, 100], the bicyclopropylidene **22a** of Moore and Ward [101], tetramethyldichlorobicyclopropylidene **28a** [46], and 1,1-dideuteriobicyclopropylidene [51]. The mechanistic and kinetic aspects of these rearrangements as well as subsequent transformations of the resulting methylenespiropentanes to dimethylenecyclobutane derivatives at higher temperatures have been reviewed [102].

Bicyclopropylidene (**1**) undergoes a clean rearrangement to methylenespiropentane (**6**) when heated to 330 °C in a flow system [103]. When heated in a closed vessel as a pure compound [1, 3] or in solution (50% in toluene) [95], a substantial fraction of **1** dimerizes to yield [4]rotane (**102**) (Scheme 19) [104].

The thermodynamic driving force for the interconversion of **1** to **6** is not very pronounced (see Table 4), and thus should be expected to be reversible. Al-

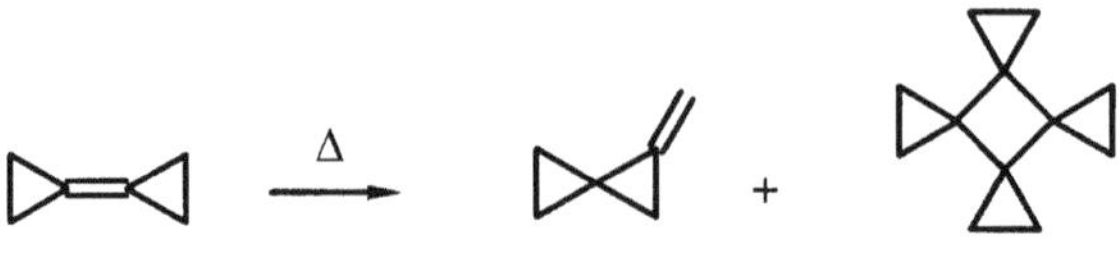

1 6 102

T [°C]	t [h]	Conditions	6 (%)	102 (%)
230	6	flow pyrolysis	no reaction	
330	6	flow pyrolysis	85	-
210	4	sealed ampoule	55	35
170	84	50% soln in toluene, sealed ampoule	18	68

Scheme 19. Thermal transformations of bicyclopropylidene (**1**) under various conditions [1, 3, 95, 103, 104]

though for a long time, this transformation was believed to proceed irreversibly [51, 101], kinetic measurements eventually revealed that this rearrangement is indeed reversible [103a].

The thermolysis of monospirocyclopropanated bicyclopropylidene **55** leads to a mixture of both possible isomeric products **103** and **104** [105], but prolonged heating of the sterically congested perspirocyclopropanated bicyclopropylidene **62** gave, along with polymeric materials, two different rearranged dimerization products **105** or **106** depending on the conditions (Scheme 20) [106].

As a rule, the thermal rearrangements of functionally substituted bicyclopropylidenes of type **107** yield mixtures of compounds (Scheme 21) [103, 107, 108]. But the methoxycarbonyl-substituted carbamate **53** underwent a clean regioselective isomerization to give **54** as a single product [55].

55 **103** (87%) **104** (13%)

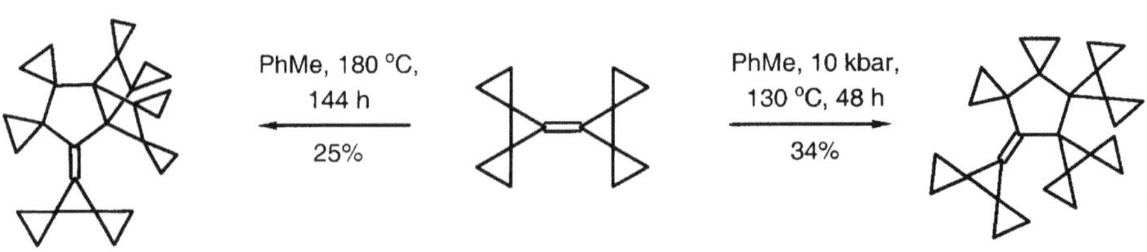

105 **62** **106**

Scheme 20. Thermal transformations of spirocyclopropanated bicyclopropylidenes **55** and **62** [105, 106]

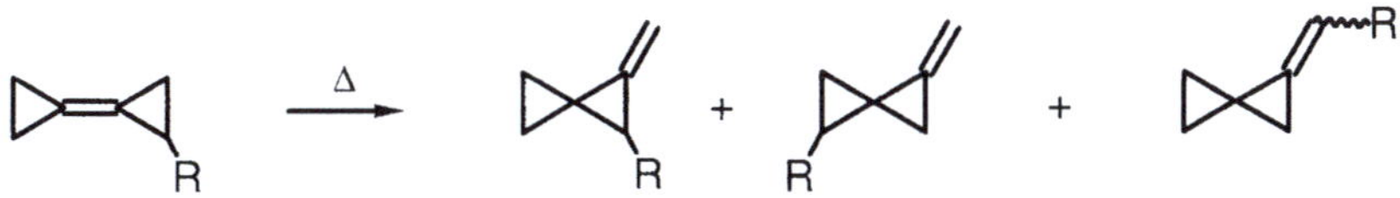

107 **108** **109** **110**

R	T [°C]	t [h]	Yield (%)		
			108	**109**	**110**
CH_2OH (**44a**)	200	5	62	28	10
COOEt (**42-Et**)	200	5	57	–[a]	43
 (**43m**)	200	3.5	47	31	17
$NHCOO(CH_2)_2TMS$ (**46b**)	80	6	10	–[a]	–[a]

[a] Not detected.

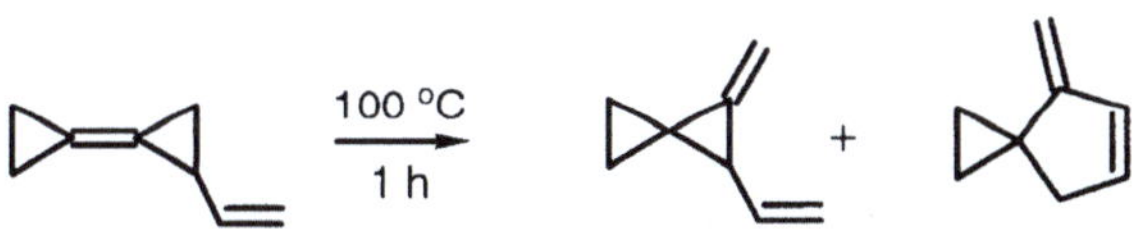

48 **111** (63%) **112** (36%)

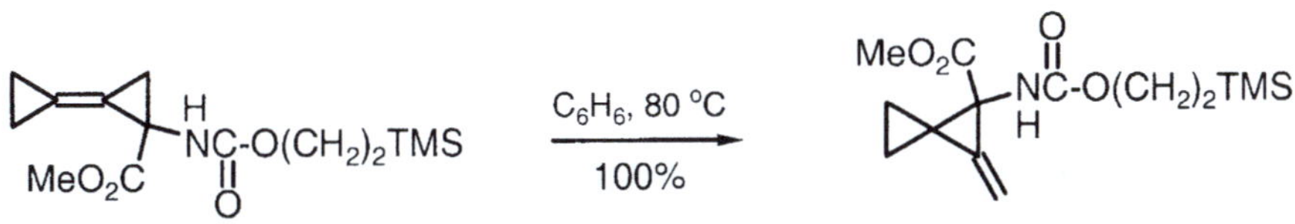

53 **54**

Scheme 21. Thermal rearrangements of substituted bicyclopropylidenes [55, 103, 107, 108]

A remarkable reorganization has been observed for 2,2-diarylmethylenespiropentanes **113** after photoinitiated electron transfer from 9,10-dicyanoanthracene (DCA) leading to the reversible formation of 2,2-diarylbicyclopropylidenes **114**, the proportion of which decreases with an increase of the electron-donating ability of the aryl substituents (Scheme 22) [109].

The biscyclohexane-annelated bicyclopropylidene **22a** has been observed to undergo a reversible *syn-anti* isomerization under ZnI_2 catalysis [110] to end up as a 1:3 mixture of *syn-* and *anti-*isomers [16] (Scheme 22).

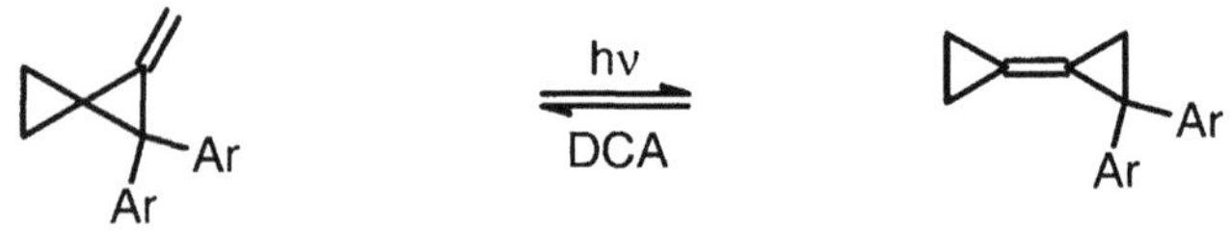

113 **114**

Ar	**113** (%)	**114** (%)
Ph	13	87
4-MeC_6H_4	23	77
4-$MeOC_6H_4$	40	60

DCA =

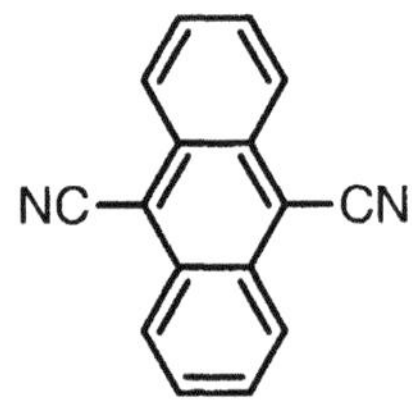

syn-**22a** 1 : 3 *anti*-**22a**

Scheme 22. Photoinitiated reversible reorganization of 2,2-diarylmethylenespiropentanes **113** and ZnI_2-promoted reversible *syn-anti* isomerization of bicyclopropylidenes **22a** [16, 109]

4.2 Reactivity Towards Electrophiles and Cycloaddends

4.2.1 *Carbene Additions Leading to So-Called Triangulanes*

The addition of different types of carbenes onto bicyclopropylidene (**1**) is a common method for the preparation of [3]triangulane derivatives as well as branched triangulanes and normally proceeds without complications (for a review see [77]). Thus, the cyclopropanation under Gaspar-Roth [60] or modified Simmons-Smith [111] conditions gave dispiro[2.0.2.1]heptane ([3]triangulane, **97**) in 80 [105] and 15% yield [5], respectively (Scheme 23). The palladium(II) acetate-catalyzed cyclopropanation of **1** with diazomethane, however, gave a number of products resulting from insertion of one or more than one methylene units into an initially formed palladacyclobutane **115** [112, 113] (Scheme 23).

(*M*)-(–)-Trispiro[2.0.0.2.1.1]nonane [(*M*)-(–)-**118**], the first enantiomerically pure unbranched [4]triangulane, has been prepared from enantiomerically pure bicyclopropylidenecarboxylic acid [(*R*)-(–)-**42**-H] starting with the cyclopropanation under Simmons-Smith conditions of its ethyl ester (*R*)-**42**-Et as a key step (Scheme 23) [59]. In spite of the fact that [4]triangulane has no chromophore which would lead to any significant absorption above 200 nm, it has a remarkably high specific rotation even at 589 nm with $[\alpha]_D^{20} = -192.7$ ($c = 1.18$, $CHCl_3$). Its outstanding rotatory power is in line with its helical arrangement of sigma bonds. Thus, it is appropriate to call [4]triangulane a "σ-[4]helicene", representing the first σ-bond analogue of the aromatic [*n*]helicenes.

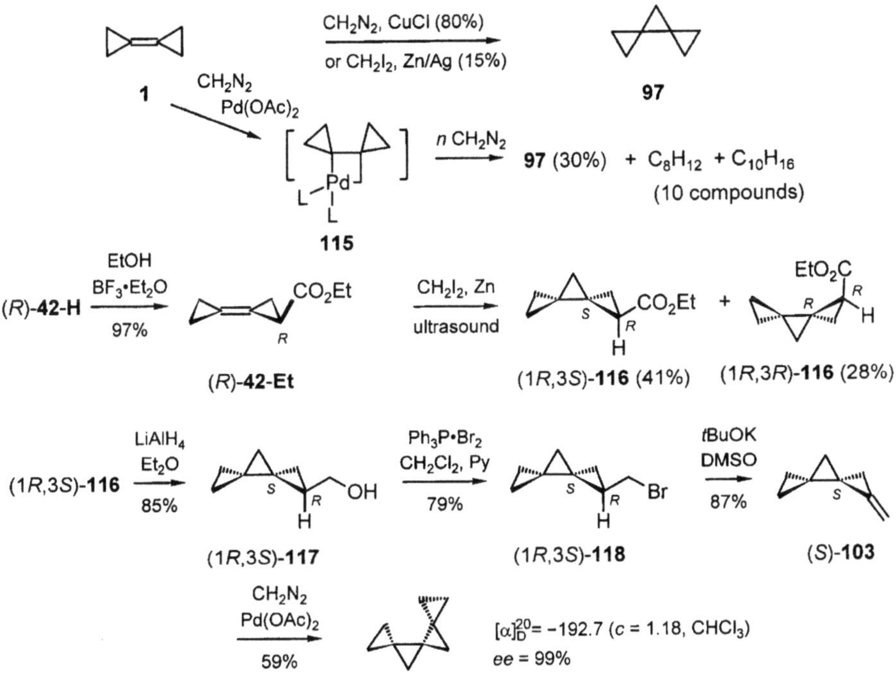

(1R,3S)-116 (41%) (1R,3R)-116 (28%)

(1R,3S)-117 (1R,3S)-118 (S)-103

(M)-(–)-119

Scheme 23. The cyclopropanation of bicyclopropylidene (1) and of enantiomerically pure ester (*R*)-42-Et as a key step in the preparation of enantiomerically pure (*M*)-[4]triangulane [(*M*)-**119**] [59, 112, 113]

The addition of dichloro- [114] and dibromocarbenes generated from Ph-$HgBr_3$ [63] or bromoform [33, 34, 115] gave the corresponding 7,7-dihalotrispiro[2.0.2.1]heptanes **120** in 86–95% yields (Scheme 24).

The additions of difluorocarbene onto the bicyclopropylidene derivative **121** (Scheme 24) [116a], of dibromocarbene onto oligospirocyclopropanated bicyclopropylidenes **56** [33, 65] (66% yield) and **62** [117] (94% yield), of dichlorocarbene onto **56** [33] (90% yield), **28a** [118] (38% yield) and **121** (99% yield) [116a] as well as of chloro(phenylthio)carbene (90% yield) [33, 119], of bromo-

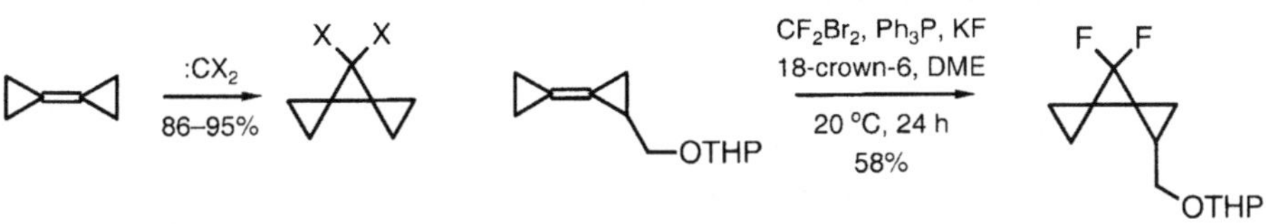

1 120a X = Cl b X = Br 121 122

Scheme 24. The addition of dihalocarbenes to bicyclopropylidene (**1**) and its derivative **121** [33, 34, 63, 115, 116a] phenylcarbene (77% yield) [119] and of phenylthiocarbene (71% yield) [119] onto **1** have also been performed. The absolute rates for the addition of chlorophenylcarbene and fluorophenylcarbene onto bicyclopropylidene (**1**) at 25 °C have been determined with rate coefficients of $k = 3.36 \times 10^8$ and 2.96×10^8 mol^{-1} · s^{-1}, respectively [116b]. For comparison, the corresponding rate coefficients for the addition of these carbenes onto tetramethylethylene were 2.8×10^8 and 1.6×10^8 mol^{-1} · s^{-1}, respectively [116b]. The cyclopropanation with dichloro- or dibromocarbenes of oligocyclic bicyclopropylidenes of type **22** was used in the preparation of "tube-like" molecules or "tunable molecular clefts" [25, 29]. However, the addition of dibromocarbene onto the bicyclopropylidene *anti*-**22a** could not be achieved, probably due to steric overcrowding [16].

The addition of an alkoxycarbonylcarbene onto **1** and oligospirocyclopropanated bicyclopropylidenes has been discussed above (see Scheme 13). It is noteworthy, however, that sterically congested bicyclopropylidenes like **24**, **62**, when treated with carbene-generating reagents, can yield by-products like **124** [120] and **126** [117, 121] (Scheme 25).

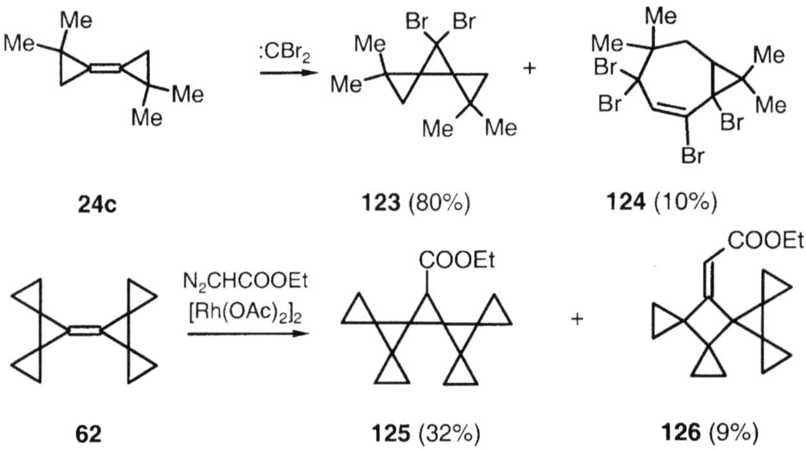

Scheme 25. Carbene cycloaddition onto sterically congested bicyclopropylidenes **24c**, **62** [117, 120, 121]

The addition of methylchlorocarbene onto bicyclopropylidene (**1**) (Scheme 26) and functionalized bicyclopropylidenes like **44a** and **121** has been used as a key step in the synthesis of branched triangulanes [53, 122–124]. This was applied

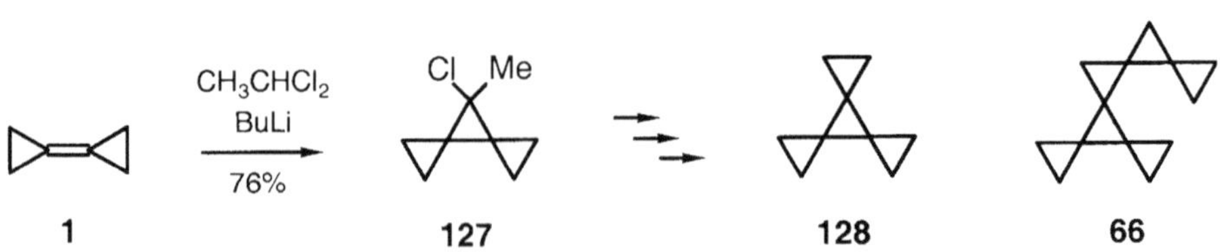

Scheme 26. Application of methylchlorocarbene addition onto **1** for the preparation of branched [*n*]triangulanes [65, 122, 123]

in the preparation of [3]rotane (**128**) [122, 123] and, after additional ubsequent transformations, compounds like **66** [65].

[3]Rotane (**128**) was also prepared by addition of cyclopropylidene liberated from in situ generated diazocyclopropane [62] onto **1** [65, 105] (Scheme 27). This carbene adds also to bicyclopropylidenes **56** and **62** producing triangulanes **129** [33, 65] (Scheme 27) and **67** [33, 64, 65] (Scheme 11).

An alternative approach to the perspirocyclopropanated spiropentane **129** includes initial addition of tetrahydropyranyloxyethylchlorocarbene [65] onto **1** (Scheme 27), but this carbene did not react with the sterically congested bicyclopropylidene **62** [117].

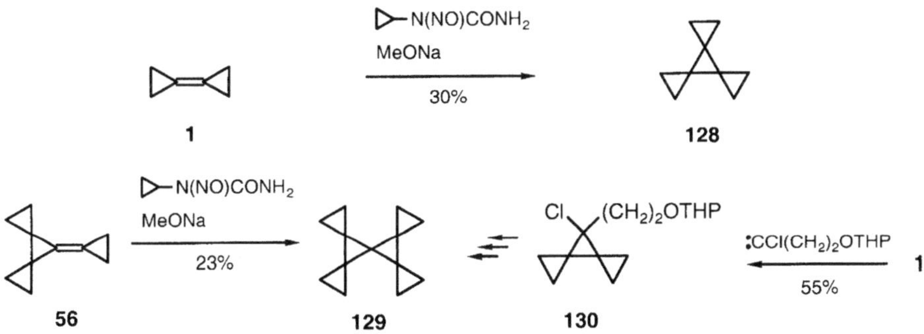

Scheme 27. Preparation of branched triangulanes **128, 129** via carbene cycloadditions [33, 65, 105]

The addition of oligospirocyclopropanated cyclopropylidenes **131** and **132** onto bicyclopropylidene (**1**) has also been probed (Scheme 28) [77, 117]. Carbenes **131** and **132** were generated from the corresponding *N*-nitrosourea compounds. The addition of such carbenes may serve as the only possible approach to compounds like the highly strained perspirocyclopropanated [3]rotane **133** [117].

Bicyclopropylidene (**1**) reacts with dimethoxycarbene generated thermally from the diazirine at 25 °C to give a complex mixture of products [116b]. The cycloaddition of several dialkoxycarbenes generated in situ from the corresponding 2,2-dialkoxy-Δ^3-1,3,4-oxadiazolines of type **134** with bicyclopropylidene (**1**) affords the dialkyl acetals of dispiro[2.0.2.1]heptanone **135** (Scheme 29) [125].

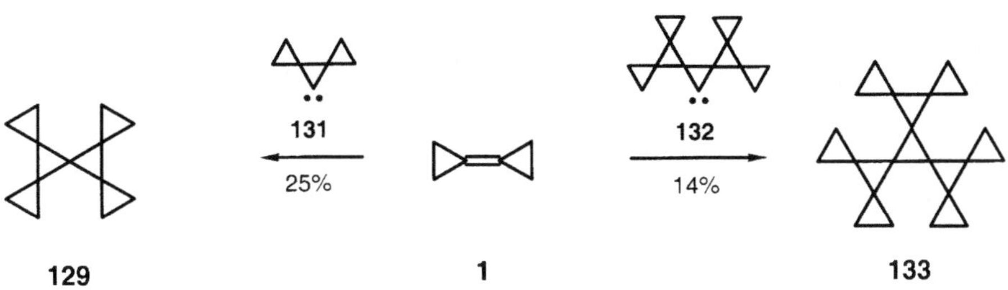

Scheme 28. Cycloaddition of oligospirocyclopropanated cyclopropylidenes onto bicyclopropylidene (**1**) [77, 117]

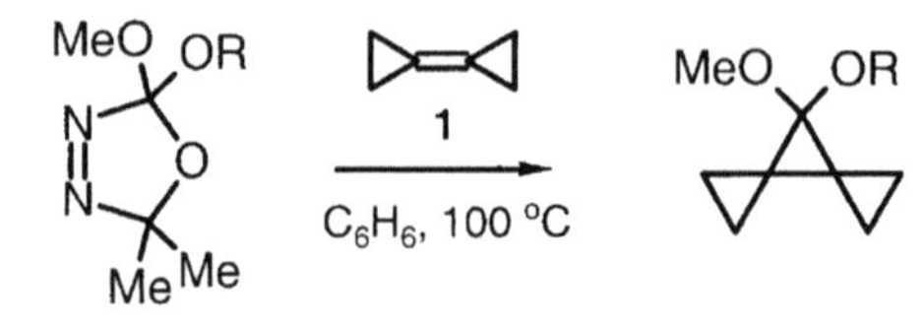

135a, R = Me (71%)
135b, R = *t*Bu (37%)
135c, R = $CH_2CH_2SiMe_3$ (91%)
135d, R = CH_2CH_2Cl (72%)
135e, R = CH_2CH_2Br (56%)

134

Scheme 29. Cycloaddition of dialkoxycarbenes onto bicyclopropylidene (**1**) [125]

The attempted transacetalization of the dimethyl acetal **135a** was partially successful only when it was boiled in excess ethanol in the presence of boron trifluoride etherate. The formation of the corresponding diethyl acetal **137** indicates that dialkyl acetals of dispiro [2.0.2.1]heptan-7-one can, in principle, enter reactions without ring opening and thus be applied as a synthetic equivalent of [3]triangulanone (Scheme 30) [125].

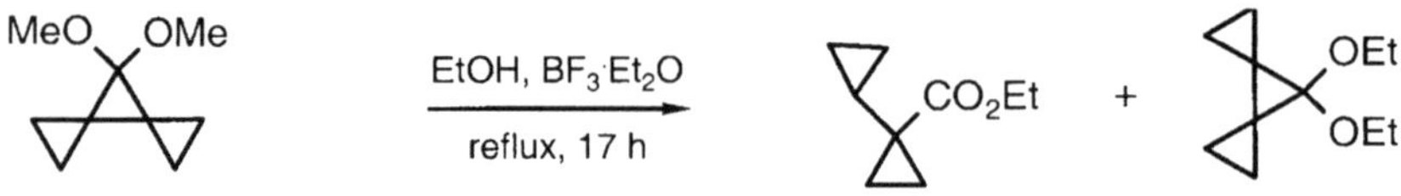

135a **136** (85%) **137** (9.5%)

Scheme 30. Attempted deprotection of acetal **135a** [125]

The attempted deprotection of acetals **135** usually gave the products of ring opening. Only in two reactions was the triangulane skeleton conserved. Treatment of **135d** with *tert*-butyllithium afforded in high yield 7-*tert*-butyldispiro [2.0.2.1]heptanol (**138**). Reaction of **135e** with *n*BuLi gave only 10% of dispiroheptanol **140**, and the three products **139**, **141**, and **142** formed with opening of the central cyclopropane ring (Scheme 31) [125]. Several of these products

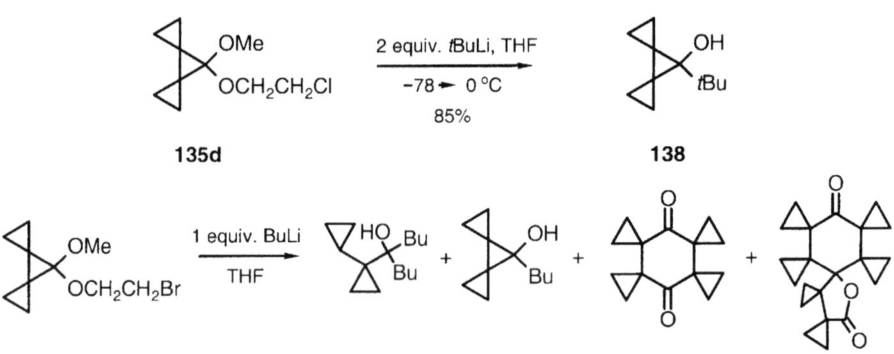

135e **139** **140** **141** **142**

	139	140	141	142
−78 °C	15%	10%	8%	0%
−110 °C	7%	10%	15%	26%

Scheme 31. Attempted deprotection of acetals **135d, e** with alkyllithiums [125]

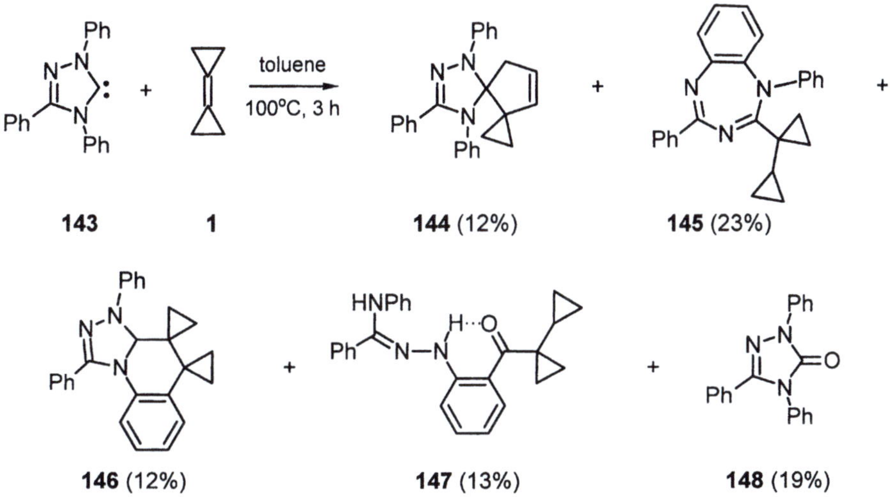

Scheme 32. Reaction of bicyclopropylidene (**1**) with the stable carbene 1,3,4-triphenyl-4,5-dihydro-1*H*-1,2,4-triazol-5-ylidene (**143**) [126]

138–142 apparently arise by intermediate formation of dispiro[2.0.2.1]heptan-7-one.

The stable carbene 1,3,4-triphenyl-4,5-dihydro-1*H*-1,2,4-triazol-5-ylidene (**143**) reacts with bicyclopropylidene (**1**) to yield the four unexpected products **144–147**, none of which resembles the typical mode of [2+1] cycloaddition observed for reactions of **1** with other carbenes (Scheme 32) [126].

4.2.2
Diels-Alder Reactions and Cycloadditions onto Electron-Deficient Cycloaddends

Bicyclopropylidene (**1**) is capable of undergoing cycloadditions by different modes, depending on the nature of the cycloaddend. Whereas cyclopentadiene (**149a**) gave the [4+2] cycloadduct **150a** only, the reactions with 1,3-cyclohexadiene (**149b**) and 1,3-butadiene (**83**) led to mixtures of the [4+2] cycloadducts **150b**, **152** and the [2+2] cycloadducts **151**, **153**, with an increasing proportion of the latter in this order (Scheme 33) [104, 127].

Only [4+2] cycloadducts **155a, b** were formed and isolated in 48 and 43% yield, respectively, in the reaction of bicyclopropylidene (**1**) with thiophene *S*-oxide derivatives **154a, b** under high pressure (Scheme 33) [128].

The reaction of bicyclopropylidene (**1**) with 1,2,4,5-tetrazine (**156**), a diene with inverse electron demand, afforded two stereoisomeric products of type **159**, trimers of the 8,9-diazadispiro[2.0.2.4]deca-7,9-diene (**158**), obviously formed via the normal [4+2] cycloadduct **157** (Scheme 34) [104].

Bicyclopropylidene (**1**) does not undergo an intermolecular Diels-Alder reaction with furan and 2-methoxyfuran even under high pressure. Intramolecular cycloadditions of compounds **160** with a furan tethered to bicyclopropylidene, however, were easily brought about under high pressure (10 kbar) and gave cycloadducts **161** stereoselectively in yields ranging from 32 to 95% (Scheme 35) [58].

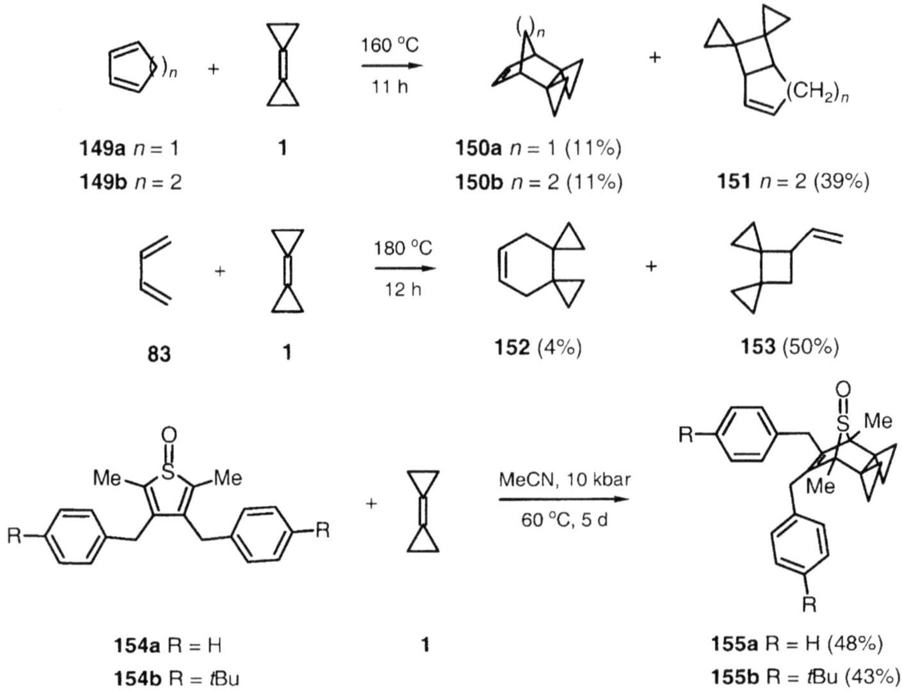

Scheme 33. Two modes of addition of acyclic and cyclic dienes onto bicyclopropylidene (**1**) [104, 127, 128]

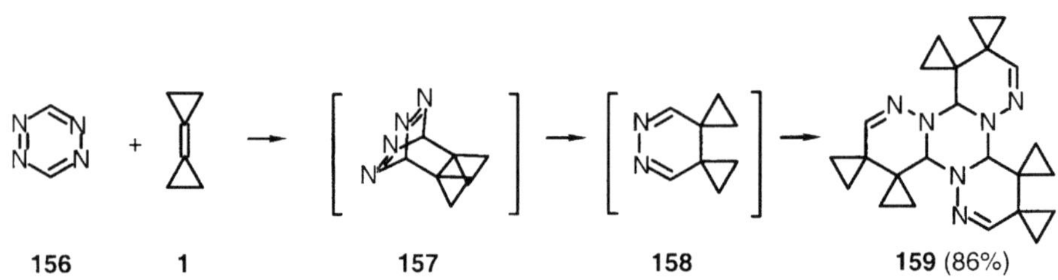

Scheme 34. The reaction of bicyclopropylidene (**1**) and 1,2,4,5-tetrazine (**156**) [104]

The kinetics of these reactions in comparison with those for methylenecyclopropane analogs of compounds **160** have been studied by following the progress at pressures up to 3 kbar by on-line FT-IR spectroscopy [129]. The rate-enhancing influence of the additional strain in **160** overcompensates the expected retarding effect of the increased steric shielding by the second cyclopropane unit in **1** compared to methylenecyclopropane, and the cyclization rates for compounds **160** were faster by a factor of 6.8 to 8.1 in comparison with the corresponding methylenecyclopropane derivatives.

Bicyclopropylidene (**1**) reacts with electron-deficient cycloaddends in different ways, depending on the partner. Trichloroethylene and acrylonitrile under-

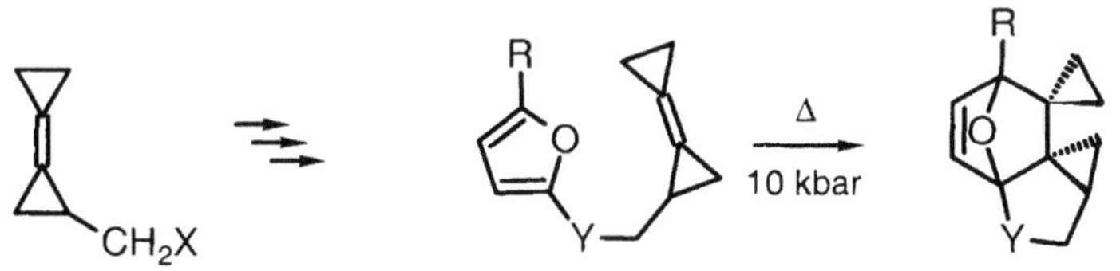

44a X = OH, **44c** X = OSO_2Ph **160** **161**
43i X = CH_2OH, **47c** X = CH_2I

R	H	MeO	H	H
Y	CH_2	CH_2	CH_2O	Me_2SiO
Yield of **161** (%)	32	95	46	83

Scheme 35. Intramolecular Diels-Alder reactions of bicyclopropylidene derivatives **160** [58]

went cycloaddition onto **1** at elevated temperatures yielding the corresponding [2+2] adducts **162** and **163**, respectively. Fumaronitrile upon reaction with **1** gave rise to both *trans*-(**164**) and *cis*-7,8-dicyanodispiro[2.0.2.2]octanes (**165**). The vinyldispiro[2.0.2.2]octane derivative **168** was isolated from the reaction of **1** with 1,2-dicyanocyclobutene (**166**). Apparently, **166** underwent ring-opening to **167**, and then **1** was added across one of the double bond of **167** to give **168** (Scheme 36) [104, 130].

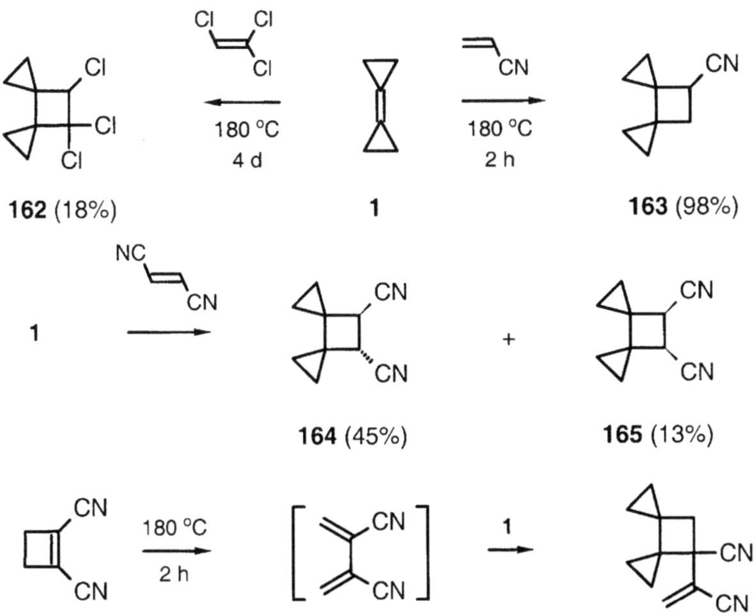

Scheme 36. Cycloaddition reactions of bicyclopropylidene (**1**) with substituted ethenes [104, 130]

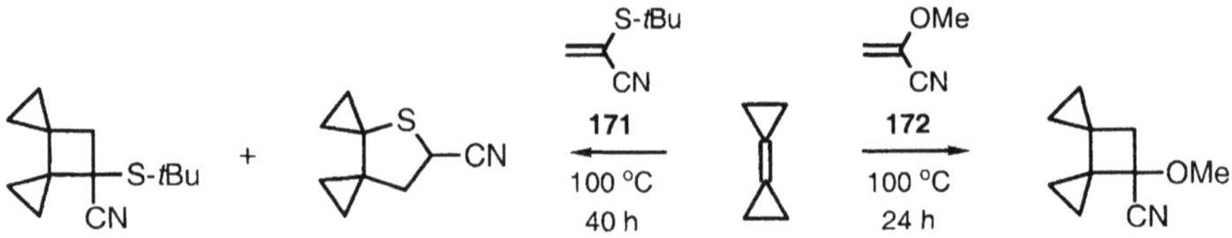

169 (34%) 170 (24%) 1 173 (59%)

Scheme 37. Reactions of capto-dative olefins **171** and **172** with bicyclopropylidene (**1**) [131]

Capto-dative olefins **171** and **172** also underwent formal [2+2] cycloadditions with **1** to yield compounds **169**, **170**, and **173**, respectively (Scheme 37) [131].

All of the above-mentioned [2+2] cycloadducts presumably arise via intermediate 1,4-diradicals in a stepwise manner [104, 131]. However, cycloadducts **174** and **175** of **1** to chloro- and dichloroketene most likely arise from a concerted $[_{\pi}2_s + _{\pi}2_a]$ cycloaddition mode (Scheme 38) [104, 130].

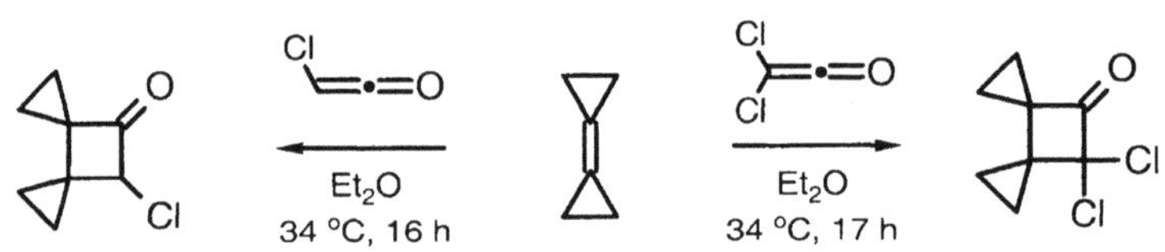

174 (63%) 1 175 (59%)

Scheme 38. [2+2] Cycloadditions of bicyclopropylidene (**1**) to chloro- and dichloroketene [104, 130]

Sulfuryl chloride isocyanate with **1** gave the expected β-lactam **177** only as a minor product, the principal product being the γ-lactam derivative **178**. It is reasonable to assume that the 1,4-zwitterionic intermediate **176** is responsible for the formation of **177** and **178** (Scheme 39) [104, 130]. Sulfonation of **1** with SO_3 also proceeds with ring opening of one of the cyclopropyl groups to give quantitatively the spirocyclopropane-γ-sultone **179** (Scheme 39) [76, 132].

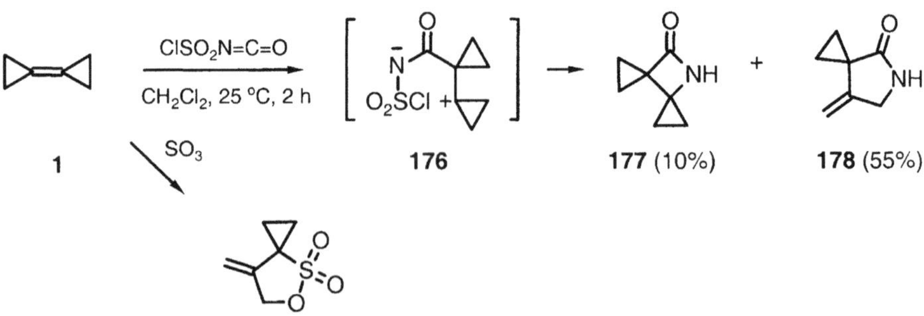

1 176 177 (10%) 178 (55%)

179 (≈ 100%)

Scheme 39. Reactions of bicyclopropylidene (**1**) with sulfuryl chloride isocyanate and SO_3 [76, 104, 130, 132]

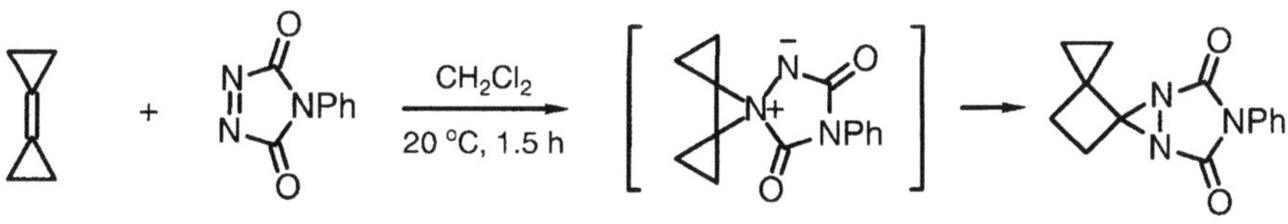

1 180 181 182 (83%)

Scheme 40. Reaction of bicyclopropylidene (**1**) with 4-phenyl-1,2,4-triazoline-3,5-dione (**180**) [104, 130]

With 4-phenyl-1,2,4-triazoline-3,5-dione (**180**), bicyclopropylidene (**1**) reacted by yet another mode and gave the spiroannelated diaziridine **182** (Scheme 40) [104, 130].

All three modes of cycloadditions described above for various electron-deficient cycloaddends were realized in the reaction of **1** with tetracyanoethylene (TCNE) which gave a mixture of cycloadducts **183–185** (Scheme 41) [104, 130].

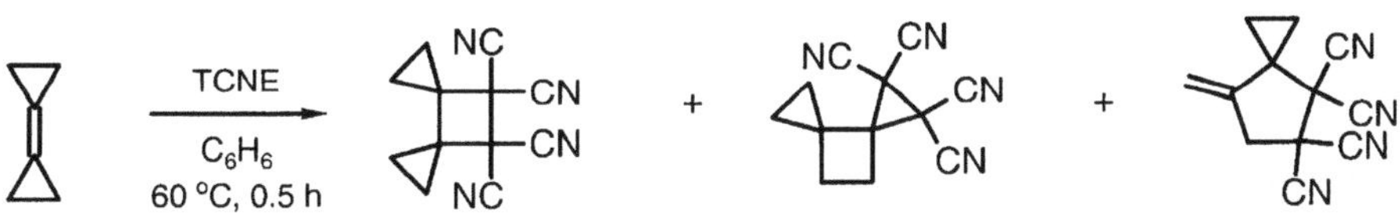

1 183 (17.5%) 184 (17.5%) 185 (45%)

Scheme 41. Reactions of bicyclopropylidene (**1**) with TCNE [104, 130]

4.2.3
Electrophilic and Radical Additions

In the investigation of electrophilic additions onto bicyclopropylidene (**1**) it was of prime interest to answer at least the two prominent questions of whether a cyclopropyl substituent can efficiently stabilize a cyclopropyl cation and prevent it from undergoing ring opening and whether the rate of such an electrophilic addition adequately correlates with the strain in the molecule. Surprisingly, the published data in this context are scarce. Thus, the addition of mercuric acetate to **1** led to a mixture of **186** and **187** in a ratio of 8:1 and a total yield of 97%. It was concluded that both products arise as a result of initial attack by the mercuric cation on the double bond of **1**. Apparently, the bicyclopropyl derivative **186** was formed by the usual 1,2-acetoxymercuration without ring opening of the intermediate 1-cyclopropylcyclopropyl cation, only the minor product **187** arose by cyclopropyl to allyl cation rearrangement (Scheme 42) [133].

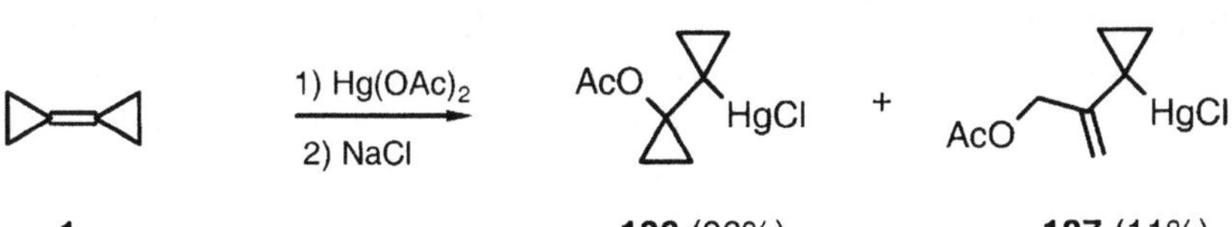

1 186 (86%) 187 (11%)

Scheme 42. Oxymercuration of bicyclopropylidene (**1**) [133]

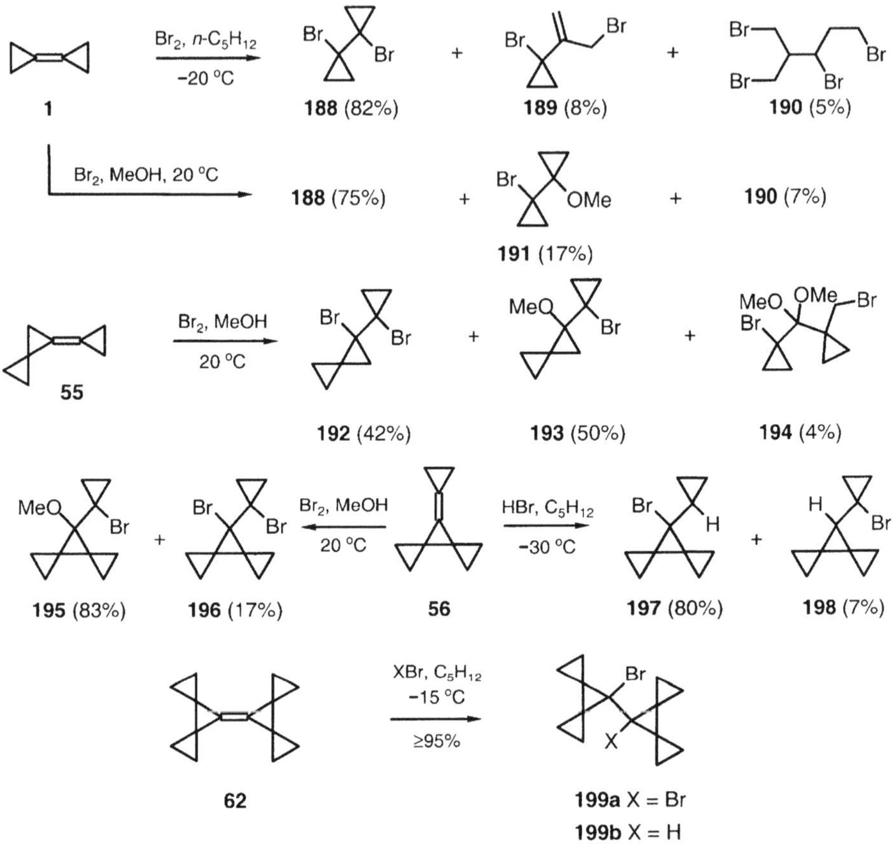

Scheme 43. Bromination and hydrobromination of bicyclopropylidene (**1**) and spirocyclopropanated bicyclopropylidenes **55, 56, 62** [33, 64, 119, 134]

Bromine additions to bicyclopropylidene (**1**) as well as spirocyclopropanated bicyclopropylidenes **55, 56** have been performed in methanol at 25 °C. An increasing number of spiroannelated three-membered rings was found to stabilize the intermediate cyclopropyl cations against ring opening (Scheme 43) [134]. Thus, the bromination as well as hydrobromination of di- and tetraspirocyclopropanated bicyclopropylidenes **56** and **62** proceeded with complete retention of all cyclopropane rings (Scheme 43) [33, 64, 119, 134].

The recently determined kinetic data for the bromination of bicyclopropylidene (**1**) and spirocyclopropanated bicyclopropylidenes **55, 56** in methanol at 25 °C disclose that the addition of Br_2 onto the double bonds in **1, 55, 56** proceeds essentially with the same rate as the bromination of corresponding oligomethylated ethylenes. The bromination rate increases with an increasing number of spiroannelated three-membered rings, and the rate of bromination correlates with the π-ionization energies of the molecules (Table 5) [134].

The addition of thiols **200a–h** onto the double bond of bicyclopropylidene (**1**) in benzene proceeds rapidly at 20 to 75 °C in the absence of catalysts or radi-

Table 5. Rate coefficients (k_{BR_2}) of bromine addition to alkenes **1, 55, 56** and π-ionization energies (π-IE$_v$) [134]

Olefin	k_{Br_2} [mol^{-1}•l•s^{-1}]	$\lg k_{Br_2}$	π-IE$_v$ [eV]
 **1**	$(4.70 \pm 0.05) \times 10^4$	4.67	8.93
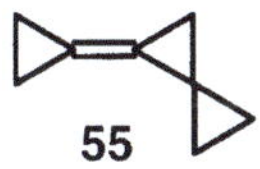 **55**	$(1.63 \pm 0.04) \times 10^6$	6.21	8.70
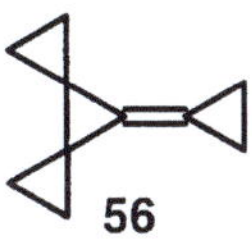 **56**	$(3.42 \pm 0.11) \times 10^7$	7.53	8.50

cal initiators to give products **201 a–h** almost quantitatively with complete retention of both three-membered rings (Scheme 44) [135]. The addition of thiols to *n*-alkylbicyclopropylidenes **43 e–g** does not proceed stereoselectively, though in all cases the mercapto function adds to the double bond with retention of the cyclopropane ring to give interesting new compounds containing bicyclopropyl fragments [135]. Apparently, the intermediate 1-(1′-alkylthiocyclopropyl)cyclopropyl radicals in this radical addition to **1** undergo ring opening far less rapidly than ordinary cyclopropylmethyl radicals.

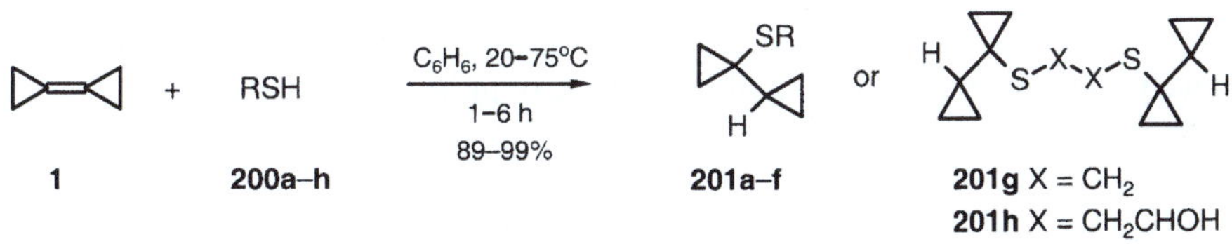

	a	b	c	d	e	f	g	h
R	Ph	Et	*n*Bu	$(CH_2)_2OH$	$(CH_2)_2COOH$	CH_2CO_2Et	$(CH_2)_2SH$	$CH_2(CHOH)_2CH_2SH$

Scheme 44. Radical addition of thiols and dithiols **200** onto bicyclopropylidene (**1**) [135]

The facile quantitative addition of mercaptoethanol **200d** onto the double bonds of bicyclopropylidene derivatives has successfully been applied to separate the difluorocarbene adduct **122** from the starting material **121** (see Scheme 24) [116].

4.2.4
Reactions with 1,3-Dipolar Cycloaddends

Ozonolysis of **1** gave rise to a mixture of the epoxide **202**, spiro[2.3]hexan-4-one (**203**), and 4-oxaspiro[2.4]heptan-7-one (**204**). While **202** and **203** apparently arise by 1,3-ring closure and cyclopropylcarbinyl to cyclobutyl ring enlargement, respectively, of an intermediate 1,3-zwitterion formed by oxygen split off from a 1,5-zwitterion en route to a primary ozonide, the major product **204** must be formed via an oxy-analog of a cyclopropylmethyl to homoallyl cation rearrangement of an intermediate 1,5-zwitterion formed by heterolytic ring opening of the primary ozonide between two oxygen atoms (Scheme 45) [104, 130].

1 **202** (22%) **203** (11%) **204** (67%)

Scheme 45. Ozonolysis of bicyclopropylidene (**1**) [104, 130]

Nitrones **205** react at ambient or slightly elevated temperature (60 °C) with bicyclopropylidene (**1**) to give the bis(spirocyclopropane)-annelated isoxazolidines **206**. Heating of compounds **206** in xylene solution at 110–125 °C led to a clean rearrangement with radical ring opening of the spirocyclopropane ring in the 5-position to give spirocyclopropane-annelated piperidones **207** after ring closure. The same sequence of cycloaddition and rearrangement can be achieved in a single operation with considerable benefit for the reaction yield by heating a nitrone **205** and **1** in xylene solution at 120 °C (Scheme 46) [136–138].

This reaction has recently been applied for the preparation of aza-analogs with the basic skeleton and functional groups of the extremely cytotoxic sesquiterpenes illudin (**208**) and ptaquiloside (**209**) [139]. Compounds **212**–**215** were successfully prepared by heating bicyclopropylidene (**1**) with the enantiomerically pure nitrones **210**, **211** (Scheme 47) [138]. Some of these simple aza-analogs have indeed been found to exhibit DNA-cleaving abilities [138b, c].

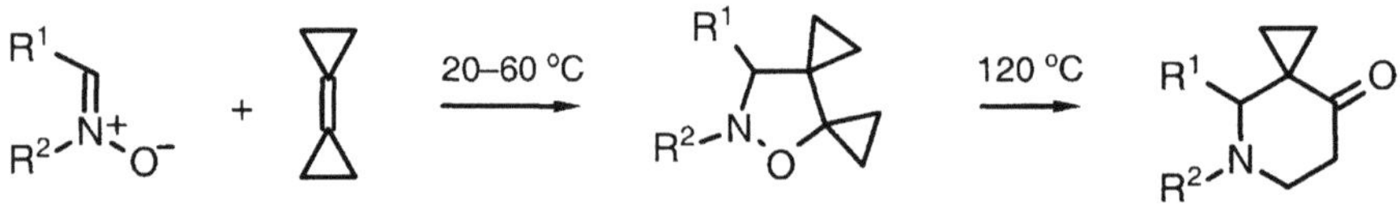

205 **1** **206** **207**

R^1	R^2	Yield of **206** (%)	Yield of **207** (%) (one-pot yields in parentheses)
Me	Ph	93	63 (61)
Me	CO_2Et	34	61 (54)
$-(CH_2)_3-$		42	60 (25)
$-(CH_2)_2-CMe_2-$		80	76 (80)
$-(CH_2)_4-$		37	68 (48)
o-$CH_2CH_2C_6H_4-$		73	50 (73)

Scheme 46. 1,3-Dipolar cycloadditions of nitrones **205** onto bicyclopropylidene (**1**) [136–138]

208a, R = OH
208b, R = H

209

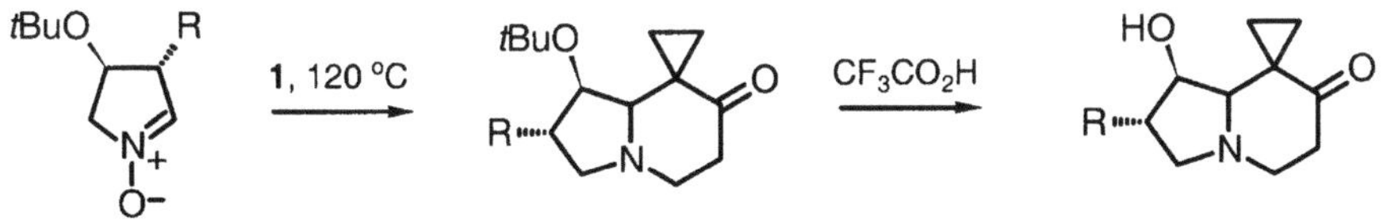

210, R = H
211, R = *t*BuO

212, R = H (64%)
213, R = *t*BuO (61%)

214, R = H (84%)
215, R = OH (93%)

Scheme 47. Preparation of aza-analogs **212**–**215** with the basic skeleton and functional groups of the sesquiterpenes illudin (**208**) and ptaquiloside (**209**) [138]

1,3-Dipolar cycloadditions of nitrile oxides **216** onto **1** gave much poorer yields of cycloadducts **217** than those of nitrones **205**. The cycloadditions of **216** to **1** require higher temperatures and unfavorably compete with their dimerization to furoxanes. However, stable nitrile oxides **216** with bulky substituents R that hamper dimerization, can be used. The thermal rearrangements of 5-spirocyclopropane-annelated isoxazolines **217** always required higher temperatures than the isoxazolidine counterparts. Under these conditions the second cyclopropane ring was also cleaved to give furopyridines **218** (Scheme 48) [136, 137].

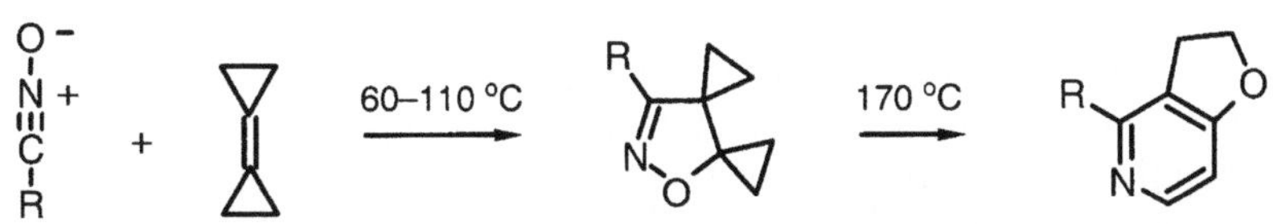

216 **1** **217** **218**

R	Yield of **217** (%)	Yield of **218** (%)
Me	10	0
Ph	40	44
2,4,6-$Me_3C_6H_2$	67	44
Ph_3C	17	21

Scheme 48. 1,3-Dipolar cycloadditions of nitrile oxides **216** onto bicyclopropylidene (**1**) [136, 137]

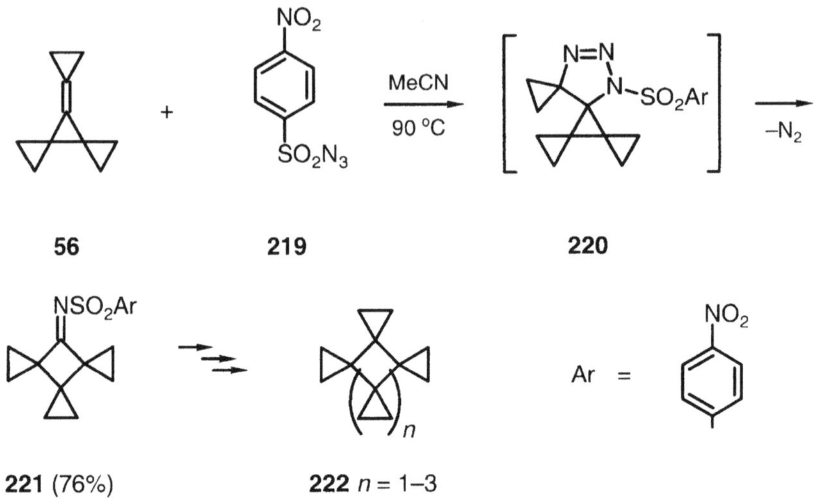

Scheme 49. 1,3-Dipolar cycloaddition of *p*-nitrobenzenesulfonyl azide (**219**) onto dispirocyclopropanated bicyclopropylidene **56** [140]

The 1,3-dipolar cycloaddition of *p*-nitrobenzenesulfonyl azide (**219**) onto dispirocyclopropanated bicyclopropylidene **56** has been used as a key step in a general repetitive procedure for the preparation of [*n*]rotanes **222** (Scheme 49) [140].

4.2.5
Miscellaneous

Bicyclopropylidene (**1**) readily reacted with photochemically generated singlet oxygen at 30–35 °C to give spiro[2.3]hexan-4-one (**203**) and 7-oxadispiro[2.0.2.1]heptane (**202**) [104, 141]. Bicyclopropylidene epoxide **202** can also be prepared by epoxidation of **1** with *m*-chloroperbenzoic acid in the presence of Na_2CO_3 [141] or with $KHSO_5$/acetone [74]. Bicyclopropylidene (**1**) was found to quench the fluorescence of 9,10-dicyanoanthracene [81].

The outstanding chemical properties of bicyclopropylidene (**1**) were once again exemplified by its unprecedented addition to tetrachlorocyclopropene (TCCP). Reaction between **1** and TCCP occurred at 80 °C and led to **223** which was readily hydrolyzed to give **224**. Adduct **223** apparently arises from a nucleophilic attack of **1**, most probably in a S_N2' fashion, on TCCP which corresponds to a chloroene reaction (Scheme 50) [142].

Treatment of bicyclopropylidene (**1**) with lithium powder led to ring opening and subsequent treatment of the reaction mixture with dimethyl sulfate afforded the methylenecyclopropane derivative **225** and 4-octyne (**226**) in variable ratios (Scheme 51) [143].

In protic solvents, alkylbicyclopropylidenes **43e, f** can be smoothly reduced with lithium to give a mixture of bicyclopropyl derivatives **227a, b** almost quantitatively [56a]. The stereoselectivity of these reductions with dissolving lithium

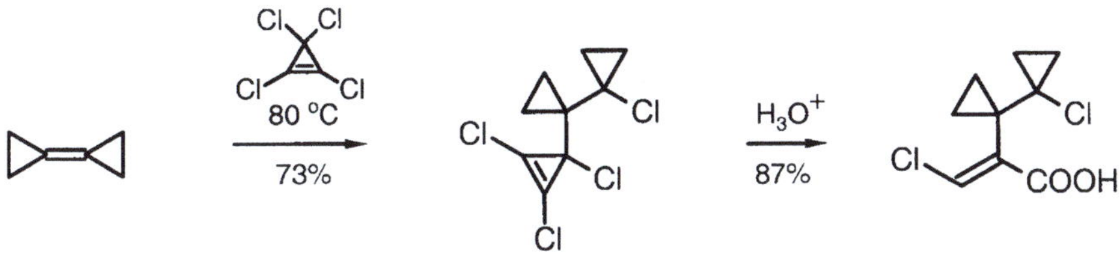

1 223 224

Scheme 50. Reaction between bicyclopropylidene (**1**) and tetrachlorocyclopropene [142]

is strongly affected by the solvent and the temperature, it is highly *trans*-stereoselective in liquid ammonia at –35 °C (Scheme 51). Along this route, the THP-protected (bicyclopropylidenyl) alcohols **43p–v** were reduced to the *trans*-bicyclopropyl derivatives **228p–v** (Scheme 52) [57].

The stereoselective reduction of *meso*-bis(bicyclopropylidenyl) (*meso*-**87**) with lithium in liquid ammonia gave an almost quantitative yield of the two diastereomeric quatercyclopropyls *trans,trans*-**229** and *cis,trans*-**230** in a ratio of 4.4:1 (Scheme 53) [56]. On the other hand, reduction of *meso*-**87** with diimine generated from 2-nitrobenzenesulfonyl hydrazide gave the *cis,cis*-quatercyclopropyl (**231**) (Fig. 6) as the main product (isolated by chromatography) along with the *cis,trans*-diastereomer **230** (Scheme 53) [56].

Thus, three of the six possible diastereomeric quatercyclopropyls were obtained from *meso*-**87**, the other three may be prepared analogously from *d,l*-bis(bicyclopropylidenyl) (*d,l*-**87**).

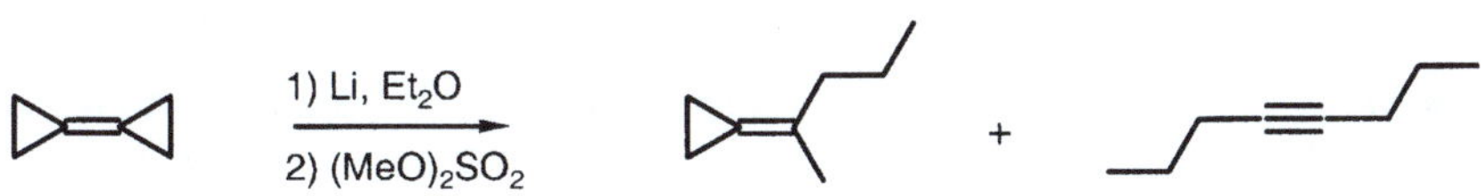

1 225 226

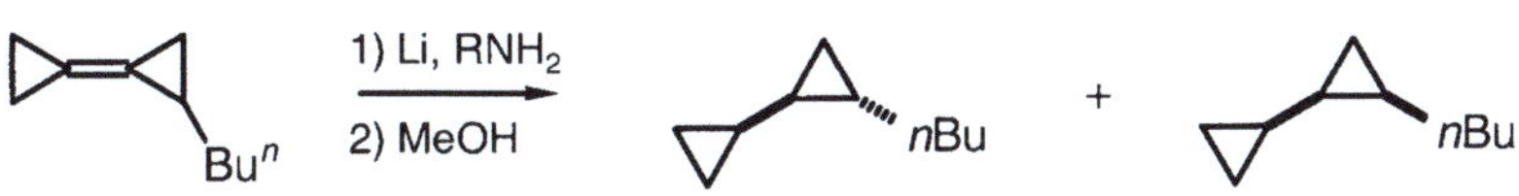

43e 227a 227b

R	T [°C]	Time [h]	Ratio (**227a/227b**)
Et	-78	3	2:1
Et/Me 1:1	-95	3	4.5:1
H	-35	1.5	95:5

Scheme 51. Reduction of the double bonds in bicyclopropylidene (**1**) and *n*-butylbicyclopropylidene (**43e**) with dissolving lithium under various conditions [56a, 143]

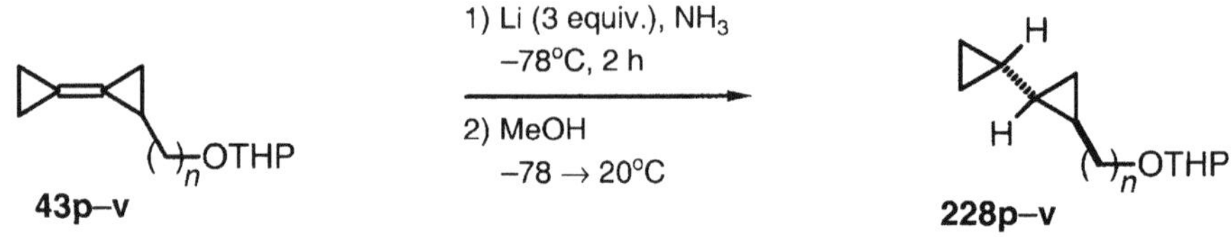

n	3	4	5	6	7	8
Product	**228p**	**228r**	**228s**	**228t**	**228u**	**228v**
Yield (%)	64	88	83	87	67	75

Scheme 52. *trans*-Stereoselective reduction of the double bonds in the THP-protected (bicyclopropylidenyl) alcohols **43p–v** [57]

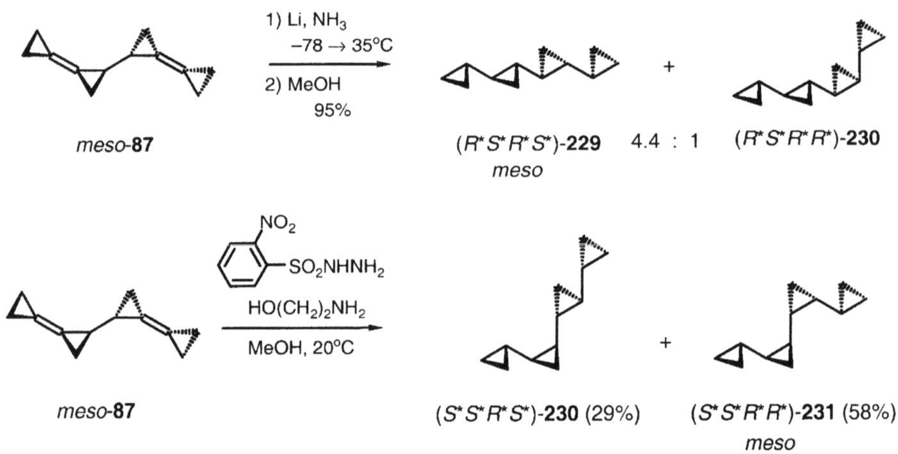

Scheme 53. Reduction of the double bonds in *meso*-bis(bicyclopropylidenyl) (*meso*-**87**) under various conditions to yield quatercyclopropyls **229–231** [56]

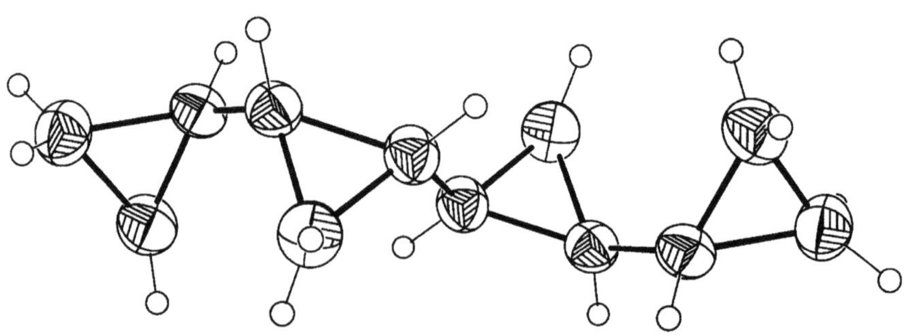

Fig. 6. Structure of ($S^*S^*R^*R^*$)-quatercyclopropyl, [($S^*S^*R^*R^*$)-**231**] in the crystal [56]

4.3
Metal-Catalyzed Reactions of Bicyclopropylidene

Nickel(0)-catalyzed [2+2] cycloaddition of cyclobutene to bicyclopropylidene (**1**) gave rise to the bis(spirocyclopropane)-annelated bicyclo[2.2.0]hexane derivative **232**, the main product was 1,5-cyclooctadiene (**233**), formed by dimerization of cyclobutene with subsequent rearrangement (Scheme 54) [8, 144].

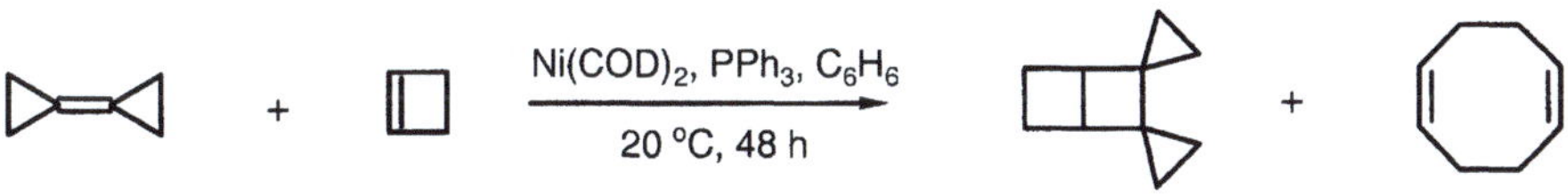

1 **232** (15%) **233** (27%)

Scheme 54. Nickel(0)-catalyzed [2+2] cycloaddition of cyclobutene to bicyclopropylidene (**1**) [8, 144]

Some of the most striking examples for an intramolecular Pauson-Khand reaction involving a tetrasubstituted double bond are the cobalt-mediated cyclizations of bicyclopropylidene derivatives **234** leading to the interesting spirocyclopropanated tricyclic products **235**. The successful cyclizations even of the trimethylsilyl-substituted enynes **234** demonstrate the unique reactivity of the strained double bond in the bicyclopropylidene moiety of these molecules (Scheme 55) [145].

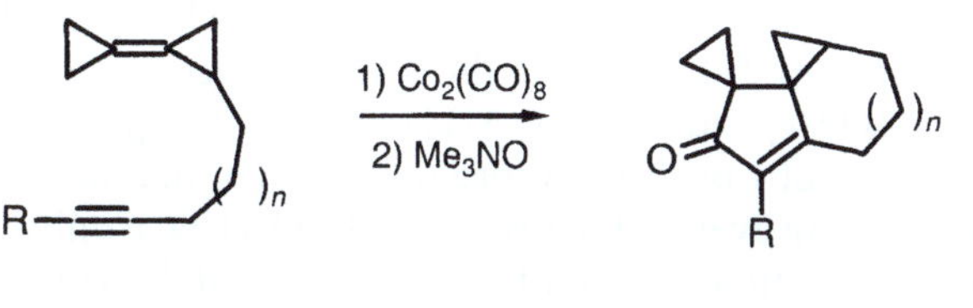

234 **235**

R	*n*	Yield (%)
$SiMe_3$	1	31
H	1	45
$SiMe_3$	2	37

Scheme 55. Intramolecular Pauson-Khand reactions of bicyclopropylidene derivatives **234** [145]

So-called domino cascade reactions have become more and more important for the efficient synthesis of complex organic molecules. In this respect the Heck reaction of aryl or alkenyl halides with bicyclopropylidene (**1**) leading to substituted 1,3-dienes or cross-conjugated trienes which subsequently undergo Diels-Alder reactions with added dienophiles, are particularly interesting [146, 147, 148]. It is noteworthy that bicyclopropylidene (**1**) itself at room temperature undergoes a slow isomerization in the presence of [$Pd(OAc)_2 \cdot 2\ PPh_3$] to form allylidenecyclopropane (**236**), which was trapped in a Diels-Alder reaction with diethyl fumarate to yield the adduct **237** (Scheme 56) [148]. Upon heating to 80 °C, this reaction led to a complex mixture of bicyclopropylidene dimers **238**–**241** apparently formed by consecutive Pd-catalyzed reaction of the isomer **236**.

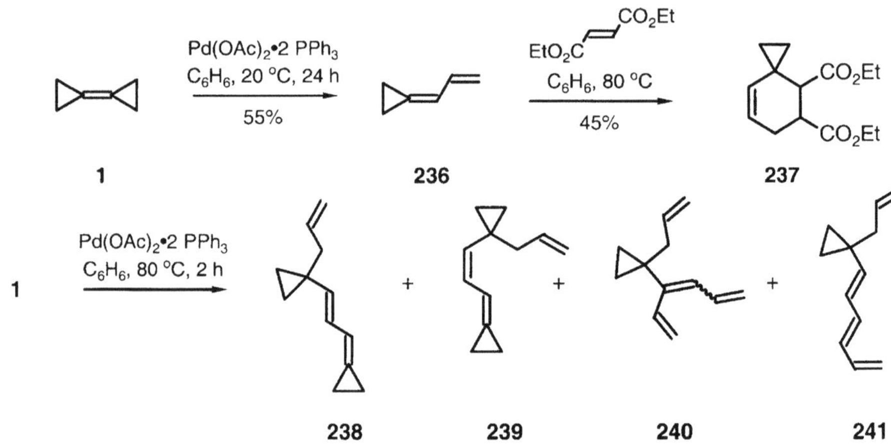

Scheme 56. Isomerization and dimerization of bicyclopropylidene (**1**) in the presence of a palladium catalyst [148]

Coupling of **1** with iodobenzene under Heck reaction conditions gave the phenyl-substituted diene **242**-Ph which was isolated in up to 78% yield [146–148]. When heated at 80 °C in DMF or MeCN with various dienophiles **244**-R (acrylate, maleate, or fumarate), **242**-Ph and its analogs **242**-Ar obtained from **1** and other haloarenes cleanly gave the spiro[2.5]octene derivatives **245**-Ar (Scheme 57).

The Heck coupling of iodoarenes and bicyclopropylidene (**1**) can be carried out in the presence of the dienophiles **244**-R to give the spiro[2.5]octenes **245**-Ar-R in a single operation in 41–100% yield (Scheme 57) [147, 148]. It is quite remarkable that the carbopalladation of **1** proceeds more rapidly than that of the methyl acrylate **244**-H, as the coupling even in the presence of **244**-H gave only a trace of methyl cinnamate resulting directly from the latter and iodobenzene. With the enantiomerically pure *N*-acryloyl-*(S)*-camphorsultam the corresponding spiro[2.5]octene derivative was obtained as a single enantiomer.

Vinyl iodide coupled to **1** gave the rather sensitive and reactive cross-conjugated diene **243** in 37–43% yield. When a dienophile **244**-R was added to the reaction mixture prior to work-up, the corresponding monoadducts of type **247** were isolated in 15–65% yield. Heating a mixture of **1**, vinyl iodide, and a dienophile **244**-R in the presence of palladium catalyst yielded the corresponding bisadducts **250** resulting from a domino Diels-Alder addition onto the cross-conjugated triene **243** (Scheme 57) [147, 148].

The interesting three-component reaction of **1**, aryl iodides, and dienophiles can be extended to difunctional aryl halides like *p*-diiodobenzene (**251**). This formal five-component reaction proceeds under ordinary Heck conditions, but gives a particularly good yield of the twofold domino adduct **252** under high pressure (10 kbar) which accelerates both the Heck coupling as well as the Diels-Alder reaction (Scheme 58) [148]. *o*-(Hydroxymethyl)iodobenzene (**253**) reacted with **1** to yield the expected diene **255** (40% yield) along with the cyclopropylidenedihydropyrane **254** (27%) apparently resulting from an intramolecular

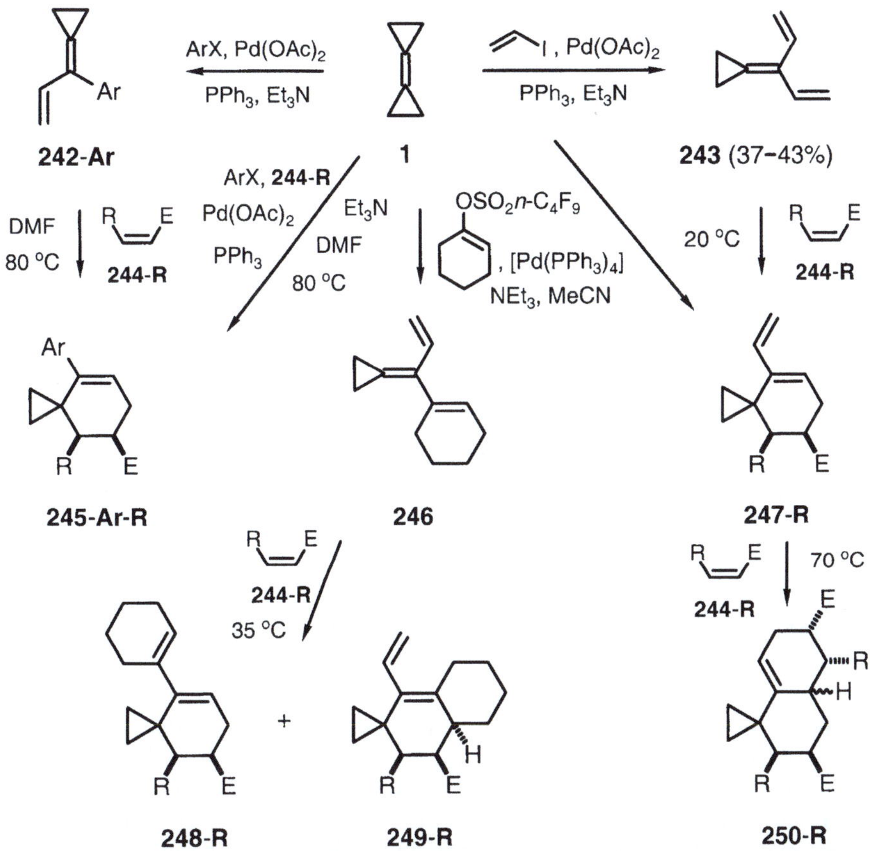

Substrate	Dienophile **244** R	Product	Yield (%)
PhI	H	**245-Ph-H**	100
PhBr	H	**245-Ph-H**	59
PhI	CO_2Me	**245-Ph-CO_2Me**	97
*p*Tol-I	CO_2Me	**245-Tol-CO_2Me**	99
4-Pyr-I	H	**245-4-Pyr-H**	81
4-Pyr-I	CO_2Me	**245-4-Pyr-CO_2Me**	60
2-Bromothiophene	H	**245-2-Thienyl-H**	88
C_6H_9ONf	CO_2Me	**248-CO_2Me, 249-CO_2Me**	41
C_2H_3I	H	**247-H**	63
C_2H_3I	H	**250-H**	59
C_2H_3I	CO_2Me	**247-CO_2Me**	60
C_2H_3I	CO_2Me	**250-CO_2Me**	49

Scheme 57. Domino Heck-Diels-Alder reactions of bicyclopropylidene (1) [147, 148]

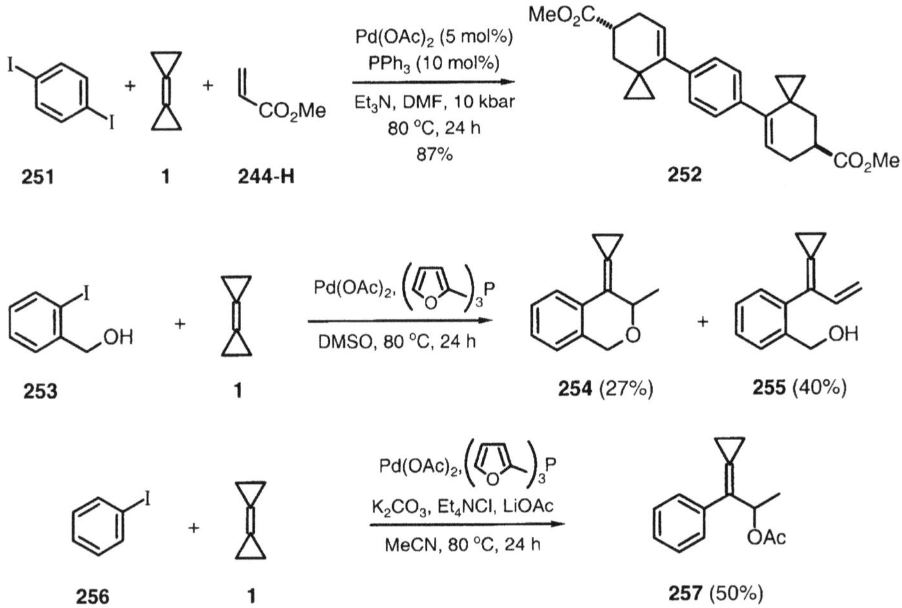

Scheme 58. Different modes of palladium-catalyzed domino reactions of bicyclopropylidene (1) [148]

trapping of an intermediate π-allylpalladium complex formed by rearrangement of the σ-homoallylpalladium intermediate initially formed after the cyclopropylcarbinyl to homoallyl rearrangement (Scheme 58) [148]. The β-hydride elimination can be retarded to favor this σ-homoallyl to π-allylpalladium rearrangement in the presence of trisfuranylphosphane as a ligand, and the π-allylpalladium intermediate then trapped with an external nucleophile such as acetate to give products of type **257** [148].

Whereas $Pd(OAc)_2 \cdot (PPh_3)_2$ decelerates the addition of thiols onto bicyclopropylidene (**1**) with the retention of the both three-membered rings by a factor of about six (see above, Scheme 44) [135], hexamethyldisilane does not react with **1** in the absence of a palladium catalyst. Under palladium catalysis, however, this addition really proceeds inter- as well as intramolecularly and, surprisingly, with retention of both cyclopropane fragments to give compounds **258** and **260**, respectively (Scheme 59) [149]. This type of addition turned out to be rather sensitive to the nature of the addend and the catalyst. For example, the addition of bis(fluorodimethylsilane) and trimethylsilyl cyanide to **1** also proceeded well, but gave ring-opened products **261** and **262**, respectively (Scheme 59). Cyclic disilanes, trimethylsilylmethyl sulfide and diphenyl disulfide do not react at all with **1** under palladium catalysis [149].

Bicyclopropylidene (**1**) also reacts with activated alkenes under transition-metal catalysis. With electron-deficient alkenes under nickel(0) catalysis, the [2+2] cycloadduct **263** was the main component in the reaction mixture [2b, 150]. Under palladium(0) catalysis, formal [3+2] cycloaddition of electron-deficient (Scheme 60) as well as strained alkenes can be achieved exclusively

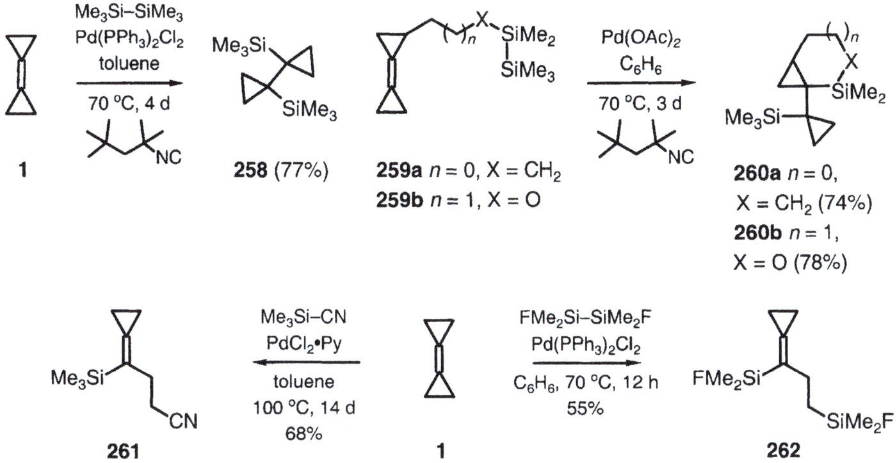

Scheme 59. Different modes of addition of Me_2Xsi-Y reagents onto the double bond in bicyclopropylidene (**1**) under palladium catalysis [149]

(Scheme 61) [2b, 150], but can give rise to a variety of products. With unsymmetrically substituted alkenes of type **266**, two regioisomeric products were obtained, but the isomer **268** bearing the ester group near the spiro atom was the minor component in all cases. Norbornadiene and norbornene react with **1** by the same mode to give formal [3+2] cycloadducts **269** and **270**, respectively, the latter as a 9:1 mixture of *exo*- and *endo*-isomers.

In the absence of another activated alkene, one molecule of bicyclopropylidene (**1**), after the opening of a distal bond, underwent formal [3+2] cycloaddition to a second bicyclopropylidene molecule to give 8-cyclopropylidenedispiro[2.0.2.3]nonane (**272**) (Scheme 61) [2b, 150].

Bromobicyclopropylidene (**43d**) reacts with the chlorozinc compound **274a** produced by metal exchange from lithiated ethyl *N*-(diphenylmethylene)glycinate [151] under $PdCl_2$(dppf) catalysis [152] to give the substituted diene **277** formed by an unusual ring opening (Scheme 62) [153].

Similar results were obtained in the couplings of 2-bromomethylenecyclopropane under $Pd(dppa)_2$ catalysis [147a, 153] and of bromobicyclopropylidene (**43d**) with organometallic derivatives of diethyl malonate **274b** under $PdCl_2$(dppf) catalysis [153]. The reversal of polarities of the reactands, i.e., treatment of bicyclopropylidenylzinc chloride (**273**) with the bromomalonate **275** or O'Donnell's acetoxyglycine derivative **276** under palladium catalysis led to the formation of the same products, but in lower yields. Although the the compounds **277** and **278** are structurally analogous to the products of the Heck reactions with **1** (see above, Scheme 57), they must be formed along a different mechanistic path [153].

Another interesting example for an unusual transformation of bicyclopropylidene (**1**) is the reaction of the higher-order cuprate **279** [154] derived from **1** with the electrophilic glycine cation equivalent **276** [155] which produces the

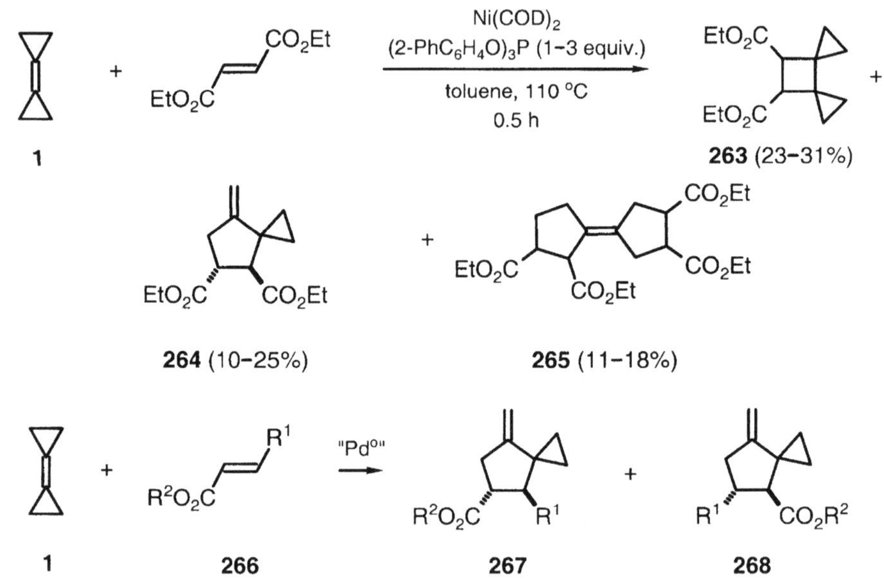

R^1	R^2	Catalyst	Conditions	Ratio (**267/268**)	Yield (%)
H	Me	(π-All)PdCp iPr_3P	130–160 °C 2 h	2.8:1	54
H	Me	$Pd(dba)_2$ $tBu(iPr)_2P$	130 °C 2 h	3:1	45
Me	Me	$Pd(dba)_2$ $tBu(iPr)_2P$	110 °C 3 h	9:1	60
Ph	Me	$Pd(dba)_2$ $tBu(iPr)_2P$	110 °C 3 h	9:1	60
CO_2Et	Et	(π-All)PdCp iPr_3P	110 °C 3 h	-	77
CO_2Et	Et	$Pd(dba)_2$ $tBu(iPr)_2P$	110 °C 3 h	-	83

Scheme 60. Reactions of bicyclopropylidene (**1**) with electron-deficient alkenes under nickel(0) and palladium(0) catalysis [2b, 150]

methylenetetrahydropyridine derivative **280a** (Scheme 63) [156]. In solutions, compound **280a** exhibits a tautomeric equilibrium with **280b**.

Stable complexes of titanium, **281** [157], cobalt, **282** [157], and platinum, **283**, with a bicyclopropylidene ligand [158] have been obtained (Fig. 7), showing that **1** is a remarkably good ligand in spite of being a tetrasubstituted alkene. The complex **282** has been fully characterized by an X-ray crystal structure analysis [157].

It shows that the bicyclopropylidene ligand in **282** is remarkably bent out-of-plane by 40° at both termini of the double bond, i.e., this complex can be considered as a derivative of 7-cobalta[3]triangulane.

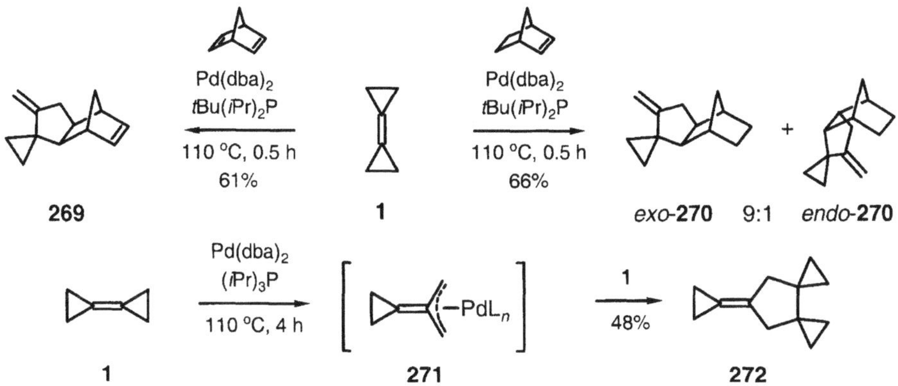

Scheme 61. Addition of bicyclopropylidene (**1**) onto strained alkenes under the palladium(0) catalysis [2b, 150]

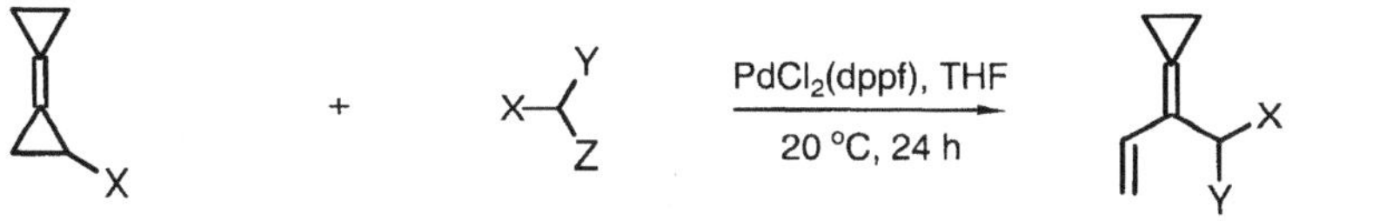

43d X = Br | **274a** X = $ZnCl_2$ | **277** Y = CO_2Et, Z = N=CPh_2 (72%)
43d X = Br | **274b** X = $ZnCl_2$ | **278** Y = Z = CO_2Et (24%)
273 X = $ZnCl_2$ | **275** X = Br | **278** Y = Z = CO_2Et (27%)
273 X = $ZnCl_2$ | **276** X = OAc | **277** Y = CO_2Et, Z = N=CPh_2 (29%)

Scheme 62. Palladium-catalyzed cross-coupling of bicyclopropylidene derivatives with metallated or brominated CH-acidic compounds accompanied by an unprecedented ring opening [153]

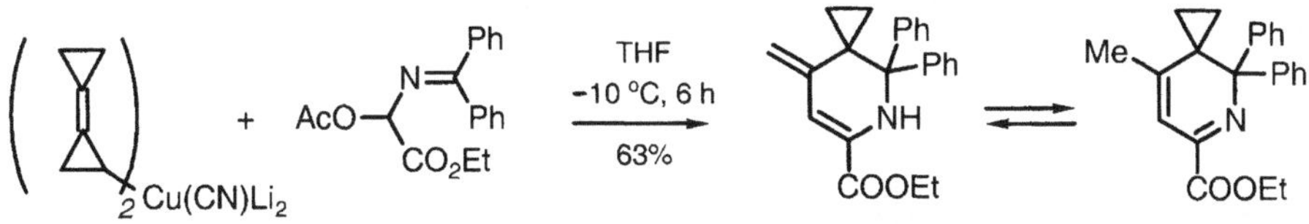

279 **276** **280a** **280b**

Scheme 63. Reaction of the higher-order cuprate **279** derived from **1** with O'Donnell's acetoxyglycine derivative **276** occurring with opening of a three-membered ring [156]

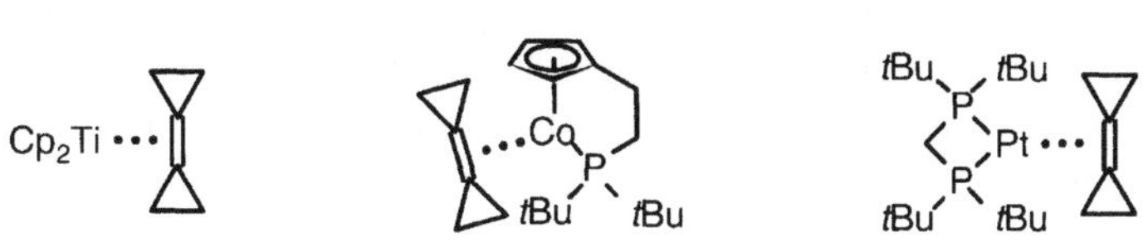

281 **282** **283**

Fig. 7. Stable complexes of bicyclopropylidene (**1**) with transition metals [157, 158]

5
Conclusion and Outlook

The presented chemistry illustrates the broad synthetic applicability of bicyclopropylidene (**1**) and its derivatives. Is there anything left to be done? Yes, indeed, there is a huge unexplored area of bicyclopropylidene chemistry in view of modern organometallic chemistry, and there is a broad scope for new bicyclopropylidene derivatives. For example, applying the recently reported conditions for the dimerization of cyclopropylidenoids generated from 1,1-dibromocyclopropanes and alkyllithiums in the presence of $CuCl_2$ [40] to 1,1-dibromospiropentane **284** enabled the preparation of the unsymmetrically bisspirocyclopropanated bicyclopropylidenes (*E*)- and (*Z*)-**285** in a preparatively acceptable yield [159]. Dimerization of such carbenoids could not be realized without ring opening in the absence of $CuCl_2$ [33, 34]. It is spectacular that this new method could also successfully be applied to 7,7-dibromodispiro[2.0.2.1]heptane (**120b**), the dibromocarbene adduct of **1**, to afford the perspirocyclopropanated bicyclopropylidene **62** (isolated in 80% yield) making this exotic hydrocarbon – a superbicyclopropylidene – which had previously been prepared along a tedious 14-step sequence [33, 64] easily available in preparatively useful quantities (Scheme 64) [159].

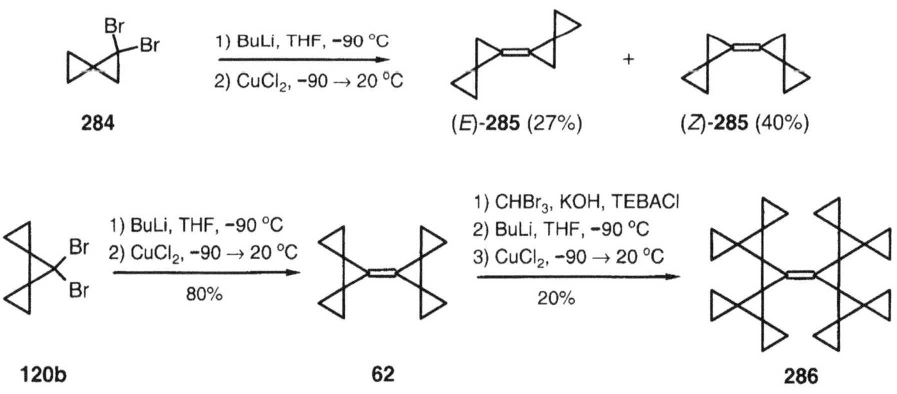

Scheme 64. Dimerization of coppercyclopropylidenoids generated from oligospirocyclopropanated dibromocyclopropanes and butyllithium in the presence of $CuCl_2$ [159]

It is even more spectacular that the dibromocarbene adduct of **62** can be dimerized again to give the "supersuperbicyclopropylidene" **286** (Fig. 8) [159]. This new methodology thus allows the construction of a whole family of dendritic molecules consisting of spiroannelated three-membered rings with a central double bond, which in turn can probably be used to prepare higher order branched triangulanes.

On another front, further progress can and will be achieved towards the preparation of bicyclopropylidene analogs of naturally occurring and biologically active methylenecyclopropane derivatives. Two particularly interesting specimens are the naturally occurring 3-(2-methylenecyclopropyl)alanine (**287**), so-

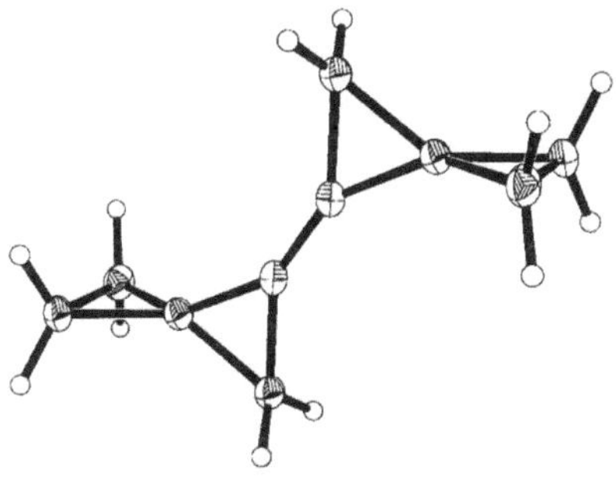

(*E*)-**285**

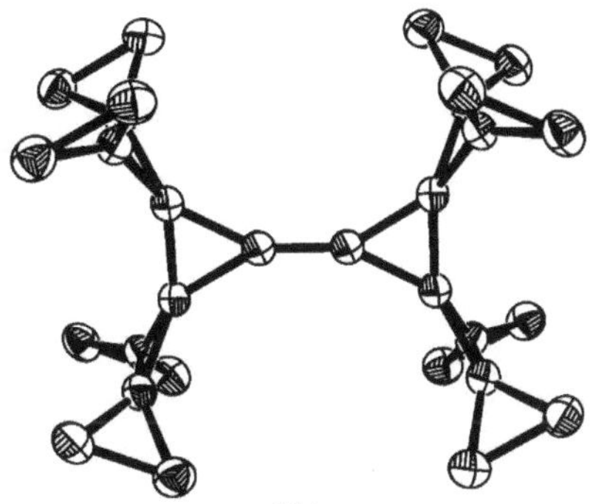

286

Fig. 8. The structures of (*E*)-dispirocyclopropanated bicyclopropylidene (*E*)-**285** and "super-superbicyclopropylidene" **286** in the crystals [159]

called hypoglycine A [160], and 1-amino-2-methylenecyclopropane-1-carboxylic acid (methylene-ACC) (**288**) [161], both show a strong hypoglycemic effect (Fig. 9). The enantioselective syntheses of individual stereoisomers of hypoglycine A (**287**) have been reported [160].

Since the more highly strained bicyclopropylidene (**1**) and methylenespiropentane (**6**), both spirocyclopropanated analogs of methylenecyclopropane (**2**), are even more reactive than the latter, it is to be expected that analogs of the amino acids **287** and **288** containing a bicyclopropylidenyl or a methylenespiropentyl moiety would also exhibit biological activities. In fact, the preparations of the amino acids **289** and **290** as well as of the methylenespiropentane amino acid **295** have been reported in the meantime (Scheme 65) [55].

Yet another interesting idea concerns a possible novel synthetic approach to analogs of the unusual natural products FR-900848 296 and U-106305 297 (Fig. 10) [162].

The key problem in the preparation of **296**, **297** undoubtedly is the stereoselective assembly of the chains of four to five contiguous cyclopropyl groups. The strategies of all total syntheses known so far rely on auxiliary-directed stereoselective cyclopropanations of double bonds [111a, 163, 164]. The chemistry of bicyclopropylidene (**1**), namely its capability to be functionalized (Schemes 8, 9), dimerized (Scheme 15), and stereoselectively reduced (Scheme 53) opens up new perspectives to approach the assembly of quatercyclopropyl derivatives.

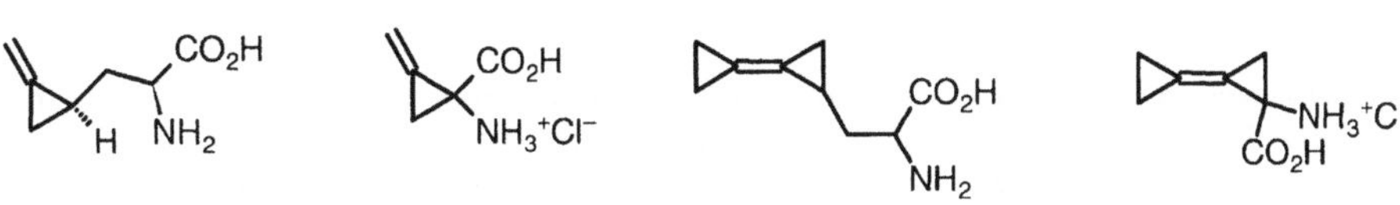

287 **288** **289** **290**

Fig. 9. Biologically active methylenecyclopropaneamino acids and their bicyclopropylidene analogs [55, 160, 161]

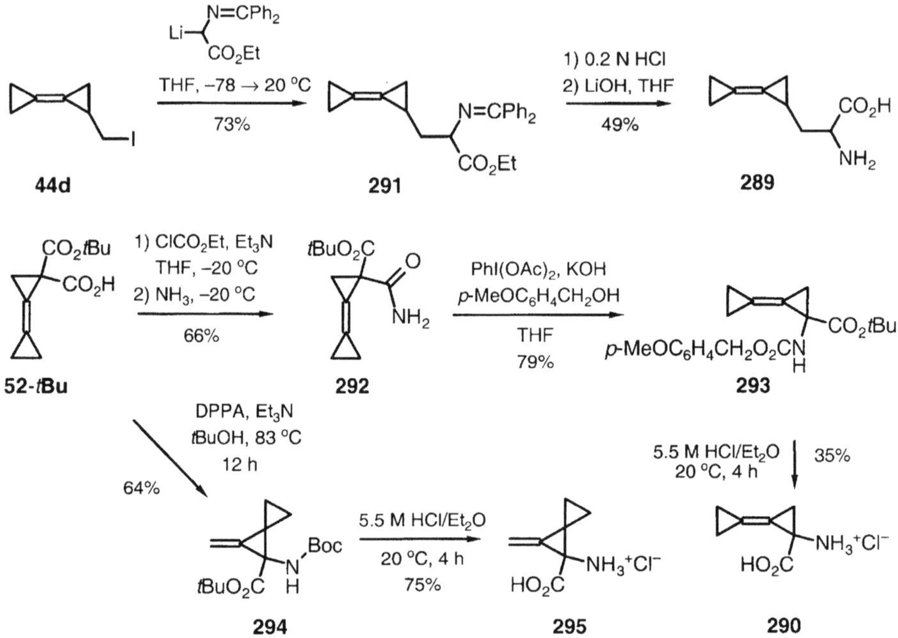

Scheme 65. Preparation of bicyclopropylidene and methylenespiropentane analogs **289**, **290**, and **295** of biologically active amino acids with a methylenecyclopropane moiety [55]

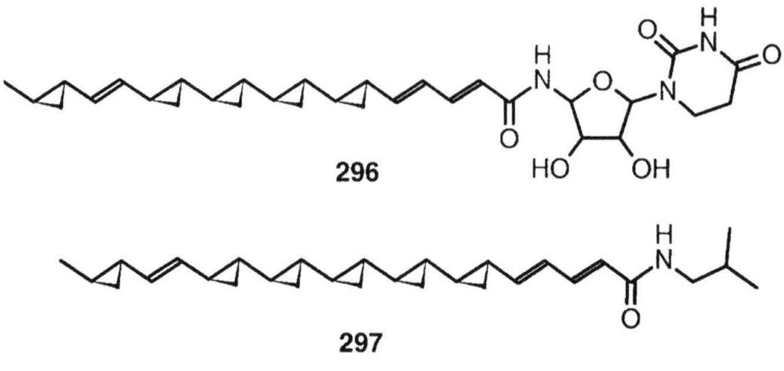

Fig. 10. Natural products FR-900848 296 and U-106305 297 containing chains of *trans*-connected three-membered rings [162, 163]

Acknowledgements. The work of our own group described herein has been supported by the Deutsche Forschungsgemeinschaft, the Volkswagen-Stiftung, and the Fonds der Chemischen Industrie as well as BAYER, BASF AG, Chemetall GmbH, Degussa, Hoechst, and Hüls AG through generous gifts of chemicals. S.I.K. is indebted to the Alexander-von-Humboldt-Foundation for a research fellowship. A.d.M. is grateful to the group of dedicated and enthusiastic young chemists who – over the years – have made this research flourish. The authors thank Dr. B. Knieriem, Göttingen, for his careful reading of the manuscript.

References

1. Le Perchec P, Conia J-M (1970) Tetrahedron Lett 1587
2. (a) Binger P, Büch HM, Top Curr Chem (1987) 135:77; (b) Binger P, Schmidt T (1997) In: de Meijere A (ed) Houben-Weyl, Thieme, Stuttgart, Vol E 17c, p 2217; (c) Brandi A, Goti A (1998) Chem Rev 98:589
3. Denis JM, Le Perchec P, Conia J-M (1977) Tetrahedron 33:1587
4. (a) de Meijere A (1971) Habilitationsarbeit, Universität Göttingen; (b) de Meijere A (1974) Chem Ber 107:1702
5. Denis JM, Girard C, Conia J-M (1972) Synthesis 549
6. Zefirov NS, Lukin KA, Politanskii SF, Margulis MA (1987) Zh Org Khim 23:1799; J Org Chem USSR (Engl Transl) 23:1603
7. (a) Schipperijn AJ (1971) Recl Trav Chim Pays-Bas 90:1110; (b) Schipperijn AJ, Smael P (1973) Recl Trav Chim Pays-Bas 92:350
8. Kaufmann D, de Meijere A (1984) Chem Ber 117:3134
9. Fitjer L, Conia J-M (1973) Angew Chem 85:347; Angew Chem Int Ed Engl 12 : 332
10. Weber W, de Meijere A (1986) Synth Commun 16:837
11. Schmidt AH, Schirmer U, Conia J-M (1976) Chem Ber 109:2588
12. Lukin KA, Kuznetsova TS, Kozhushkov SI, Piven' VA, Zefirov NS (1988) Zh Org Khim 24:1644; J Org Chem USSR (Engl Transl) 24:1483
13. Kulinkovich OG, Sviridov SV, Vasilevski DA (1991) Synthesis 234
14. (a) de Meijere A, Kozhushkov SI, Spaeth T, Zefirov NS (1993) J Org Chem 58:502; (b) de Meijere A, Kozhushkov SI, Spaeth T (1999) Org Synth accepted for checking
15. Moore WR, Ward HR (1960) J Org Chem 25:2073; for the really first, but unverified, communication about the preparation of a bicyclopropylidene derivative see: Kohler EP, Darling SF (1930) J Am Chem Soc 52:424
16. Fukuda Y, Yamamoto Y, Kimura K, Odaira Y (1979) Tetrahedron Lett 877
17. Molchanov AP, Kalyamin SA, Kostikov RR (1992) Zh Org Khim 28:122; Russ J Org Chem (Engl Transl) 28:102
18. Ando T, Muranaka T, Ichihara T (1981) Bull Chem Soc Japan 54:3227
19. Marquis ET, Gardner PD (1966) J Chem Soc Chem Commun 726
20. Köbrich G, Goyert W (1968) Tetrahedron 24:4327
21. Moore WR, Bach RD (1972) J Am Chem Soc 94:3148
22. Christ M, Moigno D, Peters E-M, Peters K, von Schnering HG (1997) Liebigs Ann/Recueil 1791
23. (a) Taylor KG, Chaney J (1972) J Am Chem Soc 94:8924; (b) Taylor KG, Chaney J, Deck JC (1976) J Am Chem Soc 98:4163
24. Pilidis G (1986) Collect Czech Chem Commun 51:2151
25. Banwell MG, Gabbe RW, Greenwood RJ, Lambert JN, McKay MF, Walter JM (1997) Synlett 953
26. Seyferth D, Massol M, Barrau J, Monteverdi S (1980) J Organomet Chem 185:307
27. Warner P, Chang S-C, Powell DR, Jacobson RA (1981) Tetrahedron Lett 22:533
28. Neidlein R, Pognee V, Kramer W, Glück C (1986) Angew Chem 98:735; Angew Chem Int Ed Engl 25:731; Such type of compounds are essentially derivatives of cycloproparenes, for the reviews see: (a) Halton B (1989) Chem Rev 89:1161, (b) Halton B, Stang PJ (1987) Acc Chem Res 20:443; (c) Müller P (1997) In: de Meijere A (ed) Houben-Weyl, Thieme, Stuttgart, Vol E 17d, p 2865; (d) Halton B, Banwell MG (1987) In Rappoport Z (ed) The Chemistry of a Cyclopropyl Group, Wiley, Chichester, p 1223
29. (a) Banwell MG, Hockless DCR, Walter JM (1996) J Chem Soc Chem Commun 1469; (b) Banwell MG, Hockless DCR, Longmore RW, Walter JM (1997) Aust J Chem 50:547
30. Osterodt J, Vögtle F (1996) J Chem Soc Chem Commun: 547; for the alternative approach to this compound see also: Fabre TS, Treleaven WD, McCarley TD, Newton CL, Landry RM, Saraiva MC, Strongin RM (1998) J Org Chem 63:3522
31. Bertrand M, Tubul A, Ghiglione C (1983) J Chem Res Miniprint 2274

32. For reviews on the transformation of dibromocyclopropanes to allenes (see so-called Doering-Skattebøl-Moore reaction) see: (a) Hopf H (1980) In: Patai S (ed) The Chemistry of Ketenes, Allenes and Related Compounds Wiley, New York, Part 2, Chapter 2, p 779; (b) Lee-Ruff E (1997) In: de Meijere A (ed) Houben-Weyl, Thieme, Stuttgart, Vol E 17c, p 2388; (c) Backes J, Brinker UH (1989) In Regitz M (ed) Houben-Weyl, Thieme, Stuttgart, Vol E 19b, p 391
33. Zöllner S (1991) Dissertation, Universität Hamburg
34. (a) Lukin KA, Zefirov NS, Yufit DS, Struchkov YuT (1992) Tetrahedron 45:9977; (b) Averina EB, Kuznetsova TS, Zefirov AN, Koposov AE, Grishin YuK, Zefirov NS (1999) Mendeleev Commun 101
35. (a) Holm KH, Mohamed EA, Skattebøl L (1993) Acta Chem Scand 47:500; (b) Brun R, Grace DSB, Holm KH, Skattebøl L (1986) Acta Chem Scand B 40:21
36. (a) Anderson HW (1972) PhD Thesis, Massachusetts Institute of Technology, Cambridge, Massachusetts; (b) Egge R (1981) Diplomarbeit, Universität Hamburg
37. Bräse S, de Meijere A (unpublished results)
38. Arct J, Skattebøl L (1982) Acta Chem Scand B 36:593
39. Tverezovsky VV, Baird MS, Bolesov IG (1997) Tetrahedron 43:14773
40. (a) Borer M, Loosli T, Neuenschwander M (1991) Chimia 45:382; (b) Loosli T, Borer M, Kulakowska I, Minder A, Neuenschwander M, Engel P (1995) Helv Chim Acta 78:1144; (c) Borer M, Loosli T, Minder A, Neuenschwander M, Engel P (1995) Helv Chim Acta 78:1311
41. Skattebøl L (1965) Tetrahedron Lett 2175
42. Kostikov RR, Vasil'eva IA, Slobodin YaM (1974) Zh Org Khim 10:2325; J Org Chem USSR (Engl Transl) 10:2339
43. (a) Jochims JC, Karich G (1974) Tetrahedron Lett 4215; (b) Karich G, Jochims JC (1977) Chem Ber 110:2680
44. Dehmlow EV, Schönefeld J (1971) Justus Liebigs Ann Chem 744:42
45. (a) du Laurens B, Bezaguet A, Davidovich G, Bertrand M, Chouteau J (1967) Bull Soc Chim France 799; (b) Bezaguet A, Bertrand M (1965) CR Hebd Seances Acad Sci 261:1695
46. Kostikov RR, Molchanov AP, Vasil'eva IA, Slobodin YaM (1977) Zh Org Khim 13:2541; J Org Chem USSR (Engl Transl) 13:2361
47. Billups WE, Haley MM, Boese R, Bläser D (1994) Tetrahedron 50:10693
48. Köbrich G, Heinemann H, Zündorf W (1967) Tetrahedron 23:565
49. Crandall JK, Paulson DR, Bunnell CA (1969) Tetrahedron Lett 4217
50. Paulson DR, Crandall JK, Bunnell CA (1970) J Org Chem 35:3708
51. Dolbier WR, Alonso JH (1973) J Am Chem Soc 95:4421
52. Lukin KA, Masunova AYu, Kozhushkov SI, Kuznetsova TS, Ugrak BI, Piven' VA, Zefirov NS (1991) Zh Org Khim 27:488; J Org Chem USSR (Engl Transl) 27:422
53. Lukin KA, Kozhushkov SI, Andrievsky AA, Ugrak BI, Zefirov NS (1991) J Org Chem 56:6176
54. de Meijere A, Kozhushkov SI, Zefirov NS (1993) Synthesis 681
55. Brandl M, Kozhushkov SI, Yufit DS, Howard JAK, de Meijere A (1998) Eur J Org Chem 2785
56. (a) von Seebach M (1997) Diplomarbeit, Universität Göttingen; (b) von Seebach M, Kozhushkov SI, Benet-Buchholz J, Boese R, de Meijere A, unpublished results
57. Löhr S (1997) Diplomarbeit, Universität Göttingen
58. Heiner T, Kozhushkov SI, Noltemeyer M, Haumann T, Boese R, de Meijere A (1996) Tetrahedron 52:12185
59. de Meijere A, Khlebnikov AF, Kostikov RR, Kozhushkov SI, Schreiner PR, Wittkopp A, Yufit DS (1999) Angew Chem in press; Angew Chem Int Ed Engl in press
60. Doering WvE, Roth WR (1963) Tetrahedron 19:715
61. Furukawa J, Kawabata N, Nishimura J (1968) Tetrahedron 24:53
62. Kirmse W, Schütte H (1968) Chem Ber 101:1674
63. Fitjer L, Conia J-M (1973) Angew Chem 85:832; Angew Chem Int Ed Engl 12:761
64. Zöllner S, Buchholz H, Boese R, Gleiter R, de Meijere A (1991) Angew Chem 103:1544; Angew Chem Int Ed Engl 30:1518

65. Zefirov NS, Kozhushkov SI, Ugrak BI, Lukin KA, Kokoreva OV, Yufit DS, Struchkov YuT, Zoellner S, Boese R, de Meijere A (1992) J Org Chem 57:701
66. Fitjer L (1976) Angew Chem 88:803; Angew Chem Int Ed Engl 15:763
67. Fitjer L (1982) Chem Ber 116:1035
68. Donskaya NA, Akhachinskaya TV, Leonova TV, Shulishov EV, Shabarov YuS (1980) Zh Org Khim 16:563; J Org Chem USSR (Engl Transl) 16:487
69. Kuznetsova TS, Averina EB, Kokoreva OV, Zefirov AN, Grishin YuK, Zefirov NS (1999) Zh Org Khim in press
70. Averina EB, Kuznetsova TS, Zefirov AN, Grishin YuK, Zefirov NS (1999) Dokl Akad Nauk 364:61; Dokl Chem (Engl Transl) 364:1
71. Garratt PJ, Tsotinis A (1990) J Org Chem 55:84
72. Kuznetsova TS, Eremenko OV, Kokoreva OV, Averina EB, Zefirov AN, Zefirov NS (1997) Zh Org Khim 33:916; Russ J Org Chem (Engl Transl) 33:849
73. Lord RC, Warrey CJ (1973) Spectrochim Acta A 30:915
74. Hofland A, Steinberg H, de Boer TJ (1985) Recl Trav Chim Pays-Bas 104:350
75. Courtieu J, Gounelle Y, Jullien J, Barnier JP, Conia J-M, Kan SK (1982) Org Magn Res 19:63
76. Schonk RM, Meijer CW, Bakker BH, Zöllner S, Cerfontain H, de Meijere A (1993) Recl Trav Chim Pays-Bas 112:457
77. (a) de Meijere A, Kozhushkov S (1995) In: Halton B (ed) Advances in Strain in Organic Chemistry, JAI Press Ltd, Greenwich, Vol 4, p 225, and references cited therein; (b) de Meijere A, Kozhushkov SI (2000) Chem Rev 100: in press
78. Gerson F, de Meijere A, Qin X-Z (1989) J Am Chem Soc 111:1135
79. Gerson F, Schmidlin R, de Meijere A, Späth T (1995) J Am Chem Soc 117:8431
80. Gleiter R, Haider R, Conia J-M, Barnier J-P, de Meijere A, Weber W (1979) J Chem Soc Chem Commun 130
81. Hofland A, de Boer TJ (1987) Recl Trav Chim Pays-Bas 106:558
82. (a) Gleiter R, Spanget-Larson J (1992) In: Halton B (ed) Advances in Strain in Organic Chemistry JAI Press Ltd, London, Vol 2, p 143; (b) Gleiter R, Kozhushkov SI, de Meijere A unpublished results
83. Trætteberg M, Simon A, Peters EM, de Meijere A (1984) J Mol Struct 118:333
84. de Boer JSAM, Stam CH (1993) Recl Trav Chim Pays-Bas 112:635
85. Boese R, Haumann T, Kozhushkov SI, de Meijere, to be published
86. Boese R, Haumann T, Stellberg P (1995) In: Hargittai M, Hargittai I (eds) Advances in Molecular Structure Research JAI Press Ltd, London, Vol 1, p 202 and references therein
87. Boese R, Haumann T, Jemmis ED, Kiran B, Kozhushkov SI, de Meijere A (1996) Liebigs Ann 913
88. Randic M, Jakab L (1971) Croat Chem Acta 43:145
89. Eckert-Macsic M, Maksic ZB, Skanske A, Skanske PN (1987) J Phys Chem 91:2786
90. Randic M, Kumar L (1975) J Mol Struct 26:145
91. Boese R (1992) In Halton B (ed) Advances in Strain in Organic Chemistry JAI Press Ltd, London, Vol 2, p 191
92. Greenberg A, Liebman JF (1978) Strained Organic Molecules, Organic Chemistry Series, Vol 38, Academic Press, New York
93. (a) Cremer D, Kraka E (1988) In: Liebman JF, Greenberg A (eds) Structure and Reactivity VCH, Weinheim, p 65 and references cited therein; (b) Schulman JM, Disch RL (1985) Chem Phys Lett 113:291
94. Schleyer PvR, Williams JE, Blanchard KR (1970) J Am Chem Soc 92:2377
95. Beckhaus H-D, Rüchardt C, Kozhushkov SI, Belov VN, Verevkin SP, de Meijere A (1995) J Am Chem Soc 117:11854 and references cited therein
96. Kuznetsova TS (1991) Habilitation Thesis, Moscow State University
97. Domalski ES, Hearing ED (1988) J Phys Chem Ref Data 17:1637
98. (a) Bushby RJ, Jesudason MV, Pollard MD, Shuhaibar KF (1982) J Chem Soc Perkin Trans 1 2647; (b) Bushby RJ, Mann S, Jesudason MV (1984) J Chem Soc Perkin Trans 1 2457; c) Bushby RJ, Mann S, Jesudason MV (1983) Tetrahedron Lett 24:4743
99. Dolbier WR, Akiba K, Bertrand M, Bezaguet A, Santelli M (1970) J Chem Soc D717

100. Dolbier WR, Akiba K, Riemann JM, Harmon CA, Bertrand M, Bezaguet A, Santelli M (1971) J Am Chem Soc 93:3933
101. Dolbier WR (1968) Tetrahedron Lett 393
102. Gajewski JJ (1981) In Hydrocarbon Thermal Isomerizations, Academic Press, New York, p 149; for the mechanism of this reaction see also ref. [104]
103. (a) Faber D (1996) Dissertation, Universität Göttingen; (b) Faber D, de Meijere A unpublished results, (c) Kozhushkov SI, de Meijere A (1997) In: de Meijere A (ed) Houben-Weyl, Vol E 17b, Thieme, Stuttgart, p 1697
104. de Meijere A, Erden I, Weber W, Kaufmann D (1988) J Org Chem 53:152
105. Fitjer L, Conia J-M (1973) Angew Chem 85:349; Angew Chem Int Ed Engl 12:334
106. Belov VN, Kozhushkov SI, Yufit DS, de Meijere A unpublished results
107. Lukin KA, Zefirov NS (1991) Zh Org Khim 27:1358; J Org Chem USSR (Engl Transl) 27:1187
108. de Meijere A, Kozhushkov SI, Haumann T, Boese R unpublished results
109. Miyashi T, Takahashi Y, Ohaku H, Ikeda H, Morishima S (1991) Pure Appl Chem 63:223
110. Fitjer L (1975) Angew Chem 87:381; Angew Chem Int Ed Engl 14:360
111. Reviews: (a) Subramanian LR, Zeller K-P (1997) In: de Meijere A (ed) Houben-Weyl, Vol E 17a Thieme, Stuttgart, p 256; (b) Simmons HE, Cairns TL, Vladuchik SA, Hoiness CM (1973) Org React 20:1
112. Lukin KA, Kuznetsova TS, Kozhushkov SI, Piven' VA, Zefirov NS (1988) Zh Org Khim 24:1644; J Org Chem USSR (Engl Transl) 24:1483
113. Lukin KA, Zefirov NS (1989) Dokl Akad Nauk SSSR 305:631; Dokl Chem (Engl Transl) 305:104
114. Lukin KA, Andrievskii AA, Zefirov NS (1991) Dokl Akad Nauk SSSR 321:521; Dokl Chem (Engl Transl) 321:353
115. Lukin KA, Zefirov NS (1987) Zh Org Khim 23:2548; J Org Chem USSR (Engl Transl) 23:2249
116. (a) Miyazawa K, de Meijere A unpublished results; (b) Moss R, de Meijere A unpublished results
117. Kozhushkov SI, Haumann T, Boese R, de Meijere A (1993) Angew Chem 105:426; Angew Chem Int Ed Engl 32:401
118. Kostikov RR, Molchanov AP, Komendantov MI (1979) Zh Org Khim 15:1437; J Org Chem USSR (Engl Transl) 15:1282
119. Späth T (1995) Dissertation, Universität Göttingen
120. Tubul A, Bertrand M (1984) Tetrahedron Lett 25:2219
121. Yufit DS, Struchkov YuT, Kozhushkov SI, de Meijere A (1995) Acta Crystallogr C 51:1325
122. Zefirov NS, Lukin KA, Kozhushkov SI, Kuznetsova TS, Domarev AM, Sosonkin IM (1989) Zh Org Khim 25:312; J Org Chem USSR (Engl Transl) 25:278
123. Erden I (1986) Synth Commun 16:117
124. Lukin KA, Andrievskii AA, Zefirov NS, Masunova AYu (1991) Dokl Akad Nauk SSSR 316:379; Dokl Chem (Engl Transl) 316:19
125. de Meijere A, Kozhushkov SI, Yufit DS, Boese R, Haumann T, Pole DL, Sharma PK, Warkentin J (1996) Liebigs Ann 601
126. de Meijere A, Kozhushkov SI, Yufit DS, Howard JAK (1999) Mendeleev Commun 5
127. Kaufmann D, de Meijere A (1973) Angew Chem 85:151; Angew Chem Int Ed Engl. 12:159
128. Thiemann T, Rauch K, de Meijere A unpublished results
129. Buback M, Heiner T, Hermans B, Kowollik C, Kozhushkov SI, de Meijere A (1998) Eur J Org Chem 107
130. Weber W, Erden I, de Meijere A (1980) Angew Chem 92:387; Angew Chem Int Ed Engl 19:387
131. de Meijere A, Wenck H, Seyed-Mahdavi F, Viehe HG, Gallez V (1986) Tetrahedron 42:1291
132. Schonk RM, Bakker BH, Cerfontain H (1991) Phosph Sulf Sil Relat Elem 59:467
133. Kuznetsova TS, Kozhushkov SI, Lukin KA, Grishin YuK, Bazhenov DV, Koz'min AS, Zefirov NS (1991) Zh Org Khim 27:78; J Org Chem USSR (Engl Transl) 27:67
134. Späth T, Kozhushkov SI, Fiebig T, Ruasse M-F, Dodin M, Albrecht K, Apeloig Y, de Meijere A (1999) J Org Chem submitted

135. Kozhushkov SI, Brandl M, de Meijere A (1998) Eur J Org Chem 1535 and references cited therein
136. Brandi A, Goti A, Kozhushkov SI, de Meijere A (1994) J Chem Soc Chem Commun 2185
137. Goti A, Anichini B, Brandi A, Kozhushkov SI, Gratkowski C, de Meijere A (1996) J Org Chem 61:1665
138. (a) Anichini B, Goti A, Brandi A, Kozhushkov SI, de Meijere A (1997) Synlett 25; (b) Goti A, Anichini B, Brandi A, de Meijere A, Citti L, Nevischi S (1995) Tetrahedron Lett 36:5811; (c) Zorn C, Anichini B, Goti A, Brandi A, Kozhushkov SI, de Meijere A, Citti L (1999) J Org Chem 64: in press
139. Review: Yamada K, Ojika M, Kigoshi H (1998) Angew Chem 110:1918; Angew Chem Int Ed Engl 37:1818
140. Fitjer L (1982) Chem Ber 116:1047
141. Erden I, de Meijere A, Rousseau G, Conia J-M (1980) Tetrahedron Lett 21:2501
142. Weber W, Behrens U, de Meijere A (1980) Chem Ber 114:1196
143. Maercker A, Klein K-D (1991) J Organomet Chem 410:C35
144. Kaufmann D, de Meijere A (1979) Tetrahedron Lett 779
145. (a) Becker H (1996) Dissertation, Universität Göttingen; (b) Becker H, de Meijere A unpublished results
146. Bräse S (1996) Dissertation, Universität Göttingen
147. Bräse S, de Meijere A (1995) Angew Chem 107:2741; Angew Chem Int Ed Engl 34:2545
148. (a) de Meijere A, Nüske H, Es-Sayed M, Schroen M, Bräse S (1999) Angew Chem 111: in press; Angew Chem Int Ed Engl 38: in press; (b) Nüske H, Bräse S, Kozhushkov SI, de Meijere A (1999) Chem Eur J in preparation
149. Pohlmann T, de Meijere A (unpublished results)
150. Binger P, Wedemann P, Kozhushkov SI, de Meijere A (1998) Eur J Org Chem 113
151. O'Donnell MJ, Polt RL (1982) J Org Chem 47:2633
152. Hayashi T, Konishi M, Kobori Y, Kumada M, Higuchi T, Hirotsu K (1984) J Am Chem Soc 106:158
153. Brandl M, Kozhushkov SI, Bräse S, de Meijere A (1998) Eur J Org Chem 453
154. Review: Lipshutz BH (1990) Synlett 119
155. O'Donnell MJ, Bennett WD, Polt RL (1985) Tetrahedron Lett 26:695
156. Kozhushkov SI, Brandl M, Yufit DS, Machinek R, de Meijere A (1997) Liebigs Ann/Recueil 2179
157. Foerstner J, Kozhushkov SI, Binger P, Wedemann P, de Meijere A, Butenschön H (1998) J Chem Soc Chem Commun 239
158. Hofmann P, de Meijere A unpublished results
159. von Seebach M, Kozhushkov SI, Boese R, Yufit DS, de Meijere A unpublished results
160. Baldwin JE, Adlington RM, Bebbington D, Russel AT (1994) Tetrahedron 50:12015 and references cited therein
161. Li K, Du W, Que NLS, Liu H-w (1996) J Am Chem Soc 118:8763
162. (a) Yoshida M, Ezaki M, Hashimoto M, Yamashita M, Shigematsu N, Okuhara M, Kohsaka M, Horikoshi K (1990) J Antibiotics 748; (b) Kou MS, Zielinski RJ, Cialdella JI, Maschke CK (1995) J Am Chem Soc 117:10629
163. (a) Barrett AGM, Tustin GJ (1995) J Chem Soc Chem Commun 355; (b) Barrett AGM, Kasdorf K, Tustin GJ, Williams DJ (1995) J Chem Soc Chem Commun 1143 and references cited therein; (c) Barrett AGM, Kasdorf K (1996) J Am Chem Soc 118:11030; (d) Barrett AGM, Doubleday WW, Kasdorf K, Tustin GJ (1996) J Org Chem 61:3280; (e) Barrett AGM, Gross T, Hamprecht D, Ohkubo M, White AJP, Williams DJ (1998) Synthesis 490; (f) Barrett AGM, Hamprecht D, White AJP, Williams DJ (1997) J Am Chem Soc 119:8608
164. (a) Charette AB, Juteau H (1994) J Am Chem Soc 116:2651; (b) Charette AB, Prescott S, Brochu C (1995) J Org Chem 60:1081

Alkyl 2-Chloro-2-cyclopropylideneacetates – Remarkably Versatile Building Blocks for Organic Synthesis

Armin de Meijere[1] · Sergei I. Kozhushkov[1] · Lazaros P. Hadjiarapoglou[2]

[1] Institut für Organische Chemie der Georg-August-Universität Göttingen, Tammannstr. 2, D-37077 Göttingen, Germany. *E-mail: ameijer1@uni-goettingen.de*

[2] Section of Organic Chemistry and Biochemistry, Department of Chemistry, University of Ioannina, 45110 Ioannina, Greece

Strain in small rings has evolved as one of the principles used to control reactivity and chemoselectivity in transformations of organic compounds. The combination of small rings with multiple bonds and functional groups establishes composite functionalities which demonstrate unique multiple reactivity and thereby potentially high synthetic utility. This survey concentrates on a family of compounds which combines the chemistry of methylenecyclopropanes and that of electron-acceptor-activated alkenes, namely alkyl 2-chloro-2-cyclopropylideneacetates of types **1**–**3**. This special feature makes the compounds **1**–**3** multifunctional as well as highly reactive, and thus extremely versatile building blocks for organic synthesis. Not only is the general synthetic access to methylenecyclopropanes **1**–**3** presented here, but particularly their rich chemistry as highly reactive Michael acceptors, dienophiles, dipolarophiles and general cyclophiles which leads to a wide range of different types of functionally substituted cyclopropane derivatives, spirocyclopropane-annelated hetero- and carbocycles, mono- and oligocondensed cycles, natural and unnatural amino acids and peptidomimetics, and more. Finally, the first results obtained with polymer-bound substrates of types **1**–**3** in a combinatorial approach to libraries of potentially biologically active compounds are presented.

Keywords: Absolute configuration, Amines, Amino acids, Carbenes, Cascade reactions, 2-chloro-2-cyclopropylideneacetates, Combinatorial libraries, Cycloadditions, Cyclobutenes, Cyclopropanes, Diels-Alder reactions, Heterocycles, Michael additions, Nitrones, Nucleophilic substitutions, Peptidomimetics, Palladium catalysis, Polycycles, Solid phase synthesis, Spiro compounds, Thiols

1 Introduction

The cyclopropane ring can be regarded as a unique three-carbon functional group which can undergo transformations that are more difficult or impossible with any of the more conventional functional groups. Simple cyclopropanes are nucleophilic, and like that of a C=C double bond, the nucleophilicity of a cyclopropane ring can be altered by the choice of substituents. Electron-donating substituents tend to increase the nucleophilicity, while electron-withdrawing substituents make the cyclopropane ring susceptible to nucleophilic attack. Furthermore, due to its ring strain and electronic properties [1,2] the combination of the cyclopropane unit with multiple bonds and other functional groups establishes composite functional groups [3], which demonstrate unique reactivities and thereby have a wide potential as building blocks for organic synthesis. In this respect, the chemistry of alkyl 2-chloro-2-cyclopropylideneacetates of types **1**–**3** (Fig. 1) is of particular interest, as the a priori high reactivity of their methylenecyclopropane units is enhanced by the electron-withdrawing substituents. The preparation and chemical transformations of methylenecyclopropane itself [4a–c] and its close relative bicyclopropylidene [4d] have recently been reviewed extensively. The two different functionalities on the methylene group of compounds **1**–**3** make them oligo-functional and thereby lead to very special chemical behavior.

The synthetic methodology based on the reactions of these highly functionalized methylenecyclopropanes, which can be regarded as allenecarboxylate homologues, has steadily been growing over the past ten years [5]; it thus appeared timely to report the considerable progress in this field.

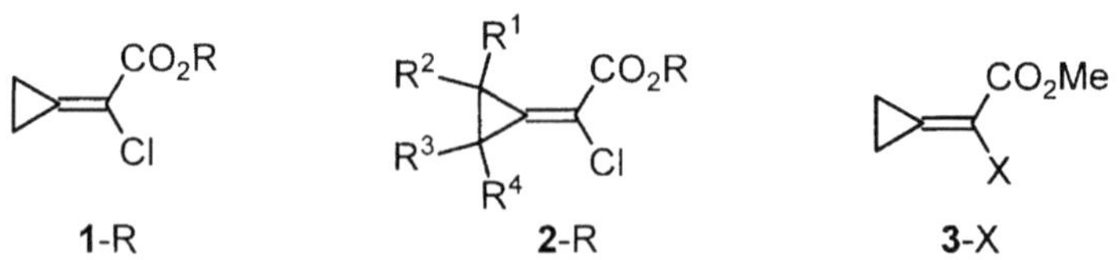

Fig. 1. Alkyl 2-chloro-2-cyclopropylideneacetates **1**, **2** and their analogs **3**

2
The Syntheses of Alkyl 2-Chloro-2-cyclopropylideneacetates and Their Analogs

It is well established [6–13] that tetrachlorocyclopropene (**4**), upon heating to temperatures above 150°C in the presence of an alkene **6**, preferentially yields 1-chloro-1-(trichloroethenyl)cyclopropanes **7** (Scheme 1) [7–11]. Apparently, tetrachlorocyclopropene (**4**) [12] reversibly undergoes ring opening to form perchloroethenylcarbene **5**, which is efficiently trapped by a large variety of alkenes [7, 13]. The yields usually range from 60 to 80%, when starting with equimolar mixtures of **4** and **6**, and can easily be raised to up to 95% with a tenfold excess of the alkene. Numerous examples show that this is a general reaction (Scheme 1); selected from more than 40 examples, only those compounds of type **7** are listed, which have been converted into alkyl 2-chloro-2-cyclopropylideneacetates **2**.

The cycloaddition of the carbene **5** to **6** occurs *stereospecifically*, i.e. with retention of the alkene configuration, and *diastereoselectively*, i.e. the more bulky trichloroethenyl group ends up predominantly *trans* to the most bulky substituent(s) on the original alkene **6** [7–10, 13]. The high efficiency with which this vinylcarbene can be intercepted intermolecularly to give vinylcyclopropanes is most surprising [14].

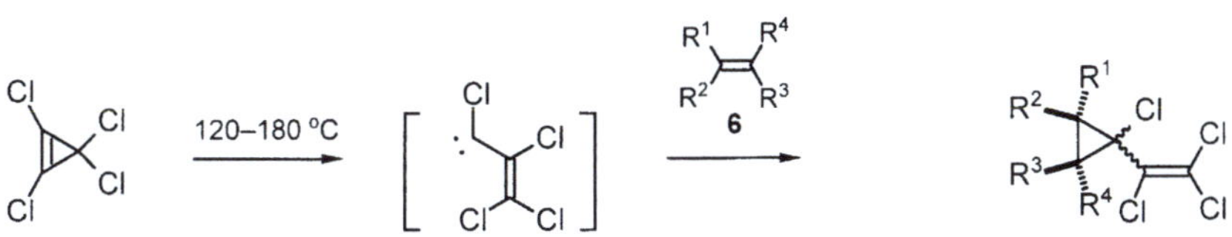

4 **5** **7**

Alkene	R^1	R^2	R^3	R^4	T [°C]	t [h]	Product	Yield (%) [Ref.]
6a	H	H	H	H	170	20	**7a**	76 [7f], 84 [7e]
6b	Et	H	H	H	170	18	**7b**	66 [7e,10]
6c	$-CH_2CH_2-$		H	H	120	336	**7c**	21 [7d], 64 [8]
6d	H	Me	Me	H	180	18	**7d**	78 [7d]
6e	H	$-(CH_2)_4-$		H	150	72	**7e**	66 [7d], 91 [7j]
6f	H	Me	H	Me	180	18	**7f**	76 [7d]
6g	Me	Me	Me	H	180	14	**7g**	88 [7d]
6h	Me	Me	Me	Me	180	12	**7h**	82 [7d], 96 [7e]
6i	Me	H	H	H	180	18	**7i**	87 [7b,10]
6j	CH_2OBn	H	H	H	170	24	**7j**	38 [9,10]
6k	$(CH_2)_2OBn$	H	H	H	170	24	**7k**	60 [7k], 59 [10c]
6l	$(CH_2)_4OBn$	H	H	H	160	15	**7l**	54 [11]
6m	CH_2OMe	H	H	H	170	24	**7m**	71 [10]
6n	CH_2OSiMe_3	H	H	H	170	24	**7n**	68 [25]
6o	CH_2Br	H	H	H	170	18	**7o**	38 [10]
6p	$-(CH_2-C)-$ (C = cyclopropane)		H	H	120	168	**7p**	55 [7l]

Scheme 1. Preparation of 1-chloro-1-(trichloroethenyl)cyclopropanes **7** by the addition of thermally ring-opened tetrachlorocyclopropene **5** onto alkenes **6** [7–10, 25]

1-Chloro-1-(trichloroethenyl)cyclopropanes **7** were transformed into methyl 2-chloro-2-cyclopropylideneacetates **1**-Me and **2**-Me by heating with sodium methoxide in methanol and subsequent treatment with acid [8–11, 15, 16]. This remarkable transformation starts with an attack of methoxide anion at the terminal vinylic carbon of **7**, replacing the chlorine substituent on the cyclopropane ring in an S_N2'-type reaction, and this is followed by substitution of the two chlorines at the newly formed terminal sp^3 carbon [17]. The orthoesters **8**, thus formed, can be isolated, but for better overall yields are directly hydrolyzed under acidic conditions. Usually, the overall yields in this conversion are good to very good, and they have been as high as 85%, even for the unsubstituted parent **1**-Me (Scheme 2).

Several derivatives of type **7** have never been treated in a way to produce the corresponding derivatives of type **2**, and in a few cases as for example compounds **7** with $R^1=Me_3Si$ or $CH_2CH(OH)CH_3$, $R^2-R^4=H$ gave neither the orthoester **8** nor the ester **2** [15b].

The approach to alkyl 2-cyclopropylideneacetates **3** (X=H) first developed by Salaün et al. [18a] and later improved by Spitzner et al. [18b], namely the Hor-

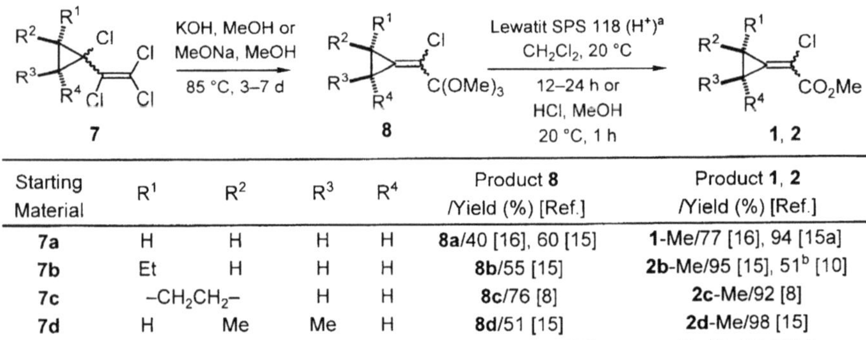

7 **8** **1, 2**

Starting Material	R^1	R^2	R^3	R^4	Product **8** /Yield (%) [Ref.]	Product **1**, **2** /Yield (%) [Ref.]
7a	H	H	H	H	**8a**/40 [16], 60 [15]	**1**-Me/77 [16], 94 [15a]
7b	Et	H	H	H	**8b**/55 [15]	**2b**-Me/95 [15], 51ᵇ [10]
7c	$-CH_2CH_2-$		H	H	**8c**/76 [8]	**2c**-Me/92 [8]
7d	H	Me	Me	H	**8d**/51 [15]	**2d**-Me/98 [15]
7e	H	$-(CH_2)_4-$		H	**8e**/24 [15], 70 [7j]	**2e**-Me/94 [15b]
7f	H	Me	H	Me	**8f**/59 [15]	**2f**-Me/98 [15]
7g	Me	Me	Me	H	**8g**/40 [15b]	**2g**-Me/74 [15b]
7h	Me	Me	Me	Me	**8h**/71 [15b]	**2h**-Me/86 [15b]
7i	Me	H	H	H	**8i**/60 [15b], 49 [16]	**2i**-Me/95 [15a], 78 [15b], 52ᵇ [10]
7j	CH_2OBn	H	H	H	**8j**/n. r. [9,10]	**2j**-Me/50ᵇ [9,10]
7k	$(CH_2)_2OBn$	H	H	H	**8k**/n. r. [7k]	**2k**-Me/47ᵇ [7k]
7l	$(CH_2)_4OBn$	H	H	H	**8l**/n. r. [11]	**2l**-Me/50ᵇ [11]
7m	CH_2OMe	H	H	H	**8k**/n. r. [10]	**2m**-Me/50ᵇ [10]
7n	CH_2OSiMe_3	H	H	H	**8n**/n. r. [25]	**2n**-Me/46ᵇ [25]
7p	$-(CH_2-C)-$		H	H	**8p**/55. [7l]	**2p**-Me/83 [7l]
7r	CH_2SCH_3	H	H	H	**8r**/n. r. [10]	**2r**-Me/50ᵇ [10]
7s	CH_2N_3	H	H	H	**8s**/n. r. [10]	**2s**-Me/52ᵇ [10]

ᵃ Lewatit SPS 118 is a strongly acidic ion exchange resin. – ᵇ Yield over two steps, n. r. = not reported

Scheme 2. Preparation of methyl 2-chloro-2-cyclopropylideneacetates **1**-Me and **2**-Me [8–11, 15, 16, 25]

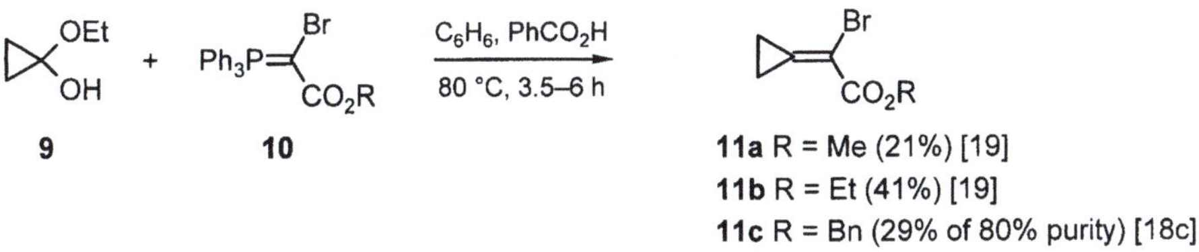

Scheme 3. Preparation of alkyl 2-bromo-2-cyclopropylideneacetates **11a–c** by olefination of the cyclopropanone hemiacetal **9** [18c, 19]

ner-Wadsworth-Emmons olefination of cyclopropanone hemiacetal, has also been applied for the preparation of alkyl 2-bromo-2-cyclopropylideneacetates **11a–c**, but in all the cases reported gave only moderate yields at best [18c, 19] (Scheme 3).

Other esters of 2-chloro-2-cyclopropylideneacetic acid can easily be prepared from the methyl esters **1**-Me, **2**-Me by transesterification with the appropriate alcohol component under titanium tetraisopropoxide catalysis [20], as has been demonstrated for a number of benzyl and ethyl esters (Scheme 4) [9, 10, 21, 22a].

Alternatively, the free acids **1**-H, **2**-H can be prepared by hydrolysis of **1**-Me, **2**-Me under standard conditions [15a, 23] or better, with ultrasonication (Fig. 2) [11b, 19b, 24, 25], and then esterified with the appropriate alcohol applying any of the known methods. The orthoesters **8** can also be directly hydrolyzed to **2**-H (74% yield for **2h**-H) [15b]. The esterification of **1**-H and **2**-H (Scheme 4) has e.g. been used to prepare *tert*-butyl (**1**-*t*Bu, **2**-*t*Bu) [25, 26], *p*-substituted benzyl (**12a–d**) [22b] and enantiomerically pure menthyl [**1**-(1′*R*)-Ment] [23, 24] esters.

Other 2-substituted 2-cyclopropylideneacetates of type **3**-X have been prepared from methyl 2-chloro-2-cyclopropylideneacetate (**1**-Me) in a sequence starting with the Michael addition of an appropriate nucleophile to **1**-Me, followed by nucleophilic substitution of the chlorine in **13** to obtain **14** (Fig. 3 and

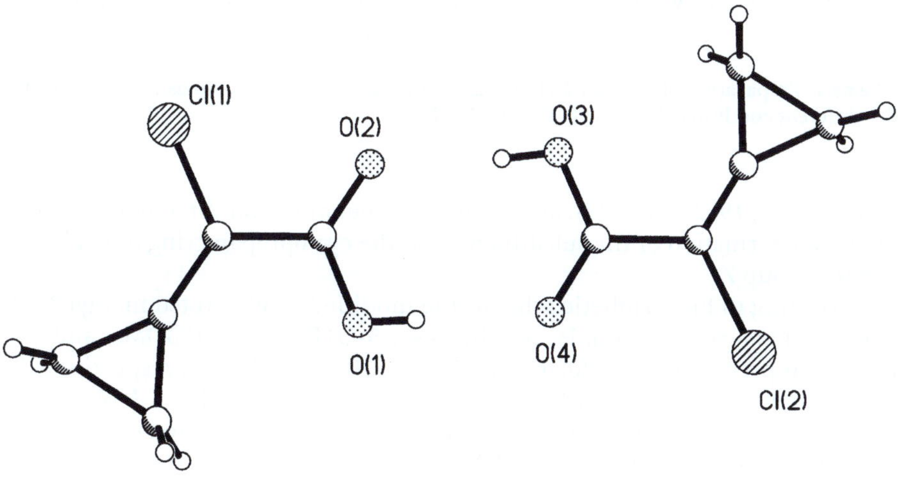

Fig. 2. The structure of 2-chloro-2-cyclopropylideneacetic acid (**1**-H) in the crystal [11b, c]

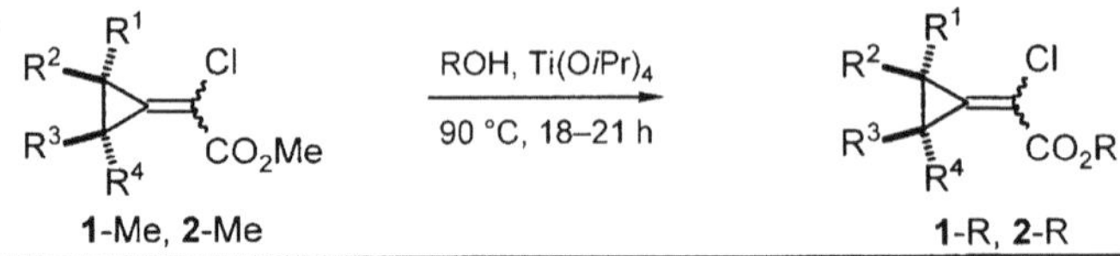

Starting Material	R^1	R^2	R^3	R^4	R	Product	Yield (%)	Ref.
1-Me	H	H	H	H	Bn	**1**-Bn	73	[9]
2b-Me	Et	H	H	H	Bn	**2b**-Bn	91	[10]
2c-Me	$-CH_2CH_2-$		H	H	Bn	**2c**-Bn	92	[21]
2e-Me	H	$-(CH_2)_4-$		H	Et	**2e**-Et	93	[22a]
2i-Me	Me	H	H	H	Bn	**2i**-Bn	70	[10]
·**2m**-Me	CH_3OCH_2	H	H	H	Bn	**2m**-Bn	73	[10]

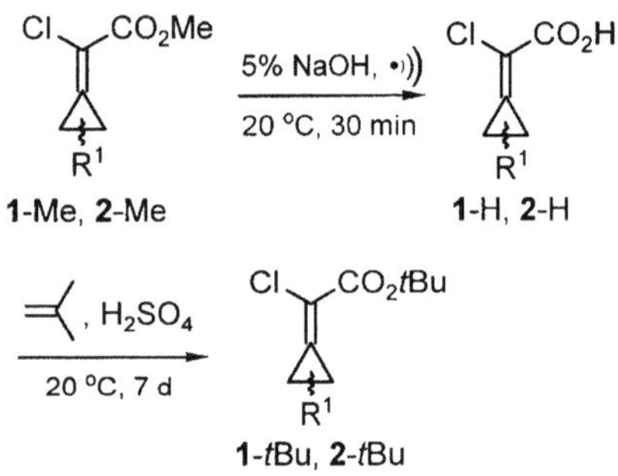

Starting Material	R^1	Product/ Yield (%)	Product/ Yield (%)
1-Me	H	**1**-H/96	**1**-*t*Bu/69
2b-Me	Et	**2b**-H/83	**2b**-*t*Bu/91
2i-Me	Me	**2i**-H/95	**2i**-*t*Bu/95
2n-Me	CH_2OSiMe_3	**2n**-H/75	**2n**-*t*Bu/76

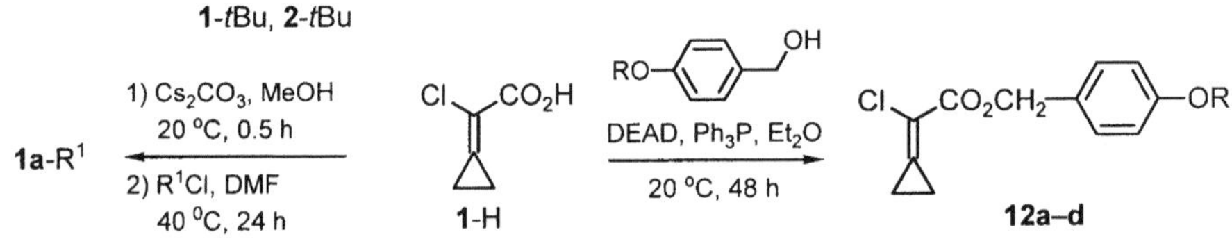

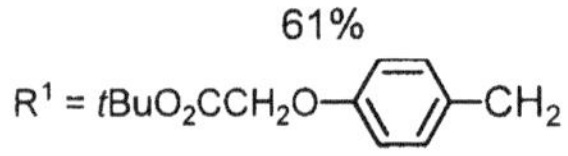

R	Me	CH_2CO_2Me	CH_2CO_2tBu	$CH_2CO_2C_2H_4SMe$
Product	**12a**	**12b**	**12c**	**12d**
Yield (%)	63	65	56	39

Scheme 4. Preparation of various 2-chloro-2-cyclopropylideneacetic acid esters **1**-R, **2**-R under different conditions [9, 10, 11b, 15, 19b, 21–26]

Table 1) [7h, 19, 25, 26b, 27], and eventually *β*-elimination directly from **14** or after transformation of the substituent Y on the cyclopropane ring into a better leaving group Z.

According to this synthetic scheme, the modified 2-substituted methyl 2-cyclopropylideneacetates **3**-SMe, **3**-SPh, **3**-N_3, **3**-OTBDPS, **3**-OTBDMS and **3**-F have been prepared in 25, 40, 60, 51, 37 and 53% overall yield, respectively. In spite of the overall yield of 74% for the parent compound **3**-H (≙ **1**-Me) along this route, the latter is more readily available by Wittig olefination of cyclopropanone hemiacetal **9** (70% yield) [9, 18a, b].

Table 1. Reaction conditions, yields and products for the conversion of methyl 2-chloro-2-cyclopropylideneacetate **1**-Me into 2-substituted methyl 2-cyclopropylideneacetates **3**-X

Starting material	Conditions	Y (*Z*)	X	Product	Yield (%)	Ref.
1-Me	**A**	SePh	–	**13**-SePh	96	[27]
1-Me	**B**, 24 h	$PhSO_2$	–	**13**-SO_2Ph	54	[27]
1-Me	**B**, 4 h	$TolSO_2$	–	**13**-SO_2Tol	62	[27]
1-Me	**C**	Me_2N	–	**13**-NMe_2	93	[19a]
13-SePh	**D**	SePh	H	**14**-SePh, H	83	[19a]
13-SePh	**E**	SePh	N_3	**14**-SePh, N_3	84	[19a]
13-SePh	**F**	SePh	OH	**14**-SePh, OH	90	[19a]
13-SePh	**G**, 15 min	SePh	SPh	**14**-SePh, SPh	97	[19a]
13-SO_2Ph	**G**, 10 min	$PhSO_2$	SPh	**14**-SO_2Ph, SPh	92	[27]
13-SO_2Tol	**G**, 10 min	$TolSO_2$	SPh	**14**-SO_2Tol, SPh	66	[27]
13-SO_2Tol	**G**, 30 min	$TolSO_2$	SMe	**14**-SO_2Tol, SMe	62	[27]
14-SePh,OH	**H**	SePh	F	**14**-SePh,F	87	[19a]
13-SO_2Tol	**I**	$TolSO_2$	H	**14**-SO_2Tol,H	94	[27]
13-NMe_2	**J**	NMe_2	N_3	**14**-NMe_2, N_3	73	[19a]
13-NMe_2	**K**	NMe_2	OH	**14**-NMe_2, OH	83	[19a]
14-NMe_2,OH	**L**, 26 h	NMe_2	OTBDPS	**14**-NMe_2, OTBDPS	90	[19a]
14-NMe_2,OH	**L**, 24 h	NMe_2	OTBDMS	**14**-NMe_2, OTBDMS	77	[19a]
14-NMe_2,N_3	**M**, 48 h	N^+Me_3	N_3	**15**-N^+Me_3, N_3	85	[19a]
14-NMe_2, OTBDPS	**M**, 4 d	N^+Me_3	OTBDPS	**15**-N^+Me_3, OTBDPS	98	[19a]
14-NMe_2, OTBDMS	**M**, 48 h	N^+Me_3	OTBDMS	**15**-N^+Me_3, OTBDMS	77	[19a]
14-SePh, H	**N**, –5 °C	–	H	**3**-H	93	[27]
14-SePh, SPh	**N**, –15 °C	–	SPh	**3**-SPh	92	[19a]
14-SO_2Tol, H	**O**, 20 min	–	H	**3**-H	90	[27]
14-SO_2Ph, SPh	**O**, 30 min	–	SPh	**3**-SPh	81	[27]
14-SO_2Tol, SPh	**O**, 55 min	–	SPh	**3**-SPh	81	[27]
14-SO_2Tol, SMe	**O**, 15 min	–	SMe	**3**-SMe	65	[27]
14-SePh, N_3	**P**	–	N_3	**3**-N_3	75	[19a]
15-N^+Me_3, OTBDPS	**R**, 1.5 h	–	OTBDPS	**3**-OTBDPS	75	[19a]
15-N^+Me_3, N_3	**R**, 10 min	–	OTBDMS	**3**-N_3	80	[19a]
15-N^+Me_3, OTBDMS	**R**, 6 min	–	N_3	**3**-OTBDMS	83	[19a]
14-SePh, F	**S**	–	F	**3**-F	71	[19a]

Reaction conditions: **A:** NaSePh, EtOH, –78 → –30 °C, 3 h; **B:** $ArSO_2Na$, H_2O/CH_2Cl_2, TEBACl, 5 °C; **C:** Me_2NH, MeOH, –10 °C, 10 min; **D:** *t*BuSH, *t*BuSNa, DMSO, 20 °C, 10 min; **E:** NaN_3, nBu_4NCl, Me_2CO/CH_2Cl_2, 20 °C, 28 d; **F:** $AgNO_3$, H_2O/Me_2CO, 100 °C, 1 h; **G:** Ph(Me)SH/Ph(Me)-SNa, DMSO, 20 °C; **H:** Et_2NSF_3, CH_2Cl_2, 20 °C, 3 h; **I:** Zn/Cu, THF/H_2O, 20 °C, ultrasonication, 20 min; **J:** NaN_3, NaI, DMF, 20 °C, 48 h; **K:** Ag_2CO_3, THF/H_2O, 35 °C, 48 h; **L:** $tBuR_2SiCl$, ImH, DMAP, DMF, 20 °C (R = Ph, Me); **M:** MeI, Me_2CO, 20 °C; **N:** *m*-CPBA, $CHCl_3$, 10 min, then Na_2CO_3; **O:** KOH, CH_2Cl_2, DB-18-c-6, ultrasonication; **P:** H_2O_2, CH_2Cl_2, 20 °C, 7 min; **R:** NaOH, TEBACl, $EtOH/H_2O$, 20 °C; **S:** 1) O_3, CH_2Cl_2, –78 °C; 2) Et_3N, $CHCl_3$, 55 °C, 2 h.

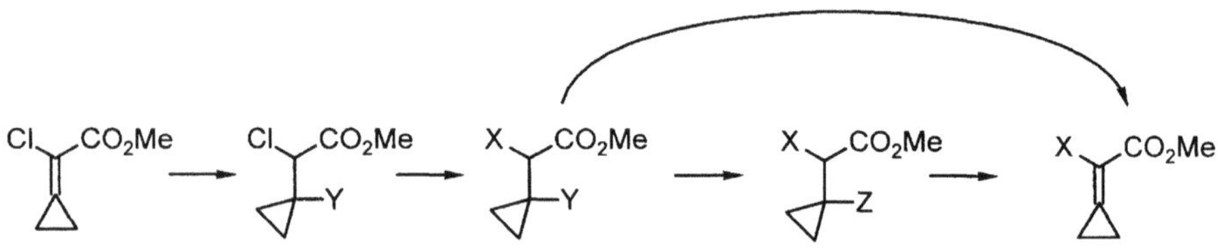

1-Me 13 14 15 3-X

Fig. 3. Synthetic strategy for the preparation of 2-substituted methyl 2-cyclopropylideneacetates 3-X from methyl 2-chloro-2-cyclopropylideneacetate (1-Me)

3 Chemical Reactivity

3.1 Cycloadditions, Carbene Additions and Diels-Alder Reactions of Alkyl 2-Chloro-2-cyclopropylideneacetates

The principles governing the outstanding reactivity of methyl 2-chloro-2-cyclopropylideneacetate (**1**-Me), have been probed by a number of physical measurements (cf. ref. [28]). However, the structural parameters of the chloro ester **1**-Me do not at all provide any explanation, as the length of the C,C-double bond is approximately the same (1.312 Å) as that in methylenecyclopropane (1.316 Å) and bicyclopropylidene (1.314 Å) [4d]. The ^{13}C-NMR chemical shifts for the vinylic carbon atoms in **1**-Me (114.4 and 138.7 ppm) indicate a difference which is characteristic for the polarization in α,β-unsaturated esters. The increased strain in the methylenecyclopropane moiety of **1**-Me compared to methyl α-chloroacrylate should influence the energy of the ground state more than that of the transition state of any reaction involving the double bond, i.e. the activation energy of any such reaction on **1**-Me should be significantly lower than that for the same reaction on methyl α-chloroacrylate. Thus, excess strain release is a significant factor determining the enhanced reactivity of **1**-Me. Moreover, the cyclopropane ring not only accelerates the majority of chemical transformations, but brings its own specificity: the capability of a three-membered ring to undergo ring opening and rearrangement reactions further enriches the chemistry of the chlorocyclopropylideneacetates **1**, **2** and makes them more versatile in comparison with simple acrylates.

Despite its high reactivity, the α-chloro ester **1**-Me does not polymerize like simple acrylates. However, in close analogy to other 1,1-disubstituted methylenecyclopropanes [29], **1**-Me slowly dimerizes in a head-to-head fashion even at room temperature to give the two diastereomeric dimethyl dispiro[2.0.2.2]octanedicarboxylates (*E*)- and (*Z*)-**16** (ratio 1:1) (Scheme 5), and at 120 °C the dimerization proceeds almost quantitatively (89% isolated yield) [15]. Apparently the dimerization of **1**-Me occurs considerably more readily than that of methylenecyclopropane [4a, b] and of bicyclopropylidene [4d]. Surprisingly, under high pressure (10 kbar) this dimerization did not proceed more efficiently (Scheme 5) [30], but more cleanly, which facilitated the separation of isomers.

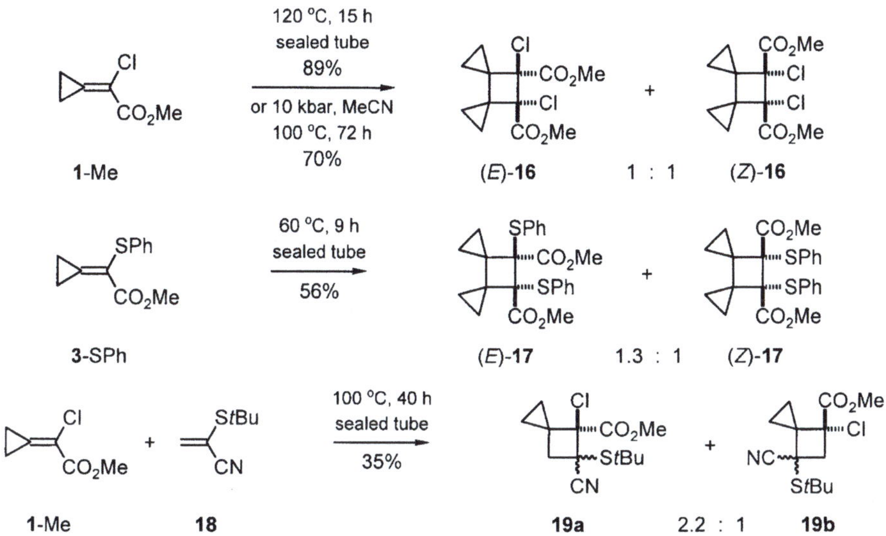

Scheme 5. Dimerizations and other [2+2] cycloadditions of methyl 2-chloro-2-cyclopropylideneacetate (**1**-Me) and its phenylthio analog **3**-SPh [7h, 15, 29, 30]

The (*E*)- and (*Z*)-isomers could be separated by chromatography or by low temperature recrystallization.

The methylenecyclopropane derivative **3**-SPh with its capto-dative substitution pattern has demonstrated essentially the same reaction mode and underwent dimerization to afford a mixture of (*E*)- and (*Z*)-**17** (ratio 1.3:1) upon attempted cycloaddition of **3**-SPh onto bicyclopropylidene [7h, 29] (Scheme 5). The assignment of these diastereomers was secured by an X-ray crystal structure analysis of (*E*)- and (*Z*)-**16** [11c, 30] as well as (*E*)-**17** [29].

These reactions of **1**-Me resemble that of (dichloromethylene)cyclopropane [31] and "radicophilic" alkenes with a capto-dative substitution pattern [32]. Thus, it is not surprising that **1**-Me reacts with *α*-*tert*-butylthioacrylonitrile (**18**), yielding the two isomeric cyclobutane derivatives **19a,b** (ratio 2.2:1) as a mixture of two diastereomers each [29] (Scheme 5), and this reaction occurs under milder conditions than the [2+2] cycloaddition of **18** onto methylenecyclopropane.

Reaction of the chloro ester **1**-Me with the enol silyl ether **23a** in the presence of dimethylaluminum chloride afforded the [2+2] cycloadduct **20** (67% yield) (Scheme 6). Deprotection of the alcohol moiety with nBu$_4$NF in THF gave an 83% yield of a mixture of the 5-keto ester **21a** and the spiropentane derivative **22** (ratio 1:90). Upon running the reaction in a mixture of THF/water (ratio 1:1) instead of anhydrous THF, the 5-keto ester **21a** was isolated exclusively [33].

However, the analogous reactions of **1**-Me with various enol silyl ethers **23** in the presence of titanium tetrachloride at –78 °C, led directly to the products **21a–c** [26a, 33] of Mukaiyama-type reactions [34] in moderate yields (Scheme 6).

The chlorine substituent in **21a,b** could be exchanged with azide (NaN_3, DMF) virtually quantitatively to yield **24a,b**, and subsequent reduction of the

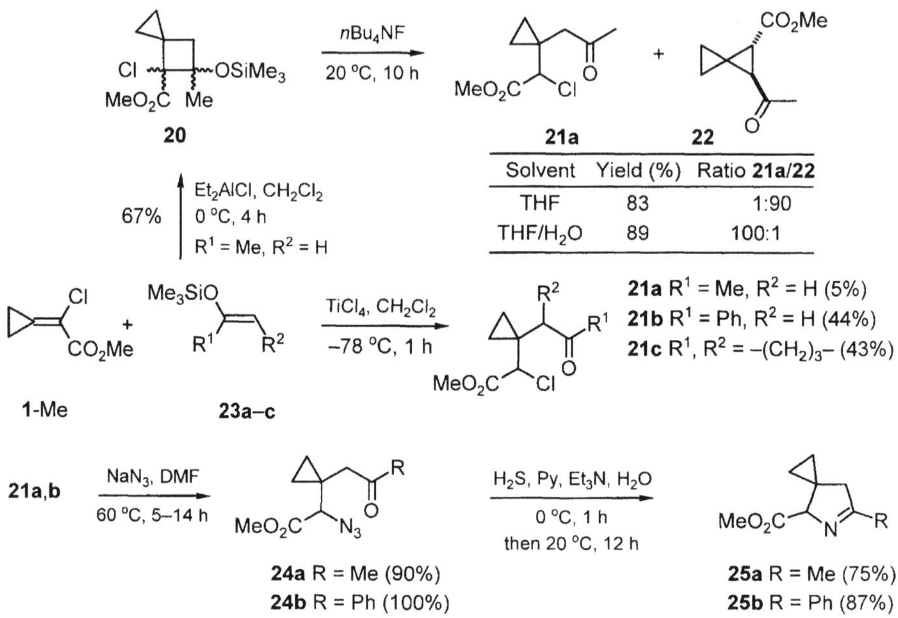

Solvent	Yield (%)	Ratio **21a/22**
THF	83	1:90
THF/H_2O	89	100:1

Scheme 6. Reactions of methyl 2-chloro-2-cyclopropylideneacetate (**1**-Me) with various enol silyl ethers **23** under different conditions [26a, 33]

azide moiety with hydrogen sulfide in pyridine/triethylamine/water [35] yielded the 3,4-dihydropyrrolidines **25a** and **25b** in 75 and 87% yield, respectively [26a, 33]. An even better yield of **25a** (91%, [33]) was obtained by the domino Staudinger-Aza-Wittig reaction upon treatment of **24a** with PPh_3 [36].

There is no published example of a cyclopropanation of the double bond in chlorocyclopropylideneacetate **1**-Me with retention of the chlorine atom. Thus, attempted cyclopropanations under Simmons-Smith [37] or Corey [38] conditions failed [25]. The treatment of the highly reactive methylenecyclopropane derivative **1**-Me with dimethoxycarbene generated by thermal decomposition of 2,2-dimethoxy-Δ^3-1,3,4-oxadiazoline **26** (1.5 equiv. of **26**, PhH, 100 °C, 24 h), gave a complex mixture of products (Scheme 7) [39], yet the "normal" cycloadduct **28** was not detected. The formation of compounds **29**–**33** was rationalized via the initially formed zwitterion **27**, resulting from the Michael addition of the highly nucleophilic dimethoxycarbene to the C,C-double bond of **1**-Me. The ring closure of **27** to the "normal" product **28** is probably reversible, and **27** can rearrange or add a second dimethoxycarbene moiety and a molecule of acetone to form **33**.

With diazoalkanes **34**, as should be expected in view of the high polarity of its double bond, the chloro ester **1**-Me demonstrated its reasonably good dipolarophilicity. Indeed, mixtures of the regioisomeric pyrazolines **35** and **36** were obtained in good yields [26a, 33], when a solution of **1**-Me was treated with any of the diazoalkanes **34a**–**e** (Scheme 8).

In the case of the diazoalkane **34b**, the ratio of **35**/**36** was observed to depend on the reaction temperature with an increasing fraction of **35b** with decreasing

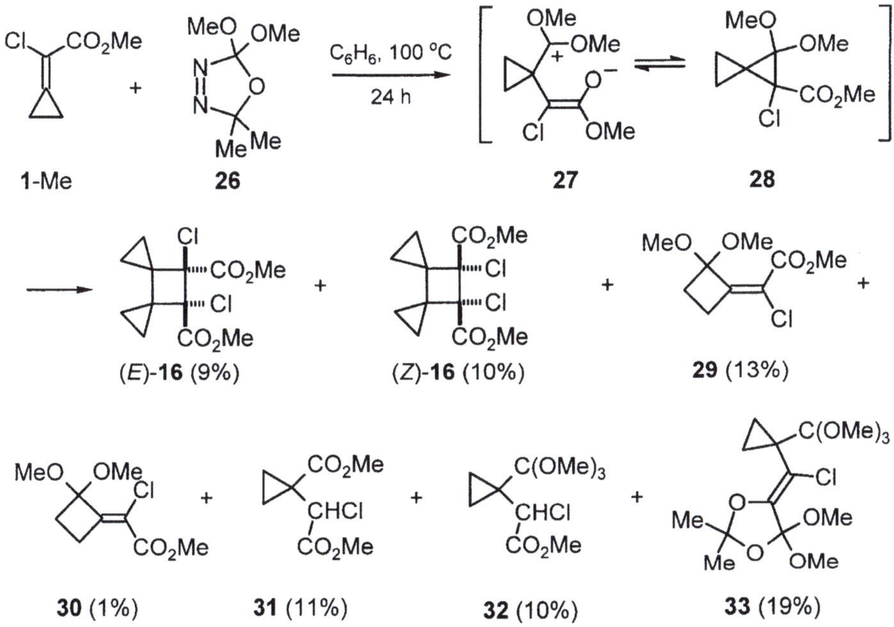

Scheme 7. Reaction of methyl 2-chloro-2-cyclopropylideneacetate (1-Me) with dimethoxycarbene [39]

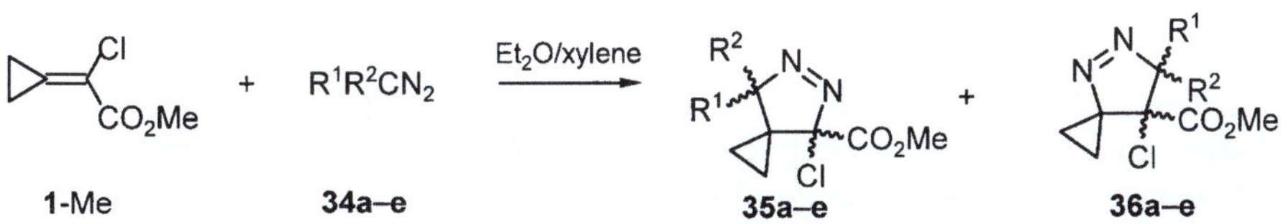

Diazoalkane	R^1	R^2	Conditions	Yield (%)	Ratio **35/36**
34a	H	H	–20 °C, 20 h	98	>98:2
34b	Ph	H	0 °C, 6 d	71	60:40
34b	Ph	H	–18 °C, 6 d	73	89:11
34b	Ph	H	–78 °C, 6 d	75	97:3
34c	Me	Me	–20 °C, 6 d	99	92:8
34d	Ph	Me	–18 °C, 6 d	75	n. r.
34e	Ph	Ph	20 °C, 3 d	64	<2:98

n. r. = not reported

Scheme 8. 1,3-Dipolar cycloaddition of diazoalkanes **34** onto methyl 2-chloro-2-cyclopropylideneacetate (1-Me) [26a, 33]

temperature (Scheme 8) [33]. The predominance of the "abnormal" [2+3] cycloadduct **36e** [40], the structure of which was unequivocally established by X-ray crystal structure analysis [26a], may arise from unfavorable steric interactions between the phenyl groups and the cyclopropane ring in the transition state that would lead to the normal product; a similar orientation had previously only been observed for the cycloaddition of diphenyldiazomethane to acetylenic dipolarophiles [41]. This inverted regioselectivity in the addition of **34e** to **1**-Me may also be caused by a difference in the electron distributions between **34a** and **34e**. The two *p*-donor phenyl substituents in **34e** may be expected to favor a larger contribution of the $N^{-}=N^{+}=CR_2$ mesomeric structure to the ground state [42].

Furthermore, the chloro ester **1**-Me also readily reacted with the nitrile ylide **37** at room temperature, however, the pyrrole derivative **40** was the only isolated product (Scheme 9) [26a]. The latter was obviously formed from the primary cycloadduct **38** by a cyclopropylcarbinyl to homoallyl rearrangement [1].

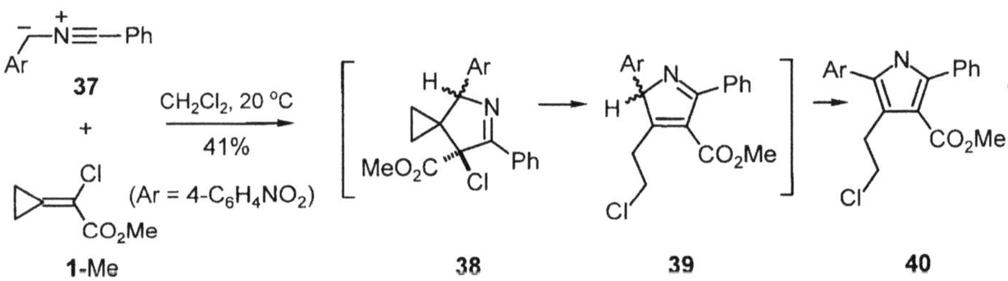

Scheme 9. 1,3-Dipolar cycloaddition of the nitrile ylid 37 onto methyl 2-chloro-2-cyclopropylideneacetate (1-Me) [26a]

Nitrones **41** readily cycloadd to methyl 2-chloro-2-cyclopropylideneacetate (**1**-Me) and its spirocyclopropanated analog **2c**-Me to give 5-spirocyclopropaneisoxazolidines **42a–f** in good yields (Scheme 10) [43].

In contrast to the cycloadducts of nitrones to methylenecyclopropane derivatives without functional substituents (cf. [4c, d]), the thermal rearrangements of the chlorocyclopropylideneacetate adducts **42** strongly depend on the polarity of the solvent as well as the nature of the substituents R^1-R^3 and they take completely different modes. Thus, only reversible isomerization on the C^4 center of the isoxazolidine ring was observed for isoxazolidinotetrahydroisoquinoline **42b** by heating in toluene (110 °C, 3 h). However, at 150 °C in xylene compounds **42a, b** underwent complete reorganization into dehydroisoquinolines **45** and their cyclic amides **44** [43], presumably by nucleophilic attack of chloride ion with ring opening of the doubly activated cyclopropane intermediate **43** as a key step [44] (Scheme 11). When the thermal rearrangement was carried out in DMSO, a completely different reaction mode of isoxazolidines **42a–d** was observed. Whereas the isomerization of **42a, b** proceeded only at 150 °C and led to considerable decomposition of the starting material and low yields of the benzoquinolizinones **47a, b**, a fast and clean reorganization of compounds **42c, d** occurred at 100 °C to give the hexahydroindolizin-5-ones **47c, d** in 83 and 73% yield, respectively (Scheme 11).

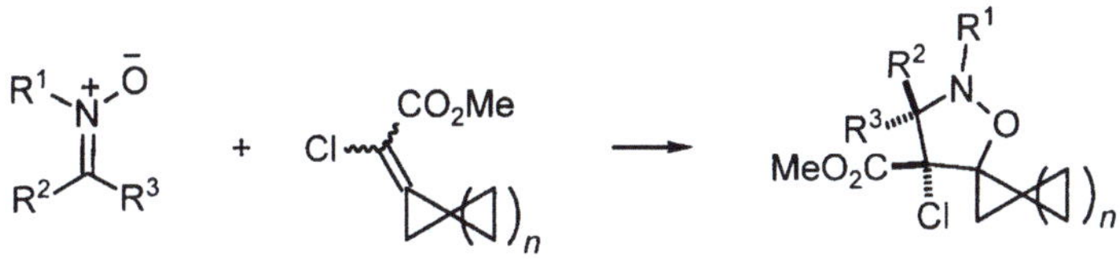

41 **1**-Me, $n = 0$ **2c**-Me, $n = 1$ **42a–f**

R^1	R^2	R^3	n	Conditions	Product	Yield of **42** (%)
o-$CH_2CH_2C_6H_4$–		H	0	CH_2Cl_2, 38 °C, 2 h	**42a**	91
o-$CH_2CH_2C_6H_4$–		H	1	CH_2Cl_2, 38 °C, 2 h	**42b**	93
–$(CH_2)_3$–		H	0	CH_2Cl_2, 25 °C, 60 h	**42c**	65
–$(CH_2)_3$–		H	1	CH_2Cl_2, 25 °C, 60 h	**42d**	58[a]
Me	H	Ph	0	toluene, 60 °C, 36 h	**42e**[b]	70[b]
Me	H	Ph	1	toluene, 60 °C, 5 h	**42f**	68

[a] Compound **46d** (12%) was also formed. – [b] Plus its diastereomer (13% yield).

Scheme 10. 1,3-Dipolar cycloadditions of nitrones **41** onto methyl 2-chloro-2-cyclopropylideneacetates (**1**-Me, **2c**-Me) [43]

The mechanism of this reaction is still not clear, but the key steps are probably a cyclopropylcarbinyl to cyclobutyl ring enlargement [45] with subsequent ring enlargement of the cyclobutane derivative **46.** In fact, such cyclobutane derivatives **46c, d** could easily be prepared in 86 and 94% yield, respectively, by stirring dichloromethane solutions of **42c, d** in the presence of Al_2O_3 at 20 °C, and **46c, d** quantitatively isomerized into **47c, d** upon heating in DMSO at 100 °C for 2 h.

The isoxazolidines **42e, f** only gave the cyclobutane derivatives of type **46** upon thermal rearrangement (isolated in 82% yield both after heating in DMSO at 100 °C for 7 and 2 h, respectively), and **46e, f** appeared to be inert to further thermal transformations.

Not only are the cyclopropylideneacetates **1**–**3** exceptional dipolarophiles, they also readily undergo [4+2] cycloadditions. The Diels-Alder reactivity of **1**–**3** is also enhanced by the additional strain in the double bond and the cyclopropane ring, any cycloaddition across the double bond leads to strain release in the ring as well. In many cases, the borderline between a domino Michael addition and a Diels-Alder cycloaddition is not well-defined [5c]. Surprisingly, the oxysubstituted open-chain dienes **48b, c** [28, 46], gave only poor yields of the corresponding cycloadducts **49** (Scheme 12), and attempts to improve these yields using Lewis acid catalysts ($TiCl_4$, BF_3, WCl_6) were unsuccessful [46]. However, 2,3-dimethylbutadiene **48a** and a variety of vicinal dialkenylidenecyclopentanes, -cyclohexanes and -heterocycles gave good to very good yields.

Good results were also obtained in the Diels-Alder reactions of methyl 2-chloro-2-cyclopropylideneacetate (**1**-Me) with cycloalkadienes. Thus, **1**-Me reacted with cyclopentadiene within 5 h at ambient temperature to give *endo/exo*-

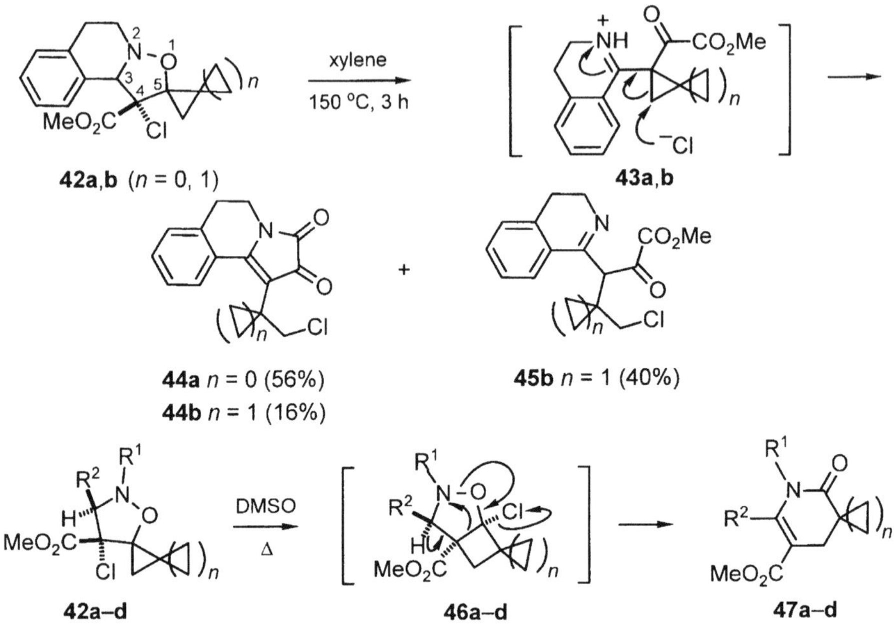

R^1 R^2	n	Conditions	Product	Yield of **47** (%)
–o-$CH_2CH_2C_6H_4$–	0	150 °C, 3 h	**47a**	15
–o-$CH_2CH_2C_6H_4$–	1	150 °C, 3 h	**47b**	21
–$(CH_2)_3$–	0	100 °C, 5 h	**47c**	83
–$(CH_2)_3$–	1	100 °C, 3 h	**47d**	73

Scheme 11. Thermal rearrangements of 5-spirocyclopropaneisoxazolidines **42a–d** [43]

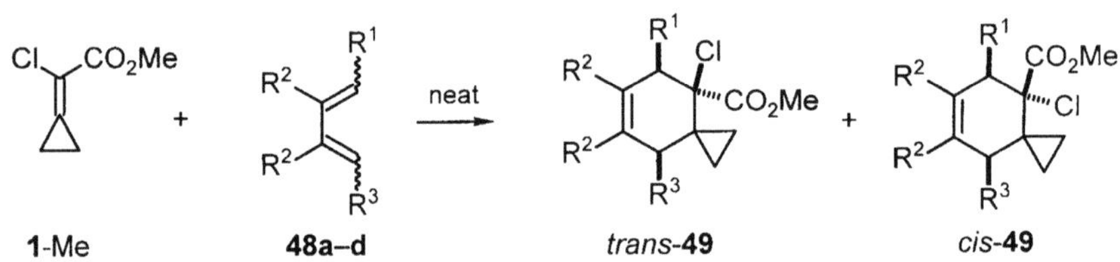

R^1	R^2	R^3	T [°C]	t [h]	Product	Yield (%)	Ratio *cis/trans*
H	Me	H	20	12 d	**49a**	79	–
OEt	H	Me	80	18	**49b**	18	0:100
$OSiMe_3$	H	Me	70	24	**49c**	12	n. r.
Me	–$(CH_2)_2$–	H	18	8	**49d**	64	3.1:1

n. r. = not reported

Scheme 12. [4+2] Cycloaddition reactions of methyl 2-chloro-2-cyclopropylideneacetate (1-Me) with open-chain dienes **48** [46]

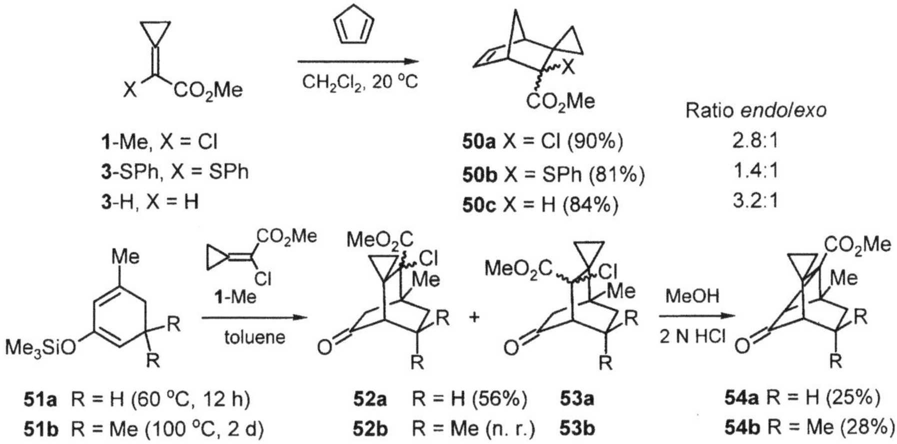

Scheme 13. [4+2] Cycloadditions of cyclopropylideneacetates **1**-Me, **3**-SPh and **3**-H with cyclic dienes [7h, 28]

50a in 90% yield (Scheme 13) [15, 28]. The α-phenylthio derivative **3**-SPh (reaction time 3 h) was similarly reactive towards cyclopentadiene, and the unsubstituted cyclopropylideneacetate **3**-H required a significantly longer reaction time (4 d) [7h].

But the trimethylsilyloxycyclohexadienes (**51 a, b**) reacted with the chloro ester **1**-Me only under more drastic conditions to give in moderate yield about equal amounts of the regioisomeric cycloadducts **52** and **53**, each as a mixture of *endo*- and *exo*-isomers (Scheme 13) [28]. Upon treatment with acid or acidic work-up of the reaction mixture, compounds **52** and **53** were converted to the tricyclic keto esters **54a** and **54b** in 25 and 28% yield, respectively (cf. Sect. 4.4).

N-substituted pyrroles **55** entered the Diels-Alder cycloaddition with **1**-Me only under high pressure to give mixtures of *endo*- and *exo*-adducts **56** (the ratio was not determined) (Scheme 14) [30].

Furan (**57**) which is less aromatic and thereby more reactive than pyrroles and especially non-aromatic 6,6-dimethylfulvene (**58**), as one would expect, underwent facile cycloaddition upon treatment with neat **1**-Me to give the corresponding bicyclic cyclohexene derivatives **59** and **60** (Scheme 15) [19, 46]. The

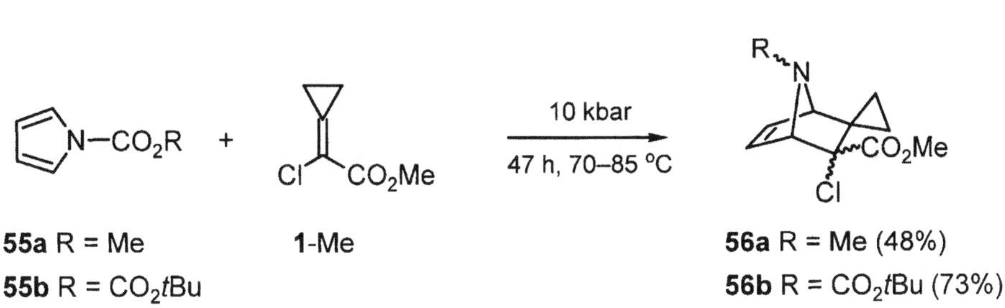

Scheme 14. Diels-Alder [4+2] cycloadditions of cyclopropylideneacetate **1**-Me onto *N*-substituted pyrroles **55** [30]

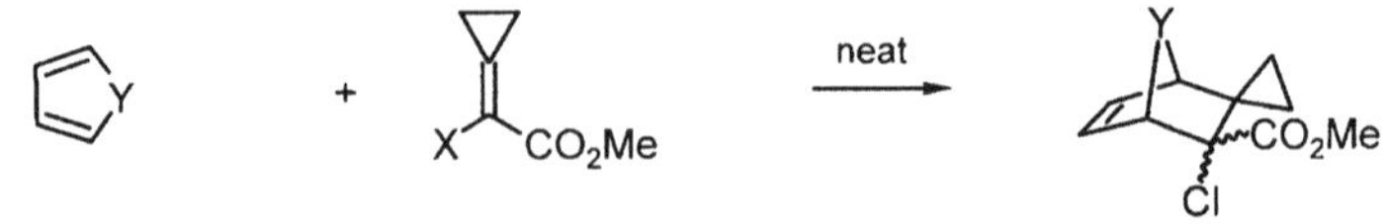

57 Y = O **1**-Me, **3**-X **59a**, **60**

58 Y = C=CMe$_2$

Starting Material	Diene	T [°C]/ t [h]	X	Y	Product	Yield (%)	Ratio *endo*/*exo*
1-Me	**57**	45/48	Cl	O	**59a**	90	1.4:1
1-Me	**58**	64/21	Cl	C=CMe$_2$	**60a**	66	1.3:1
3-H	**57**	60/4 d	H	O	**59b**	60	4.3:1
3-H	**58**	64/66	H	C=CMe$_2$	**60b**	49	1.8:1
3-Br	**57**	60/85	Br	O	**59c**	43	1:1
3-Br	**58**	64/12	Br	C=CMe$_2$	**60c**	86	1.2:1
3-F	**57**	45/2 d	F	O	**59d**	72	0.5:1
3-F	**58**	45/36 d	F	C=CMe$_2$	**60d**	85	0.5:1
3-N$_3$	**57**	22/7.5[a]	N$_3$	O	**59e**	16	1.7:1
3-N$_3$	**58**	32/10 d	N$_3$	C=CMe$_2$	**60e**	34	1.8:1

[a] Reaction in CH_2Cl_2 solution under a pressure of 10 kbar

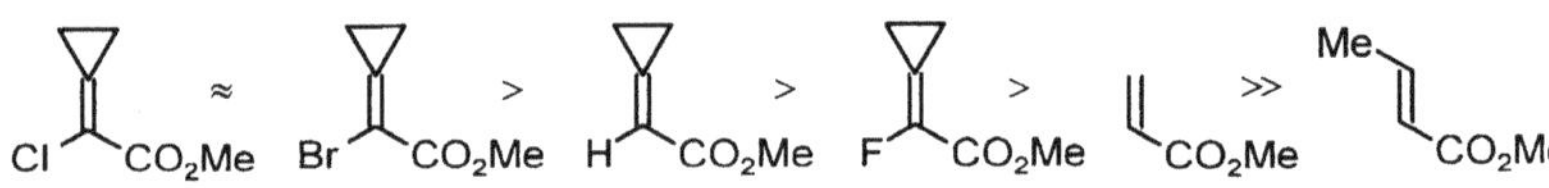

k_{rel}[furan]	16	16	4	3	1	0
k_{rel}[fulvene]	230	210	10	9	1	0

Scheme 15. Diels-Alder reactions of cyclopropylideneacetates **1**-Me, **3**-X with furan (**57**) and 6,6-dimethylfulvene (**58**) and relative reaction rates compared with methyl acrylate and methyl crotonate [19, 46]

other 2-substituted cyclopropylideneacetates of type **3**-X also entered this cycloaddition (Scheme 15) [19]. The *endo*/*exo* selectivity is low but usually still higher than that of simple acrylic acid esters. The relative Diels-Alder reactivities of dienophiles **1**-Me and **3**-X as determined by competition experiments (Scheme 15) suggest a mechanism involving either diradicals or zwitterions as intermediates [19]. Surprisingly, the 2-fluoro derivative **3**-F is less reactive than the parent compound **3**-H. The 2-chloro and 2-bromo derivatives **1**-Me and **3**-Br have similar reactivities and cycloadd to furan (**57**) about 16 times faster than methyl acrylate.

Unsymmetrically 2,5-disubstituted furan derivatives **61** reacted with **1**-Me in good to very good yields, but gave mixtures of regioisomers and diastereomers *endo*/*exo*-**62**, *endo*/*exo*-**63**, the bicyclic acetals **62c–d** showed a distinct tendency to be cleaved upon chromatographic separation. Dimethylfuran **61b**, surprisingly, is less reactive towards **1**-Me than furan (**57**) which is probably due to steric reasons (Scheme 16) [46].

The reasonably good regioselectivities observed in the [2+4] cycloadditions of **1**-Me onto unsymmetrically disubstituted furans **61d, e** can be utilized in a new synthetic approach to the cytotoxic sesquiterpenes illudin M **64a** [47] and

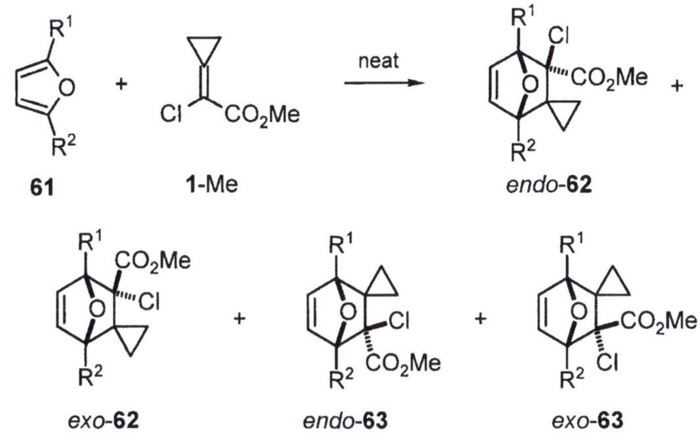

Furan 61	R^1	R^2	T [°C]/ t [h]	Yield (%)	Product Ratio endo-62	exo-62	endo-63	exo-63
a	Me	H	20/100	76	1.5	1	–	–
b	Me	Me	60/12	72	2	1	–	–
c	OMe	H	20/32	95	1.2	1	–	–
d	$OSiMe_3$	Me	20/140	81	8	2	1	0.5
e	OMe	Me	20/120	78	10	3	1	0.3

Scheme 16. Diels-Alder reactions of methyl 2-chloro-2-cyclopropylideneacetate (**1**-Me) with substituted furans **61** [46]

structurally related ptaquilosin **65** [48]. A possible synthetic strategy for **64a** [19b, 46] would rely on the [2+4] cycloaddition of **1**-Me to appropriately substituted furan **61** to produce **62**, **63** as precursors to the bicyclic cyclohexenone derivative **66**, which contains all the required functionalities of the six-membered ring in **64a** (Scheme 17). The five-membered ring would be attached at the stage of the initial Diels-Alder adducts **62**, **63** or after modification of its functionalities before opening the bicyclic acetal moiety, i.e. via **68** or an analog, or after hydrolysis of the acetal and modification of the functionalities, i.e. via the cyclohexenone **69** or an analog.

The Diels-Alder reaction of **1**-Me with 2-methoxy-5-methylfuran (**61e**) yielded the cycloadduct *endo*-**62e** as the main product which could be isolated in 48% yield (Scheme 16). Its structure was proved by X-ray crystal structure analysis [19b, 46a], and *endo*-**62e** was converted in three simple steps into the acetal protected tertiary ketoalcohol **71** (Scheme 18) [19b, 46]. A Pauson-Khand reaction [49] was envisaged as the second key step to annelate the five-membered ring to progress towards the target molecule. This cobalt-mediated cyclopentenone annelation can be brought about both with the spiroepoxide **70** and with the tertiary alcohol **71**. In each case, two regioisomers resulted from the reaction with propyne. In the case of **70**, the isomer **72**, with its carbonyl group at the correct position in the five-membered ring for the target molecule, predominated by a factor of 4.4 to 1. To prove the stereo- and regiochemical assignments, an X-

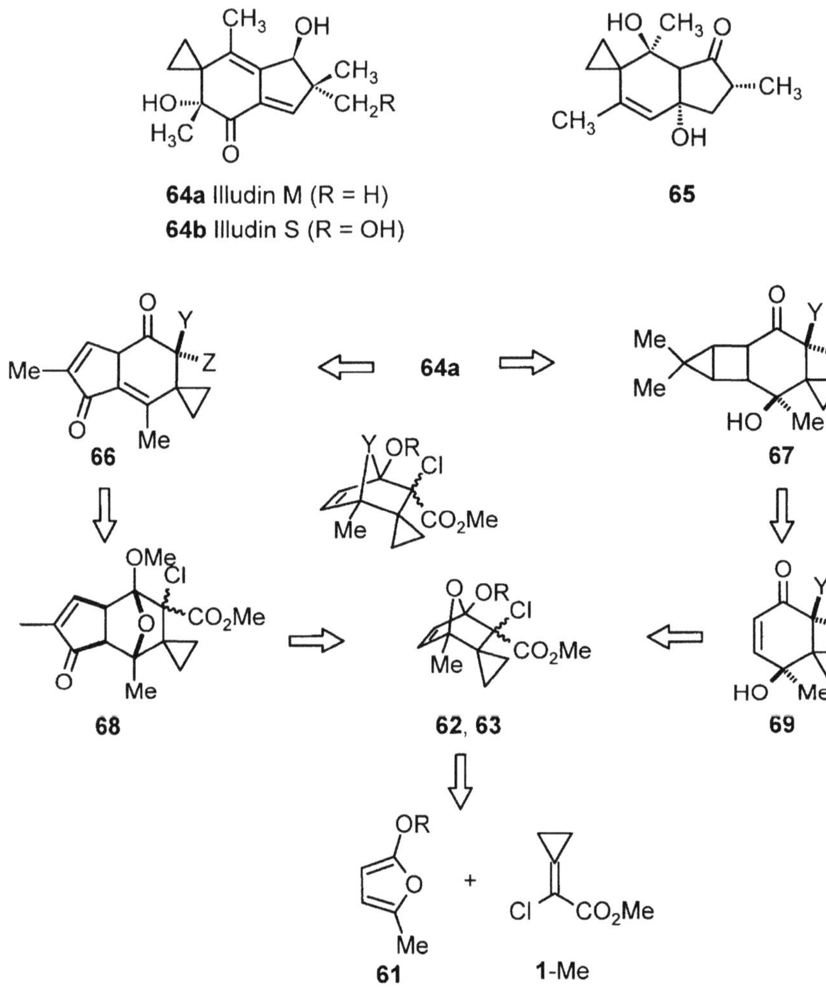

Scheme 17. Retrosynthetic analysis of (±)-illudin M (**64a**) [19, 46]

ray structure analysis of **72** was carried out [19b, 46a]. The Pauson-Khand adduct **73** from **71** was obtained in better yields, but without any predominance of one of the two possible regioisomers. Compound **72**, which can be prepared from methoxymethylfuran (**61e**) and **1**-Me in four steps with an overall yield of 18% has all the structural features of illudin M (**64a**) and has all the necessary functionalities for further elaboration.

Another useful synthetic application of a Diels-Alder adduct of the chloro ester **1**-Me is the facile preparation of the spirocyclopropanated analog **75** of so-called "naked sugar" **76** [50] (Scheme 19) [30]. The transformation of the adduct *endo*-**59a** of **1**-Me onto furan (**57**) into α-chloroamide **74** followed by fragmentation of the latter under basic conditions allowed to prepare the versatile building block **75** in 71% overall yield. Starting from isomer *exo*-**59a**, the yields were 82 and 52% for the first and the second step, respectively. The analogous trans-

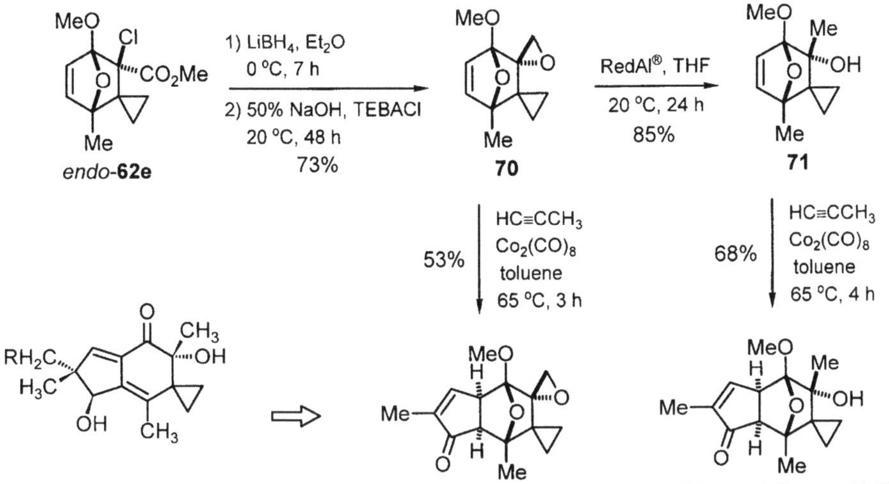

Scheme 18. Along the route to the total synthesis of (±)-illudin M (**64a**) from methyl 2-chloro-2-cyclopropylideneacetate (**1**-Me) [19b, 46a]

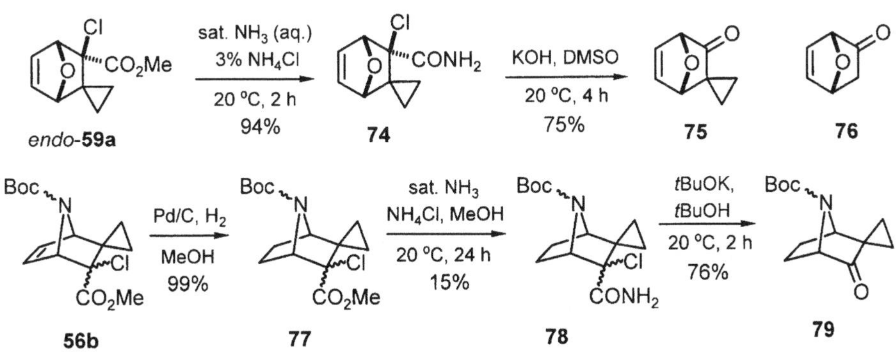

Scheme 19. Useful chemical transformations of Diels-Alder adducts *endo*-**59a** and **56b**, respectively, of the chloro ester **1**-Me onto furan (**57**) and *N*-Boc-protected pyrrole (**55b**) [30]

formation of the aza-analog **56b** succeeded only after hydrogenation of the double bond in **56b** to give **77**, and the overall yield was significantly lower (Scheme 19) [30].

3.2 Michael Addition Reactions Followed by Chemical Transformations of the Adducts

The 2-chloro-2-cyclopropylideneacetates **1**, **2** are ultimately better Michael acceptors than any other 3,3-disubstituted acrylate. The addition of most nucleophiles occurs smoothly, and it is accelerated by the high strain in the starting material which is released upon conversion of the *sp*- to *sp*^{2}-hybridized carbon

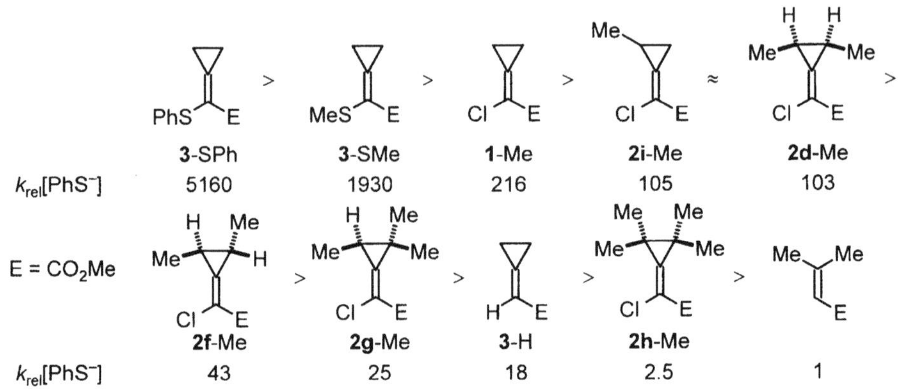

Fig. 4. Relative reactivities of cyclopropylideneacetates **1**–**3** towards thiophenolate anion compared to 3,3-dimethylacrylate [5b, c, 15b, 27]

[51]. Thus, in a competition experiment, the chloro ester **1**-Me reacted with thiophenolate anion 216 times faster than methyl 3,3-dimethylacrylate (Fig. 4). Part of this enhanced reactivity is due to the α-chloro substituent in **1**-Me, as the parent methyl cyclopropylideneacetate (**3**-H) reacts only 18 times faster than 3,3-dimethylacrylate, while **3**-SPh reacts 5160 times more rapidly [5b, c, 15b, 27].

The Michael additions to **1**–**3** of a large variety of carbon, nitrogen, oxygen, sulfur and selenium nucleophiles, as well as hydride, followed by inter- or intramolecular transformations of the chlorine substituent or/and the methoxycarbonyl fragment, offer versatile synthetic approaches to a large variety of synthetically useful and important organic molecules.

3.2.1
Monoadditions and Nucleophilic Substitution of Chlorine

The Michael additions of any nucleophile onto a cyclopropylideneacetate **2**-R with substituents on the cyclopropane ring generally occur with a high degree of diastereoselectivity, with the nucleophile attacking from the side opposite to the most bulky substituent on the cyclopropane ring [5]. A typical example is the sodium trimethoxyborohydride reduction of the 2′-substituted α-chloro esters **2**-*t*Bu to the corresponding 2′-substituted 2-chloro-2-cyclopropylacetates **80** in comparison to the parent ester **1**-*t*Bu (Scheme 20). In all cases the *cis* isomer of **80** was formed predominantly [25]. But the reduction of the methyl ester **2j**-Me (R=CH_2OBn) with LiBH(*sec*-Bu)$_3$ (L-Selectride) in THF at −78 °C which gave the corresponding derivative **80j**-Me in 42% yield was found not to be selective [9]. Compounds **80a**-Me and **80c**-Bn were also prepared by reduction of **1**-Me and **2c**-Bn with $NaBH_4$ adsorbed on silica gel (Scheme 20) [9, 21].

This transformation of alkyl 2-chloro-2-cyclopropylideneacetates has been applied for the preparation of compounds of type **83** containing a *cis*-configurated disubstituted cyclopropane moiety [25] with potential liquid crystalline properties [52]. In the first approach to such compounds, 2-chloro-2-cyclopropylideneacetic acids **2b**-H, **2i**-H were coupled with the linear diaryl components **81** fol-

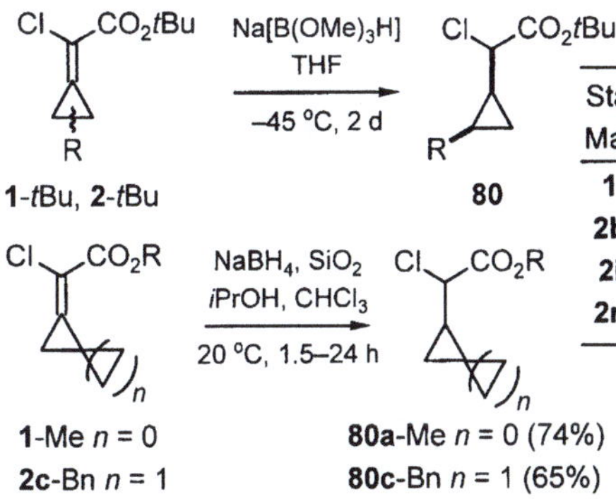

Starting Material	R	Ratio E/Z	Product 80/ Yield (%)	Ratio trans/cis
1-*t*Bu	H	–	**80a**-*t*Bu/69	–
2b-*t*Bu	Et	1:2.1	**80b**-*t*Bu/58	9:91
2i-*t*Bu	Me	1:1.6	**80i**-*t*Bu/66	5:95
2n-*t*Bu	CH_2OTMS	1:1.3	**80n**-*t*-Bu/56	2:98

Scheme 20. Stereoselective addition of hydride onto *tert*-butyl 2-chloro-2-cyclopropylideneacetates **1**-*t*Bu, **2**-*t*Bu and reduction of the double bond in **1**-Me, **2c**-Bn [9, 21, 25]

lowed by a stereoselective reduction of the double bond in the chlorocyclopropylideneacetate moiety. This approach turned out not to be very efficient (Scheme 21), the overall yields obtained in reducing the 2′-substituted cyclopropylideneacetates first and then coupling the corresponding 2′-substituted 2-chloro- 2-cyclopropylacetic acids **84** with the biaryl derivatives **81** were significantly better.

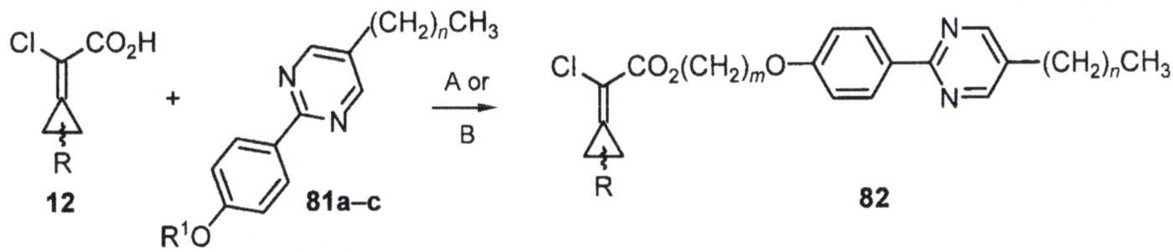

Starting Material	R	R^1	Product	Conditions	*m*	*n*	Yield (%)
12b	Et	H	**82a**	A	0	7	25
12i	Me	H	**82b**	A	0	7	27
12i	Me	$HO(CH_2)_2$	**82c**	B	2	7	39
12i	Me	$HO(CH_2)_6$	**82d**	B	6	9	37

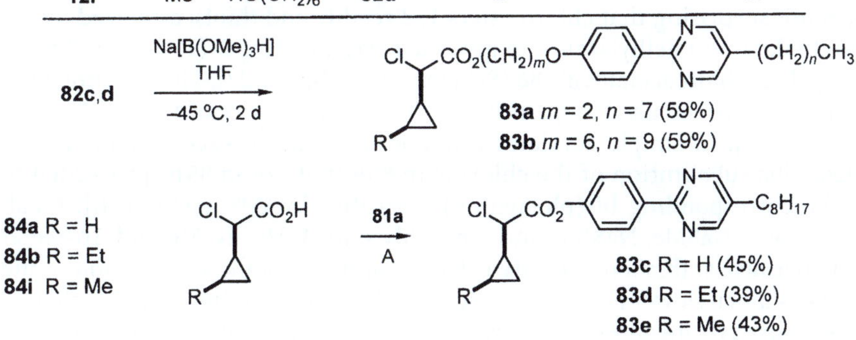

Scheme 21. Preparation of biaryl derivatives of type **83** with potential liquid crystalline properties from 2-chloro-2-cyclopropylideneacetic acid. Conditions: A: DCC, CH_2Cl_2, 0°C, 5 h. B: DEAD, Ph_3P, Et_2O, 20°C, 14 h [25]

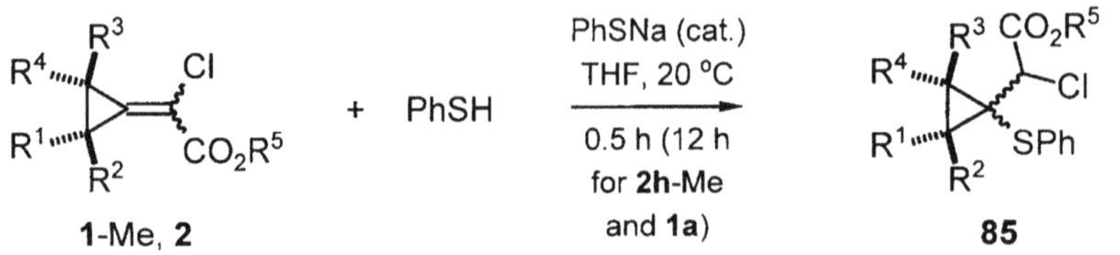

Starting Material	R^1	R^2	R^3	R^4	R^5	Product	Yield (%)
1-Me	H	H	H	H	Me	**85a**	54
2d-Me	H	Me	Me	H	Me	**85d**	59
2f-Me	H	Me	H	Me	Me	**85f**	49
2g-Me	Me	Me	Me	H	Me	**85g**	54
2h-Me	Me	Me	Me	Me	Me	**85h**	68
2i-Me	Me	H	H	H	Me	**85i**	52
1a	H	H	H	H	$4\text{-}MeOC_6H_4CH_2$	**85j**	65[a]

[a] Et_3N was used as a base in CH_2Cl_2

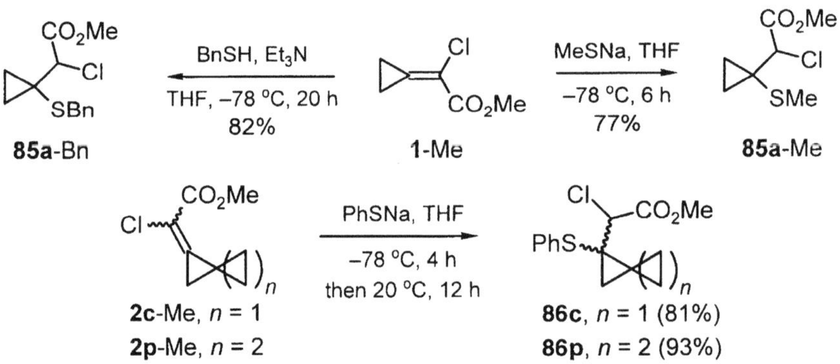

Scheme 22. Michael addition of thiols onto chlorocyclopropylideneacetates **1**-Me, **2** [71, 9, 15b, 22b, 27]

It is not surprising that chloro esters **1**, **2** readily add thiols, catalyzed by sodium thiolates or triethylamine, to give the corresponding 2-(1′-organylthiocyclopropyl)-2-chloroacetates **85**, **86** (Scheme 22) [15b, 22b, 27]. This reaction with thiophenol has been used to quantify the Michael reactivity of **1**-Me, **2**-Me, **3**-X in comparison to simple acrylates (see above). With an excess of PhSH, the nucleophilic substitution of the chlorine in **85a** (but not in **85h**) proceeded to give the corresponding bis(phenylthio) derivative in 63% yield [15b]. Alkali thiolates (e.g. NaSMe, NaSBn) add smoothly onto **1**-Me, **2c**-Me and **2p**-Me at –78 °C, because at this temperature subsequent nucleophilic substitution of the chlorine is much slower [71, 9]. The Michael additions of sodium phenylselenide and sodium arylsulfenates onto **1**-Me and their synthetic utility have been discussed above (see Table 1).

Addition of methanol onto **2** sometimes even occurs spontaneously upon attempted preparation of **2** [15b]. Under alkoxide catalysis, such alcohol additions

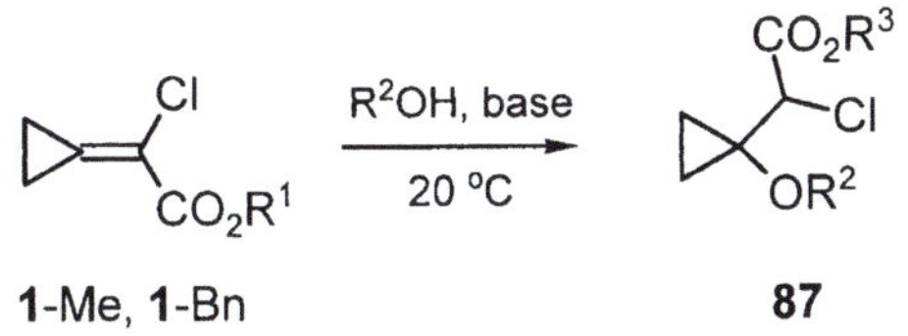

Starting Material	R^1	Product	R^2	R^3	Base	t [h]	Yield (%)
1-Me	Me	**87a**	Me	Me	n. r.	n. r.	64
1-Me	Me	**87b**	Me	Bn	Et_3N	20	60
1-Me	Me	**87c**-Me	Bn	Bn	BnOK	24	61
1-Bn	Bn	**87c**-Bn	Bn	Bn	BnONa	18	74

n. r. = not reported

Scheme 23. Addition of alcohols onto chloro esters **1**-Me and **1**-Bn [9, 15b]

were accompanied with transesterification. Although benzyl alcohol can be added without simultaneous transesterification of the methyl ester, it is easier to combine both steps (Scheme 23) [9, 15b].

The addition of alkoxides generated from secondary alcohols (for example, different ephedrine derivatives, [10b, c]) proceeded very slowly and not selectively with yields up to 53%. *tert*-Alkoxides did not add to **1**-Me, **2** at all [53].

The Michael additions of various carbon nucleophiles such as cyanide [15b, 22b], anions generated from nitromethane [27], *tert*-butyl acetate [9], malonates [27] and O'Donnell's glycine equivalent [54], cuprates [9, 15b] or Grignard reagents [53] under copper catalysis [55] have also been reported (Scheme 24).

The Michael adducts of nitrogen nucleophiles are among the most important ones for organic synthesis (their synthetic applications will be discussed in the corresponding Sections). The addition of primary amines onto **1** and **2** in methanol was normally complete after a few minutes at –10 °C and gave moderate to good yields of the monoadducts **89** (for example, 67% for *n*-butylamine and benzylamine adducts **89a**-Me and **89b**-*t*Bu [26]). But later it was found that performing these additions in THF leads to better and well reproducible yields of the adducts **89** (Scheme 25) [11b, 53, 56].

Thus obtained secondary amines **89** did not add again to a second molecule of **1**-Me, but often gave rise to side reactions either with the neighboring chloromethine group or the ester function (see Sects. 4.2.2, 4.2.3 and 4.4).

The addition of secondary amines in methanol proceeded very quickly even at –10 °C (see, for example, Table 1), but at ambient temperature, addition and substitution can also occur to form products of type **90** and even of addition with subsequent methanolysis [11b, 15b]. Therefore, these conditions have been applied mainly for the synthesis of diamines **90** (Scheme 26) [15b]. Nevertheless, the primary adducts **91** can also be obtained in methanol when working carefully (e.g. the adduct **91b** from **1**-Me and diethylamine in 78% yield; see also Table 1).

1, 2 **88**

X	Reaction Conditions	R^1	R^2	Product	Yield (%)	Ref.
CH_2CO_2tBu	$LiCH_2CO_2t$Bu, THF, HMPA −78 °C, 1 h	H	Me	**88a**	91	[9]
$CH(CO_2Et)_2$	$CH_2(CO_2Et)_2$, KF, *i*Pr_2NH C_6H_6, DB-18-c-6	H	Me	**88b**	69	[27]
CH_2NO_2	MeCN, KF, *i*Pr_2NH C_6H_6, DB-18-c-6	H	Me	**88c**	91	[27]
CH(CN)(CO_2Et)	$NCCH_2CO_2$Et, KF, *i*Pr_2NH C_6H_6, DB-18-c-6	H	Me	**88d**	42	[27]
N=CPh_2 / CO_2tBu	$CH_2(N{=}CPh_2)(CO_2t$Bu), THF BuLi, −78→−10 °C, 2 h	H	Me	**88e**	80	[53]
Me	$LiCuMe_2$, Et_2O −40 °C, 20 min	H	Me	**88f**	61	[9]
Vinyl	CH_2=CHMgBr, CuCN, THF $BF_3{\cdot}Et_2O$, −78 °C, 3 h	H	Me	**88g**	80	[53]
CN	KCN, acetone cyanohydrine C_6H_6, DB-18-c-6, 20 °C, 30 h	Me	Me	**88h**	32	[15b]
CN	K_2CO_3, acetone cyanohydrine 70 °C, 2 h	H	4-MeO-C_6H_4-CH_2	**88i**	81	[22b]
CN	K_2CO_3, acetone cyanohydrine 70 °C, 2 h	H	4-RO-C_6H_4-CH_2 (R = CH_2CO_2tBu)	**88j**	84	[22b]

Scheme 24. Addition of carbon nucleophiles onto chlorocyclopropylideneacetates **1, 2** [9, 15b, 22b, 27, 53]

The Michael adducts of dibenzylamine onto chloro esters **1**-*t*Bu, **2i**-Me, and **2k**-Me were also prepared in methanol in reasonably good yields (67, 79 and 77%, respectively) [7k, 26b]. However, like for the addition of primary amines, THF turned out to be the best solvent for the preparation of the Michael monoadducts **91** of secondary amines onto chlorocyclopropylideneacetates **1, 2** (Scheme 27) [9, 11b, 21b, 22b, 53, 56, 57].

The additions of ammonia equivalents, i.e. nitrogen nucleophiles like dibenzylamine (Scheme 27) which are essentially a protected primary amino group, are of special synthetic interest with respect to their possible subsequent chemical transformations. Poorly nucleophilic ammonia equivalents like acetamide or the classical phthalimide, did not add or originally gave low yields of **92a** (Scheme 28) [9], but later phthalimide was found to add to **1**-Me very well under mild conditions [53]. However, the α-chlorine in **92a** could neither be substituted nor could the phthalimido group be cleaved without destroying the cyclopropane ring. The potassium bis(alkoxycarbonyl)amides $(Boc)_2NK$ and (Moc)(Boc)NK add to **1**-Me in satisfactory yields, but the α-chlorine atom in

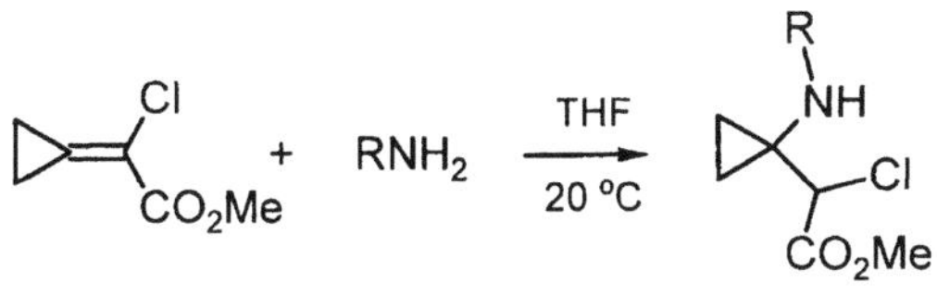

1-Me 89

R	Product	t [h]	Yield (%)
*n*Bu	**89a**	0.25	94
Bn	**89b**	1	93
CH_2CO_2*t*Bu	**89c**	3	91
4-$MeOC_6H_4CH_2$	**89d**	0.5	88
cyclo-C_6H_{11}	**89e**	3	59
2-Cl-C_6H_4	**89f**	24	43[a]
4-O_2N-C_6H_4	**89g**	53	54[b]

Additionally a base was used: [a]Et_3N, [b]NaH

Scheme 25. Addition of primary alkyl- and arylamines onto chlorocyclopropylideneacetate 1-Me in THF [11b, 53]

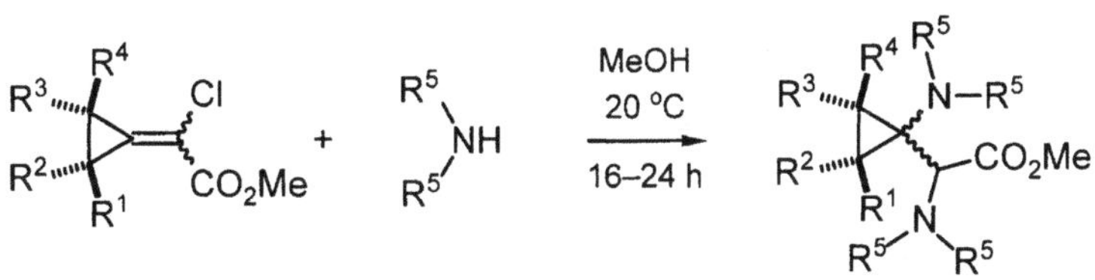

1-Me 90

Starting Material	R1	R2	R3	R4	R5	Product	Yield (%)
1-Me	H	H	H	H	$-(CH_2)_5-$	**90a-Pipy**	71
1-Me	H	H	H	H	Et	**90a-Et**	56
2d-Me	H	Me	Me	H	$-(CH_2)_5-$	**90d-Pipy**	59
2f-Me	H	Me	H	Me	$-(CH_2)_5-$	**90f-Pipy**	51
2g-Me	Me	Me	Me	H	$-(CH_2)_5-$	**90g-Pipy**	67
2h-Me	Me	Me	Me	Me	$-(CH_2)_5-$	**90h-Pipy**	57
2d-Me	H	Me	Me	H	Me	**90d-Et**	67
2h-Me	Me	Me	Me	Me	Me	**90h-Et**	64

Scheme 26. Addition-substitution reactions of chlorocyclopropylideneacetates 1-Me, 2 with secondary amines [15b]

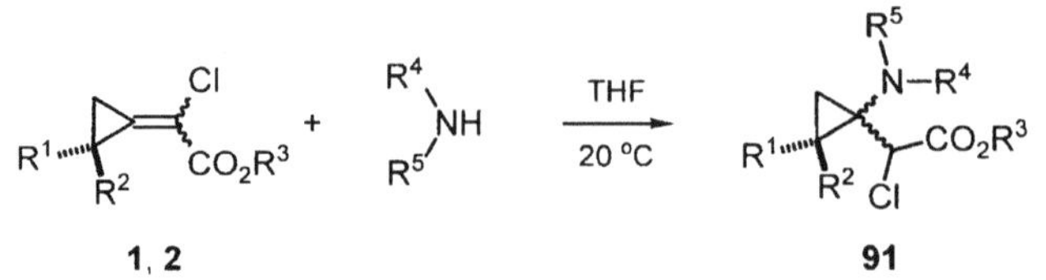

R^1	R^2	R^3	R^4	R^5	t [h]	Product	Yield (%)	Ref.
H	H	Me	Bn	Bn	14	**91a**	98	[9]
H	H	Me	Et	Et	1	**91b**	80	[57]
H	H	Me	Allyl	Allyl	1	**91c**	78	[57]
H	H	Me	$-(CH_2)_5-$		1	**91d**	90	[57]
H	H	Me	Et	*n*Bu	4	**91e**	88	[11b]
H	H	Me	*n*Pr	*n*Pr	6	**91f**	95	[11b]
H	H	Bn	Et	*n*Bu	16	**91g**	87	[11b]
H	H	Bn	Bn	Bn	22	**91h**	75	[9]
H	H	Me	Me	$CH(CH_2OH)(Bn)$	8	**91i**	74	[11b]
H	H	Me	Me	Bn	6	**91j**	94	[11b]
H	H	Me	Bn	$4-MeOC_6H_4CH_2$	8	**91k**	94	[11b]
H	H	$4-(CH_2CO_2Me)Bn$	Et	*n*Bu	14	**91l**	89	[11b]
H	Et	Me	Et	*n*Bu	16	**91m**	66	[11b]
H	$HO(CH_2)_4$	Me	Et	*n*Bu	48	**91n**	73	[11b]
H	H	Me	Me	$N_3(CH_2)_2$	2	**91o**	87	[11b]
H	H	Me	Me	$Br(CH_2)_2$	6	**91p**	83[a]	[11b]
H	H	Me	Me	CH_2CO_2Me	48	**91r**	65[b]	[11b]
H	H	Me	$-CH(CO_2Me)CH_2CH_2CH_2-$		48	**91s**	65[b]	[11b]
H	$BnOCH_2$	Me	Bn	Bn	16	**91t**	79[c]	[7k]
H	$BnO(CH_2)_2$	Me	Bn	Bn	16	**91u**	77[c]	[7k]
$-(CH_2)_2-$		Bn	Bn	Bn	22	**91v**	88	[21b]
H	H	Me	Me	CH(Me)(Ph)	24	**91w**	36	[53]
H	H	$4-(tBuO_2CCH_2O)Bn$	Bn	Bn	48	**91x**	94	[22b]
H	H	$4-(tBuO_2CCH_2O)Bn$	$-(CH_2)_4-$		14	**91y**	82	[22b]
H	H	4-(MeO)Bn	Ph	Me	24	**91z**	64	[22b]
H	H	4-(MeO)Bn	Bn	Gly-OEt	48	**91α**	56	[22b]

[a] The starting amine was generated in situ from (2-bromoethyl)methylamine hydrobromide and Et_3N. – [b] With Et_3N as a catalyst. – c Reaction in MeOH.

Scheme 27. Addition reactions of secondary amines in THF onto chlorocyclopropylideneacetates **1**-Me, **2** [7k, 9, 11b, 21b, 22b, 53, 56, 57]

92b and even in the less sterically encumbered **92c** could not be substituted, most probably due to steric hindrance [9].

Surprisingly though, the rather weakly nucleophilic (diphenylmethylene) amine (DPMA-H) and its cyclopropyl analog (cyclopropylphenylmethylene) amine (CPMA-H) **93** [8, 9, 10c, 21b, 22b, 58–60] which can also serve as ammonia equivalents, add to **1**-R and **2**-R cleanly and in most cases quantitatively in a 1,4-fashion (Scheme 29).

Methanol proved to be the best solvent for this reaction, it proceeds significantly more slowly in aprotic and less polar solvents requiring a basic catalyst

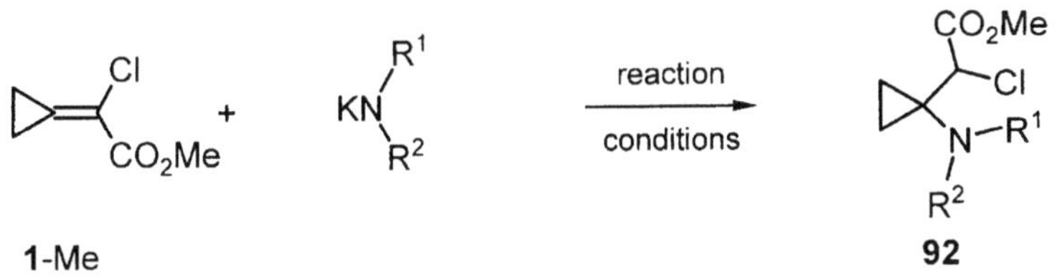

Reagent	Reaction Conditions	R^1	R^2	Product	Yield (%)	Ref.
KNPhth[a]	DMF, 100 °C, 4 h	Phth		**92a**	18	[9]
	DMF, 0 °C, 1 h	Phth		**92a**	87	[53]
$KN(Boc)_2$	CH_2Cl_2, DB-18-c-6 20 °C, 48 h	Boc	Boc	**92b**	69	[9]
KN(Boc)(Moc)	CH_2Cl_2, DB-18-c-6 20 °C, 96 h	Boc	Moc	**92c**	64	[9]

Phth = phthaloyl

Scheme 28. Addition of ammonia equivalents onto the chloro ester **1**-Me [9, 53]

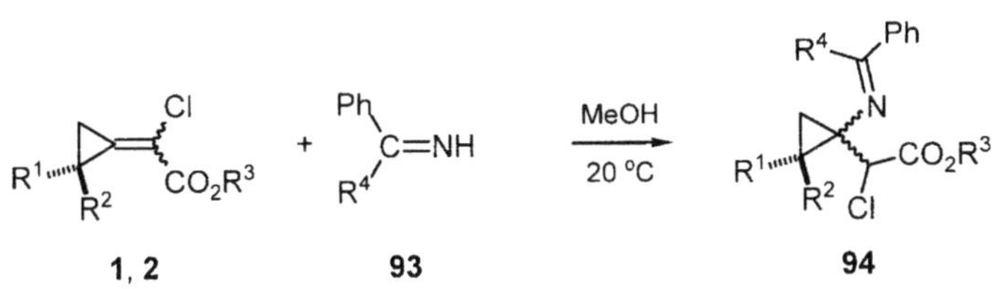

Starting Material	R^1	R^2	R^3	R^4	t [d]	Product	Yield (%)	Ref.
1-Me	H	H	Me	Ph	1	**94a**	98	[9,58]
1-Bn	H	H	Bn	Ph	2	**94b**	86	[9]
2c-Me	$-(CH_2)_2-$		Me	Ph	4	**94c**	96	[8]
2c-Bn	$-(CH_2)_2-$		Bn	Ph	4	**94d**	98	[21b]
2i-Me	Me	H	Me	Ph	2	**94e**	66	[8]
1-Me	H	H	Me	*c*Pr	6	**94f**	95(98)	[8]
1-Me	H	H	Me	1-Me-cPr	16	**94g**	87(96)	[8]
2j-Me	CH_2OBn	H	Me	Ph	2	**94h**	68[a]	[10c]
2r-Me	CH_2SMe	H	Me	Ph	2	**94i**	72[a]	[10c]
1a-R^1	H	H	4-($t$$BuO_2CCH_2O$)Bn	Ph	1	**94j**	64	[22b]

[a] Reaction in CH_2Cl_2

Scheme 29. Reaction of chlorocyclopropylideneacetates **1**, **2** with (diphenylmethylene)amine (DPMA-H) and its cyclopropylsubstituted analogs (CPMA-H) **93** [8, 9, 10c, 21b, 22b, 58]

like 1,4-diazabicyclo[2.2.2]octane (DABCO) to activate DPMA-H [58]. The spiropentane and methylcyclopropyl derivatives **94c–e** were obtained as mixtures of diastereomers in ratios of about 2.2:1. In most cases, the β-DPMA-substituted esters **94** are obtained in a sufficiently pure form to be used without further purification. Due to their hydrolytic sensitivity, substantial losses may occur upon chromatography. In the absence of any solvent, water, and oxygen, however, they are perfectly stable. They gave good results in the substitution reaction with azide ion or alkylthiolates [58, 59]. The amino group can be easily deprotected by hydrolysis under weakly acidic conditions (acetic acid, citric acid, dilute hydrochloric acid) [54, 61] or by hydrogenation [9]. Moreover, these compounds **94** are capable to undergo interesting rearrangements to give rise to different carbo- and heterocycles (see Sect. 3.2.3).

An azido group could also be introduced by 1,4-addition of azide ion onto **1**-Me and **2d**-Me in aqueous acetic acid. The corresponding azides were isolated in 56 and 62% yield, respectively [15b].

As a potential approach towards enantiomerically pure amino acids containing a cyclopropane ring, Michael additions of enantiomerically pure chiral ammonia equivalents **95–100** have been examined (Fig. 5).

The addition of **95** onto **2m**-Me in THF at ambient temperature (89% yield) was not stereoselective [53]. All attempts to add the chiral cyclic benzylamine derivative **96** and **97** onto the spirocyclopropanated chloro ester **2c**-Me were unsuccessful [21b]. Although it was possible to add heterocycles (*S*)-4-phenyloxazolidine-2-one (**98**) and its thio analog **99** onto the chloro esters **1**-Me and **2j**-Me (Scheme 30) [10b, c] and to substitute or reductively remove the chlorine atom in the adducts **101**, all attempts to deprotect the amino functions in **101** and the products of their transformations under various conditions were unsuccessful.

A solution to this problem was found by using (4*R*, 5*S*)-4,5-diphenyloxazolidine-2-one (**100**) [63] which turned out to be a suitable chiral ammonia equivalent permitting to achieve good diastereoselectivities with respect to the stereogenic center α to the alkoxycarbonyl group especially with 2′-substituted

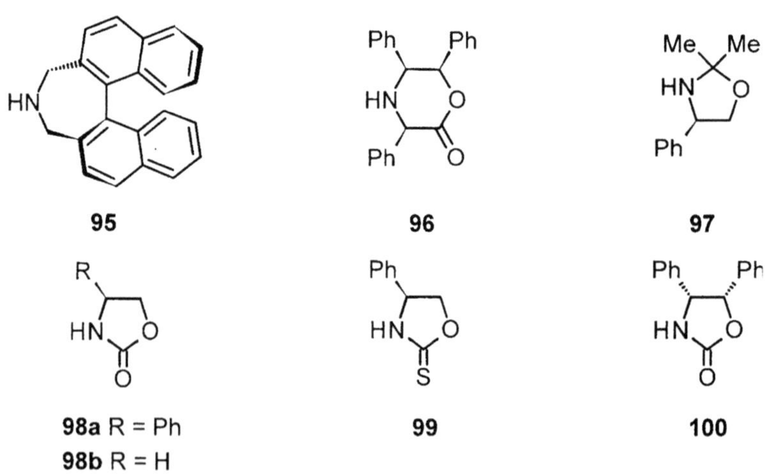

Fig. 5. Chiral cyclic nitrogen nucleophiles 95–100 as potential ammonia equivalents

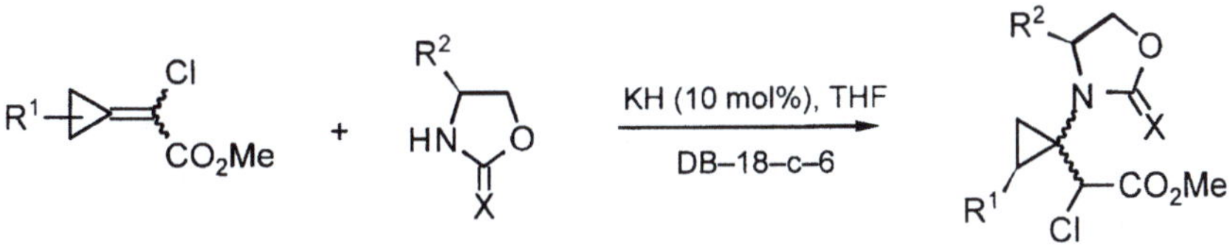

1-Me, **2j**-Me **98**, **99** **101**

Michael Donor	T [°C]	t [h]	R^1	R^2	X	Yield (%)/Ref.	Diastereomer Ratio[a]
98a	20	168	H	Ph	O	93 [10c]	7:4
98a	20	8	H	Ph	O	81 [10b]	5:1
98b	20	2.5	H	H	O	81 [10c]	–
99	–78→20	8	H	Ph	S	81 [10b]	3:1
99	–78→20	12	$BnOCH_2$	Ph	S	63 [10c]	n. r.

[a] With respect to the newly formed stereogenic center α to the methoxycarbonyl group; n. r. = not reported

Scheme 30. Substituted oxazolidines **98**, **99** as Michael donors in addition reactions onto chloro esters **1**-Me, **2j**-Me [10b, c]

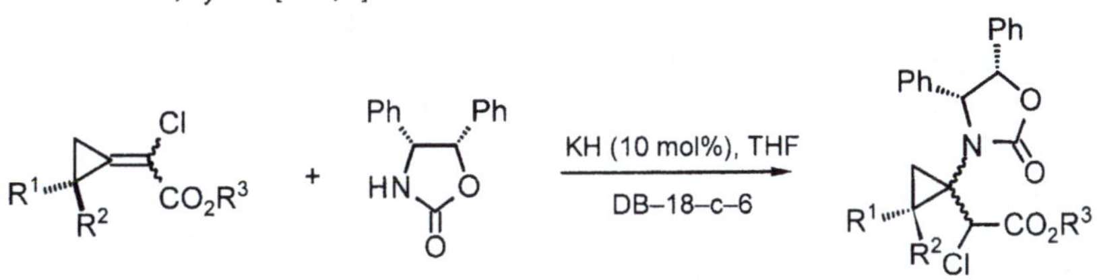

1-Me, **2** **100** **102**

Starting Material	T [°C]	t	R^1	R^2	R^3	Product	Yield[a] (%)	Ref.
1-Me	–78→20	4 d	H	H	Me	**102a**-Me	73	[10b]
2b-Me	0→20	8 h	H	Et	Me	**102b**-Me	64(58)	[10a,c]
2b-Bn	0→20	12 h	H	Et	Bn	**102b**-Bn	79(90)	[10a,c]
2c-Me	–20→20	3 h	$-(CH_2)_2-$		Me	**102c**-Me	85(74)	[21a,b]
2c-Bn	–20→20	3 h	$-(CH_2)_2-$		Bn	**102c**-Bn	91(74)	[21a,b]
2i-Me	–20→20	4 h	H	Me	Me	**102i**-Me	73	[10a,b]
2i-Bn	–78→20	12 h	H	Me	Bn	**102i**-Bn	48(53)	[10a,b]
2j-Me	20	30 min	H	CH_2OBn	Me	**102j**-Me	57(59)[b]	[10a,c]
2m-Me	20	1 h	H	CH_2OMe	Me	**102m**-Me	68[b]	[10a,c]
2m-Bn	–20→20	4 h	H	CH_2OMe	Bn	**102m**-Bn	69(73)	[10a,c]
2r-Me	20	20 min	H	CH_2SMe	Me	**102r**-Me	59(52)[b]	[10a,c]
2s-Me	–20→20	2 d	H	CH_2N_3	Me	**102s**-Me	79(58)[b]	[10b,c]

[a] For the diastereomer ratios see text. In all cases, only two diastereomers were isolated. – [b] Reaction was performed in THF/CH_2Cl_2 mixture.

Scheme 31. (4*R*,5*S*)-4,5-Diphenyloxazolidine-2-one (**100**) as a Michael donor in addition reactions onto chloro esters **1**-Me, **2** [10, 21, 62]

2-chloro-2-cyclopropylideneacetates like **2j**-Me (Scheme 31) [10, 21, 62]. Enantiomerically pure **100** was easily prepared from commercially available (1*S*,2*R*)-(+)-2-amino-1,2-diphenylethanol [10, 62, 63] and its adducts with Michael acceptors can readily be cleaved by catalytic hydrogenation over palladium on charcoal. This Michael addition onto different 2′-substituted α-chloro esters **2** [64] gave the 1,4-adducts **102**, not only with excellent *trans*-selectivities with respect to the two stereogenic centers on the three-membered ring, but also quite good diastereoselectivities with respect to the newly formed stereogenic center α to the methoxycarbonyl group except for the parent compound **1**-Me with a ratio of 3:1 (see below). In all cases of 2′-substituted chloro esters **2**-Me, only two diastereomers (approximately 1:1 ratio in most cases) corresponding to the two configurations at C-2′ in the racemic starting material were detected. Apparently, the protonation of the intermediate enolate in these cases occurs with high diastereoselectivity, in spite of the fact that protonation reactions are usually very fast [65]. These two diastereomers for each of the derivatives of type **102** can be separated by simple flash column chromatography, thus allowing the resolution of the racemate due to the stereogenic 2′-center in the starting material.

It is noteworthy that the substituent at C-2′ appears to be very important for the diastereoselectivity at C-2. The unsubstituted methyl 2-chloro-2-cyclopropylideneacetate **1**-Me and the spirocyclopropanated analogs **2c**-Me, **2c**-Bn add the oxazolidinone **100** to form **102a**-Me, **102c**-Me and **102c**-Bn only with moderate selectivities (diastereomer ratios were 3:1, 42:11:34:13 and 27:59:14, respectively) [10b, 21].

As the absolute configuration of the oxazolidinone **100** introduced into the adducts of type **102** is known, for several cases absolute configurations (compounds **102b**-Bn, **102c**-Me and **102c**-Bn) were determined by X-ray crystal structure analyses [10b, 10c, 21a].

The second synthetically useful general transformation of any of the chloro esters **1**, **2** consists of an intra- or intermolecular nucleophilic substitution of the chlorine atom in any of the Michael adducts. Several examples of such reactions have been mentioned above (Table 1 and Scheme 26), and substitutions of any specific importance (for example, with hydride or azide anion [9, 10, 62]) as well as subsequent transformations of such substitution products will be discussed in the corresponding Sections [66].

Neighboring group participation effects appear to play a crucial role in the nucleophilic substitution of chlorine in Michael adducts of **1**-R, **2**-R, **3**-X. Thus, this substitution proceeds very easily in any of the adducts formed with an electron rich nitrogen, sulfur and oxygen Michael donor. For the adducts of nitrogen nucleophiles, the facile substitution of the chlorine has been suggested to occur via formation of intermediate aziridinium ions **103** [8] (Scheme 32), and this postulate was later supported by isolation of azaspiropentane derivatives under appropriate conditions in several reactions (see Sect. 3.2.2) [11b, 53, 56]. It is most likely that alkylthio substituents in adducts of type **85** participate in the same way to first form spirocyclopropane-annelated thiiranium ion intermediates which are subsequently opened by attack of the incoming nucleophile.

Therefore, polar solvents such as methanol play a "magic" role in that they can stabilize ions of the type **103**. In methanol, substitutions of chlorine proceeded

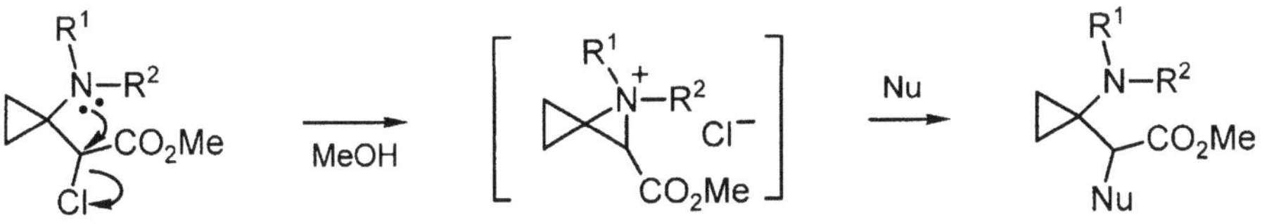

91 **103** **104–106**

Scheme 32. Possible aziridinium ion intermediate **103** in the nucleophilic substitution of chlorine in Michael adducts of type **91** [8]

under mild conditions and gave high yields of the corresponding products (Scheme 33) [11b]. For comparison, compound **104d** under commonly used conditions (NaN_3, H_2O, Aliquat 336, 60 °C [9]) was obtained in 36% yield only [53].

Of special interest is the reaction with potassium cyanate (Scheme 34) [11b] which turned out to be essentially a new method of chlorine substitution with a protected ammonia equivalent.

Under conditions favoring an S_N1 mechanism, i.e. in the presence of silver salts, a competition of nucleophiles was observed [11b, 26b]. Thus, the acetoxy derivative **104h** was obtained in 85% yield in the reaction of **91e** with *n*-butyl amine in THF in the presence of 1 equiv. of AgOAc (Scheme 33) [11b]. Incorporation of the solvent (ROH, DMF) into the product has also been detected, sometimes as the only reaction mode.

With these results available, it was not difficult to substitute the chlorine atom in compounds **91** with an additional equivalent of a primary or secondary amine in methanol [11, 57] to obtain *α*,*β*-diaminoesters **106** with different amino fragments (Scheme 35).

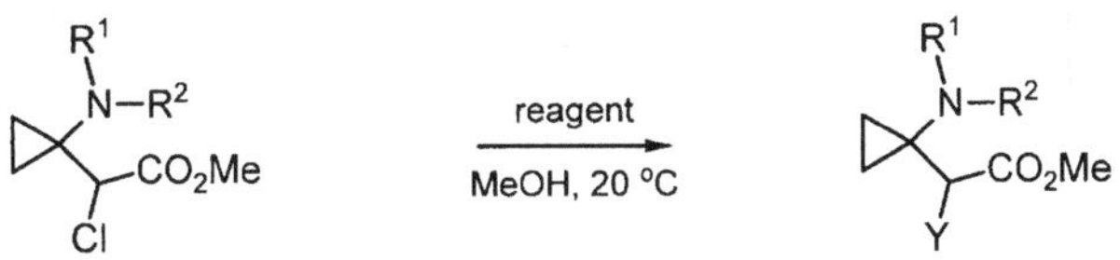

91 **104**

Starting Material	R^1	R^2	Reagent	Y	t	Product	Yield (%)
91e	Et	*n*Bu	NaN_3	N_3	4 h	**104a**	89
91r	Me	CH_2CO_2Me	NaN_3	N_3	10 d	**104b**	56
91j	Me	Bn	NaN_3	N_3	4 d	**104c**	89
91w	Me	CH(Me)(Ph)	NaN_3	N_3	12 h	**104d**	86
91e	Et	*n*Bu	R^3SH, Et_3N	EtS	6 h	**104e**	91
91e	Et	*n*Bu	R^3SH, Et_3N	*n*BuS	24 h	**104f**	42
91k	4-MeO-Bn	Bn	R^3SH, Et_3N	EtS	10 d	**104g**	88
91e	Et	*n*Bu	R^3CO_2H, Et_3N	MeCOO	12 h	**104h**	89
91e	Et	*n*Bu	R^3CO_2H, Et_3N	$CH_3C(O)(CH_2)_2COO$	12 h	**104i**	72
91e	Et	*n*Bu	*n*-$BuNH_2$, AgOAc	$MeCO_2$	12 h	**104h**	85

Scheme 33. Nucleophilic substitution of chlorine in chloroamines **91** in methanol [11b]

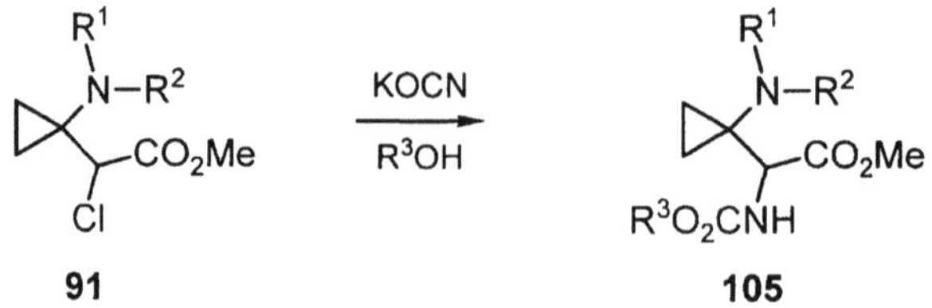

Starting Material	R^1	R^2	R^3	T [°C]	t [h]	Product	Yield (%)
91e	Et	*n*Bu	Me	20	16	**105a**	93
91e	Et	*n*Bu	Et	20	20	**105b**	76
91e	Et	*n*Bu	Bn	55	10	**105c**	61
91j	Me	Bn	Bn	55	10	**105d**	51

Scheme 34. Reaction of Michael adducts **91** with potassium cyanate in alcohols [11b]

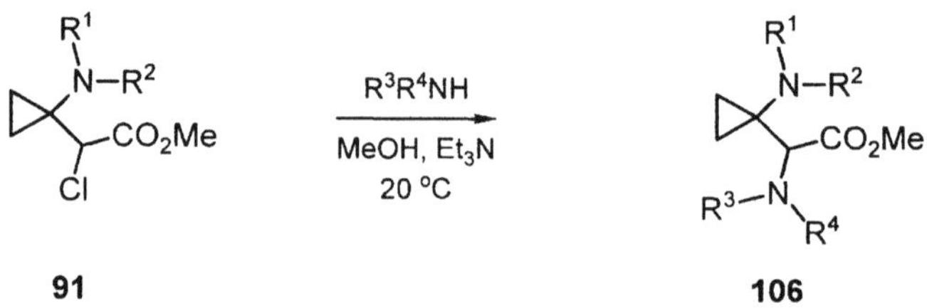

Starting Material	R^1	R^2	R^3	R^4	t	Product	Yield (%)	Ref.
91b	Et	Et	H	Allyl	7 d	**106a**	89	[57]
91b	Et	Et	Allyl	Allyl	3 d	**106b**	56	[57]
91b	Et	Et	$-(CH_2)_5-$		7 d	**106c**	89	[57]
91b	Et	Et	$-(CH_2)_2O(CH_2)_2-$		7 d	**106d**	86	[57]
91c	Allyl	Allyl	H	*n*Pr	7d	**106e**	91	[57]
91c	Allyl	Allyl	Et	Et	3d	**106f**	42	[57]
91c	Allyl	Allyl	$-(CH_2)_2O(CH_2)_2-$		3d	**106g**	88	[57]
91d	$-(CH_2)_5-$		H	Allyl	7d	**106h**	89	[57]
91d	$-(CH_2)_5-$		Et	Et	7d	**106i**	72	[57]
91d	$-(CH_2)_5-$		$-(CH_2)_2O(CH_2)_2-$		7d	**106j**	77	[11b]
91e	Et	*n*Bu	H	*n*Bu	12 h	**106k**	98	[11b]
91e	Et	*n*Bu	H	*n*Pr	6 h	**106l**	89	[11b]
91e	Et	*n*Bu	H	Bn	3 d	**106m**	93	[11b]
91e	Et	*n*Bu	H	CH(Me)(Ph)	3 d	**106n**	90	[11b]
91e	Et	*n*Bu	Me	CH_2CO_2Me	2 d	**106o**	91	[11b]
91e	Et	*n*Bu	Me	Bn	12 h	**106p**	94	[11b]
91e	Et	*n*Bu	H	$CH(Ph)(CH_2OCH_3)$	20 h	**106r**	95	[11b]
91l	Et	*n*Bu	H	*n*Bu	2 d	**106s**	86	[11b]

Scheme 35. Preparation of diaminoacid derivatives **106** via nucleophilic substitution of chlorine in β-amino-α-chloro esters **91** [11, 57]

The simplicity of these transformations and high yields in both steps indicate that the preparation of compounds of type **106** from **1**, **2** can easily be automated and performed as liquid phase parallel synthesis. The starting materials **1**, **2** can also be put on a polymer support to obtain combinatorial libraries of 1,1′-di-, 1,1′,2′-tri-, 1,1′,2′,2′-tetra-, 1,1′,2′,3′-tetra-, 1,1′,2′,2′,3′-penta- and 1,1′,2′,2′,3′,3′-hexasubstituted cyclopropylacetic acid derivatives (see Sect. 6).

3.2.2
Preparation of Carbo- and Heterocycles

The high Michael acceptor reactivity of chloro esters **1**, **2** opens wide possibilities for the preparation of various carbo- and heterocycles with different ring sizes and substituents. This may be achieved either by chemical transformations of primary Michael adducts or by addition of a bidentate nucleophile onto **1** or **2**. The adduct of benzylamine to the chloro ester **1**-*t*Bu (**89b**-*t*Bu) in a three-step sequence yielded the *β*-lactam **107** and the isomeric iminolactone **108** in approximately equal amounts (Scheme 36) [26]. The mechanism for the formation of **108** remains unknown; presumably it was formed from **107** via a 1,3-shift, in fact **107** and **108** are stable in pure form, but would be equilibrating under the reaction conditions.

An efficient synthesis of 2-[(phenylalkylmethylene)amino]cyclobutenecarboxylates **109**, **110** from primary Michael adducts **94** has been developed (Scheme 37) [8]. The key step of this dehydrochlorinative rearrangement is believed to be the lithium iodide-induced reorganization of the azaspiropentane intermediate **103**, in close analogy to the well documented rearrangement of oxaspiropentanes to cyclobutanones [67].

Compounds of type **94** with substituted and spirocyclopropane-annelated cyclopropane rings showed an increased rate of rearrangement, and the corresponding dimethyl-substituted starting material **94k** was never isolated, its transformation into **109k** proceeded even in the absence of LiI upon attempted preparation (36% yield). This isomerization apparently occurs regioselectively and ste-

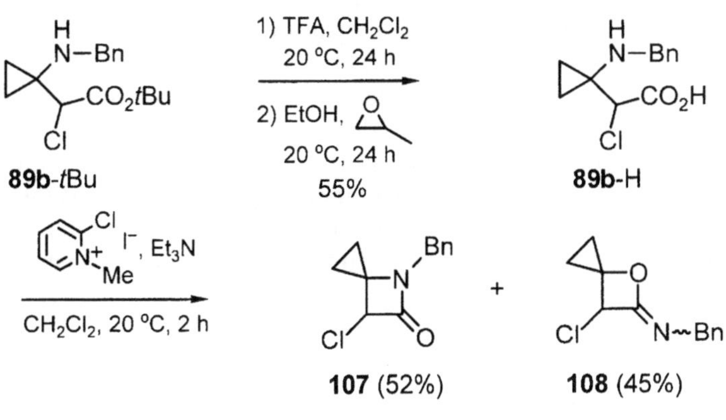

Scheme 36. Preparation of *β*-lactam **107** and isomeric iminolactone **108** from the *β*-amino-*α*-chloroester **89b**-*t*Bu [26]

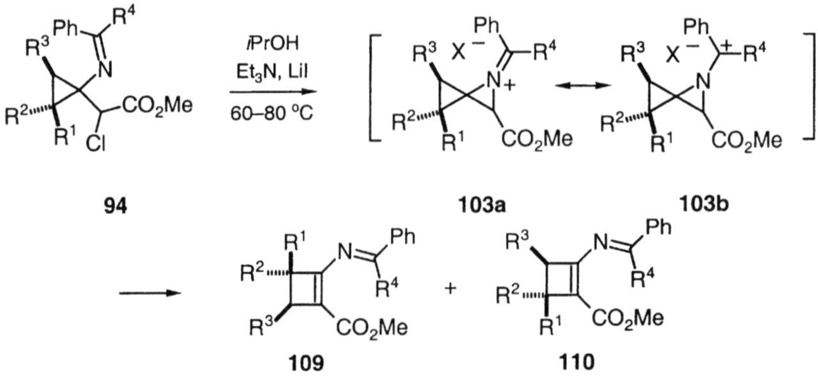

Starting Material	R^1	R^2	R^3	R^4	t [d]	Yield of **109** (%)	Yield of **110** (%)
94a	H	H	H	Ph	7	65	–
94c[a]	–$(CH_2)_2$–		H	Ph	5	20	37
94e[b]	Me	H	H	Ph	5	54	34
94f	H	H	H	*c*Pr	4	20	–
94g	H	H	H	1-Me-*c*Pr	5	65	–
[**94k**]	H	Me	Me	Ph	2	76	–

[a]A 2.1:1 mixture of diastereomers. – [b]A 2.3:1 mixture of diastereomers

Scheme 37. Lithium iodide-promoted isomerization of Michael adducts **94** into cyclobutene-carboxylates **109, 110** [8]

reoselectively with the more highly substituted fragment of the three-membered ring migrating preferentially with the retention of its configuration [8].

As was later observed [11b, 53, 56] this rearrangement is quite general for any Michael adduct of type **89, 91** of a primary or a secondary amine onto a chlorocyclopropylideneacetate **1, 2**. The nature of the amino group, elevated temperature and an appropriate choice of the solvent are more important than the presence of lithium iodide. Thus, the β-amino-α-chloroester **89b** slowly rearranges to **111b** even at −20 °C, and the dibenzylamino derivative **91a** forms the dibenzylaminocyclobutenecarboxylate **111a** in DMF almost quantitatively (Scheme 38) [56]. Moreover, in the "magic" solvent methanol which may stabilize intermediate **103** (see above), the adducts of primary amines **89** did not give any of the corresponding cyclobutene derivative **111** or the tautomeric cyclobutanoneimine, but the azaspiropentanecarboxylates **113b–d** instead, albeit in low yields (4–14%, Scheme 38). This is the first experimental evidence for the intermediacy of aziridinium ions **103** in the chemical transformations of Michael adducts of types **89, 91, 94**.

The low yields of the esters **113** may be attributed to their limited stability. The corresponding amides **112** turned out to be more stable and could be isolated after treatment of compounds **89** with a second equivalent of a primary amine or even ammonia in moderate yields (Scheme 38).The structure of the primary amide **112f** was unequivocally proved by X-ray crystal structure analy-

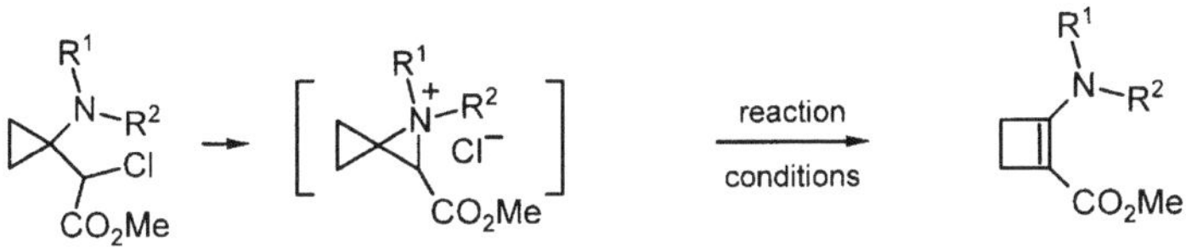

89, 91 103 111

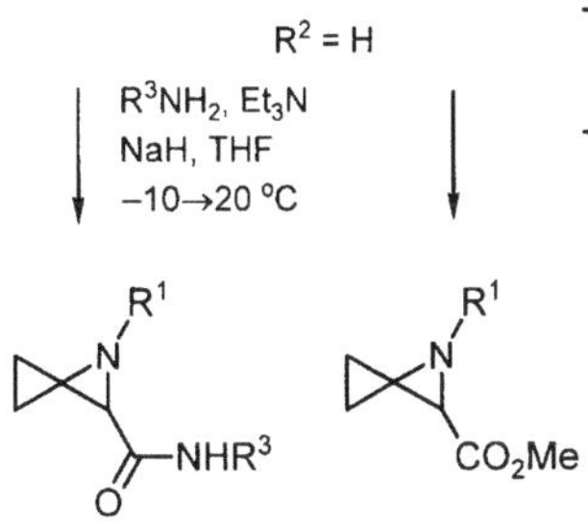

112 113b–d

Starting Material	R^1	R^2	Reaction Conditions	Yields of 111/113 (%)	Ref.
91a	Bn	Bn	DMF, 80 °C 48 h	95/–	[56]
89b	Bn	H	DMF, 20 °C 48 h	51/0	[56]
89b	Bn	H	Et_3N, 10 kbar THF, 20 °C, 4 d	42/0	[56]
89b	Bn	H	MeOH, Et_3N 20 °C, 24 h	0/14	[56]
89c	CH_2CO_2tBu	H	MeOH, Et_3N 20 °C, 9 d	0/4	[56]
89d	4-$MeOC_6H_4CH_2$	H	MeOH, Et_3N 20 °C, 20 h	0/13	[11b]

Starting Material	R^1	R^3	t [h]	Product	Yield (%)	Ref.
89a	*n*Bu	*n*Pr	65	**112a**	22	[11b]
89a	*n*Bu	H	29	**112b**	16	[11b]
89b	Bn	Bn	48	**112c**	59	[56]
89b	Bn	4-MeO-$C_6H_4CH_2$	18	**112d**	47	[56]
89b	Bn	(Ph)(Me)CH	48	**112e**	47	[56]
89b	Bn	H	24	**112f**	48	[56]
89d	4-MeO-$C_6H_4CH_2$	4-MeO-$C_6H_4CH_2$	36	**112g**	27	[56]
[a]	(Ph)(Me)CH	Bn	36	**112h**	36	[56]
[a]	(Ph)(Me)CH	(Ph)(Me)CH	96	**112i**	27	[53]
[a]	$PhCH_2CH_2$	H	50	**112j**	33	[56]
[a]	*n*Pr	H	48	**112k**	12	[11b]

[a]One-pot reaction without characterization of **89**

Scheme 38. Two reaction pathways of β-amino-α-chloro esters **89, 91** [11b, 53, 56]

sis. On one side, compounds **112, 113** are new examples of the rare class of aza-spiropentane derivatives [68]. On the other side, they are essentially protected derivatives of a spirocyclopropanated analog of the aziridinecarboxylic acid [69] which has been applied as a building block in interesting peptidomimetics. Unfortunately, up to now all attempted deprotections have been unsuccessful, and only a limited number of transformations have been achieved without ring opening [11b, 53, 56].

Probably because of its conjugation with the cyclobutene moiety, the DPMA group in the protected amino acids **109** demonstrated an increased stability towards hydrolysis. Nevertheless, deprotection was achieved under acidic conditions [oxalic acid/water in methanol/ether or aluminum trichloride/water (1:2

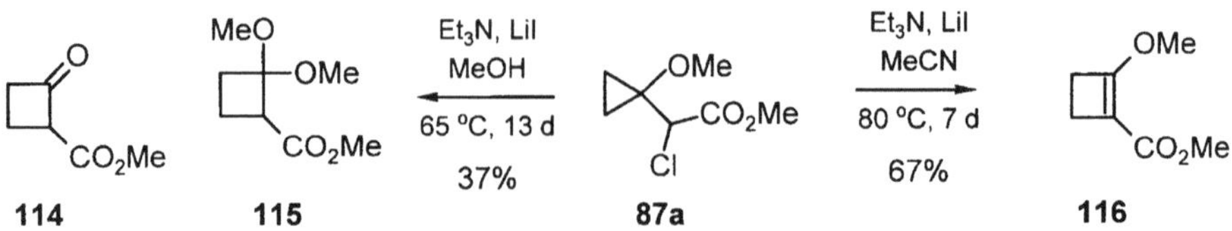

Scheme 39. Ring enlargements of the Michael adduct **87a** of methanol to **1**-Me [8]

equiv.) in chloroform] to give the known methyl 2-oxocyclobutanecarboxylate (**114**) [70] in 77% yield [8]. Two synthetically interesting derivatives of **114** – the dimethylacetal **115** and the enol methyl ether **116** [70] – were obtained by treatment of the Michael adduct of methanol **87a** onto **1**-Me with Et_3N in the presence of LiI in methanol or acetonitrile, respectively (Scheme 39) [8]. It has not been proved whether this ring enlargement occurred at the stage of a cyclopropylcarbene formed by α-dehydrochlorination of **87a** [1, 71] or as a cyclopropylcarbinyl to cyclobutyl cation rearrangement with subsequent trapping by methanol to give **115** or deprotonation to yield **116**.

The benefits of the cumulated functionalities in chlorocyclopropylideneacetates **1**, **2** is demonstrated by the facile syntheses of different types of five- and six-membered heterocyclic compounds (Fig. 6), one-pot or stepwise. The preparation of the compounds of the types **E**, **F**, **J**, **K** has been discussed above (see Sect. 3.1).

Thus, treatment of the DPMA-H Michael adducts **94a, b** with sodium azide in DMF yielded the α-azidoesters **117** (Scheme 40) [10c, 59]. Because of the hydrolytic sensitivity of the DPMA groups in the esters **117**, they were immediately hydrolyzed with dilute hydrochloric acid, and the amino group was protected with an alkoxycarbonyl group. The β-alkoxycarbonylamino-α-azidoesters **118** were reduced to the α-amino esters **119** with hydrogen sulfide in pyridine [74], and they in turn were treated with *N,N*-dimethylformamide dimethylacetal to give the formamidines **120**. The latter were cyclized without purification to give the spirocyclopropanated 2-imidazolinedicarboxylates **121** upon heating in xylene at 150 °C in the presence of ammonium sulfate.

Bicyclic β-lactams of the penam type are still a very important class of antibiotics. But like many other antibiotics its applicability is narrowing continuously due to increasing numbers of resistant strains of bacteria and a shift of the pathogen spectrum from the gram-positive into the gram-negative range. For this reason new analogs of the classical penams are constantly being developed [75]. Thus, a new synthetic approach [10c, 26, 33, 59] to azapenams containing a nitrogen in the five-membered ring with a spirocyclopropane linkage [76, 77] from heterocycles **25**, **121** has been designed (Scheme 41).

The azapenam derivatives **122** were isolated as single diastereomers from the irradiation of (dibenzylaminocarbene)pentacarbonylchromium [78] in the presence of imidazolines **121**. The relative configuration of the new stereocenters at C-3 and C-4 was predicted as being *exo*, *exo* from MMX calculations, and this configuration was proven by X-ray structure analysis [59]. High stereoselectivity was also observed in the analogous transformation of pyrazoline **25b**, but

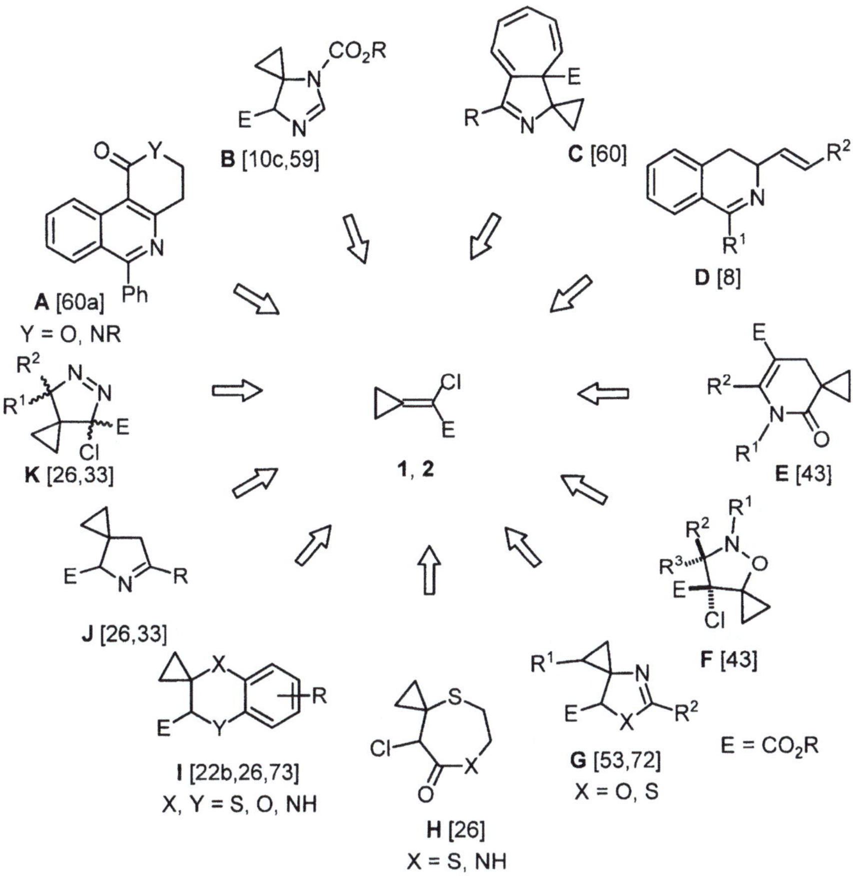

Fig. 6. Different types of heterocyclic compounds prepared from chlorocyclopropylideneacetates **1, 2**

the yields were moderate or low which may well have been due to an influence of the phenyl substituent in **25b**. Compounds **25a, c** did not react stereoselectively and gave 1:1 mixtures of both diastereomers **123** and **124** (Scheme 41) [26, 33].

Contrary to the normal Michael reaction of the chloro ester **1**-Me with (diphenylmethylene)amine (DPMA-H) in methanol (Scheme 29), reaction of DMPA-H with **1**-Me in methylene chloride or tetrachloromethane containing sodium hydride proceeded much more slowly and gave the isoquinoline **126** in 26 and 39% yield, respectively (Scheme 42) [60a]. The structure of **126** was ascertained by X-ray crystallography [60a]. Taking into consideration the stability of the Michael adducts **94** in refluxing CCl_4, the reaction mechanism with intermediacy of a formal [4+2] cycloadduct **125** which may be formed in either a concerted or, more probably, a two-step domino Michael reaction, has been postulated [60a].

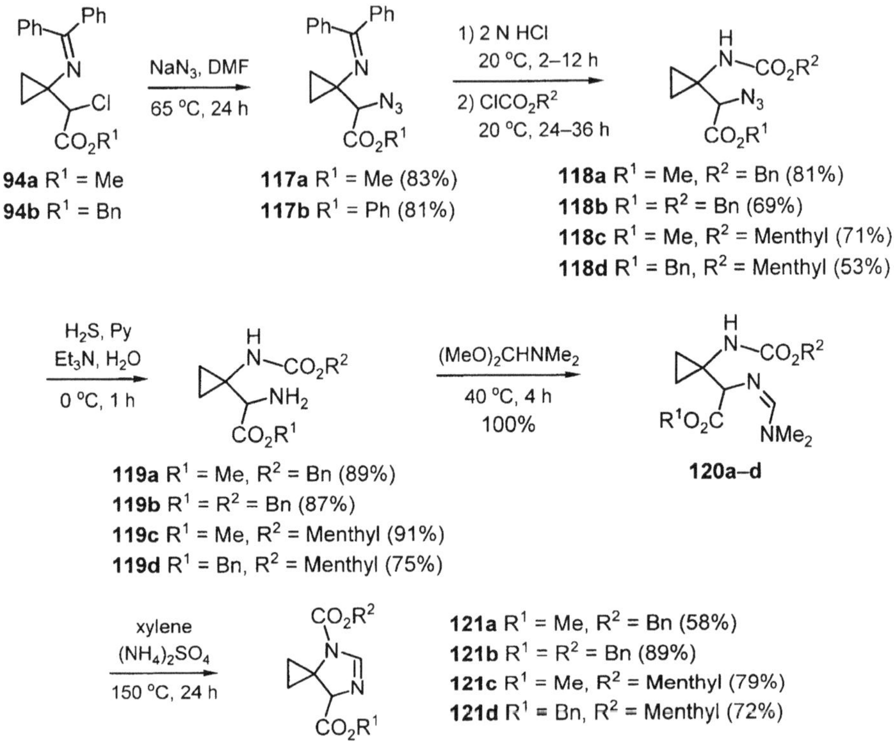

Scheme 40. Stepwise preparation of imidazoles **121** from Michael adducts **94** of (diphenylmethylene)amine to 1-Me [10c, 59]

Further transformation of the ester group in **126** was easily achieved. Thus, **126** with methylamine and 2-dimethylaminoethylamine gave the lactams **127a** and **127b**. Reaction of the free acid derived from **126** with thionyl chloride yielded the lactone **127c** (Scheme 42) [7m, 60a].

Treatment of the Michael adducts **94** with a strong base such as potassium *tert*-butoxide yielded the 2-azaazulenes **131** in 48–70% yield, respectively (Scheme 43) [60].

Presumably the ester enolate **128**, which is a carbenoid and easily formed upon deprotonation of the α-chloroester **94**, undergoes α-elimination to the carbene **129** [79]; either the carbenoid **128** or the carbene **129** intramolecularly cycloadds onto the phenyl group, and the norcaradiene **130** thus formed isomerizes to **131**. If the cyclopropylcarbene **129** is really formed as an intermediate, it is noteworthy that it does not undergo the facile rearrangement to the corresponding cyclobutene (cf. Scheme 39). With lithium diisopropylamide as a base, the yield of the azaazulene **131a** was only 47% [60a]. Although compounds **131f, g** were obtained exclusively without any by-products being detected, isolated yields were always in the range between 48 and 58%. It can only be speculated that the decreased yields are due to side reactions involving the additional cyclopropyl substituent in **94f, g** or attributed to a decreased stability of the products **131f, g** [60b].

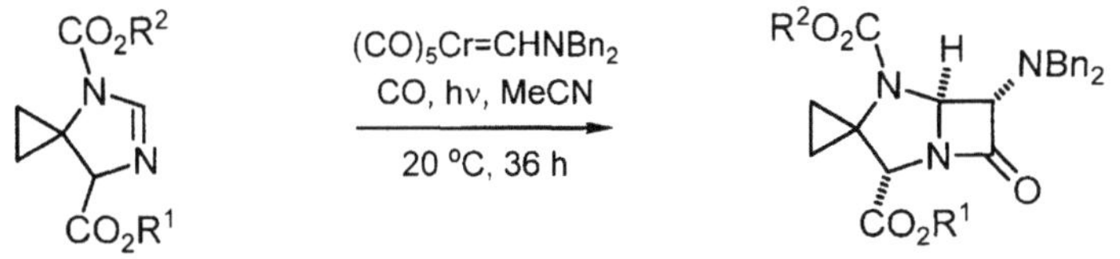

121a R^1 = Me, R^2 = Bn
121d R^1 = Bn, R^2 = Menthyl

122a R^1 = Me, R^2 = Bn (80%)
122d R^1 = Bn, R^2 = Menthyl (50%)

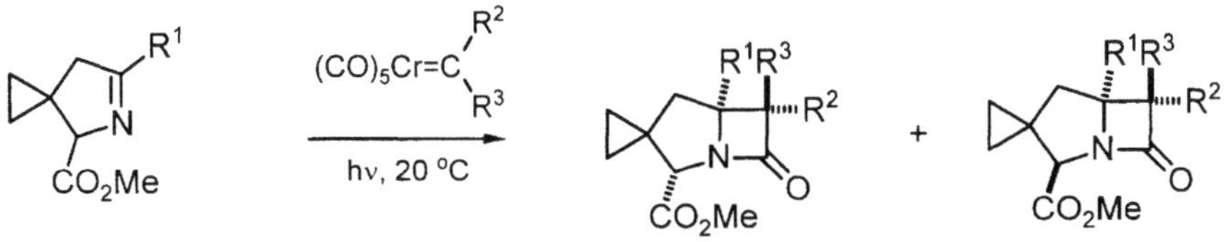

25a–c **123** **124**

Starting Material	R^1	R^2	R^3	Solvent	t [d]	Yield of **123** (%)	Yield of **124** (%)
25a	Me	NBn_2	H	MeCN	2	33	33
25b	Ph	OEt	Me	Et_2O	6	10	0
25b	Ph	NBn_2	H	MeCN	0.7	48	0
25c	H	NBn_2	H	MeCN	7	34	34

Scheme 41. Reactions of heterocycles **25**, **121** with Fischer carbene complexes as a synthetic approach to bicyclic *β*-lactams **122**–**124** [26, 33, 59]

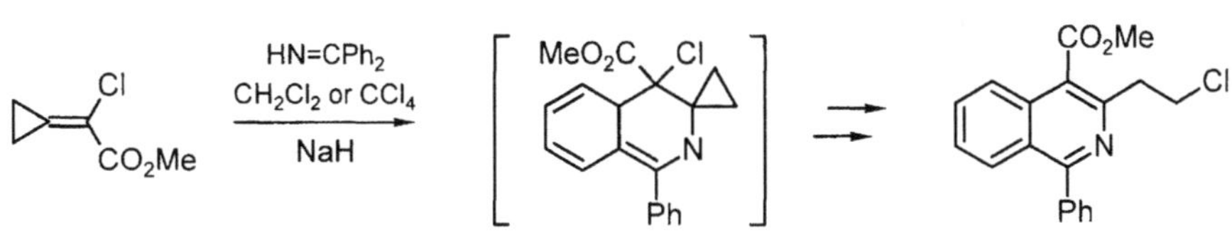

1-Me **125** **126**

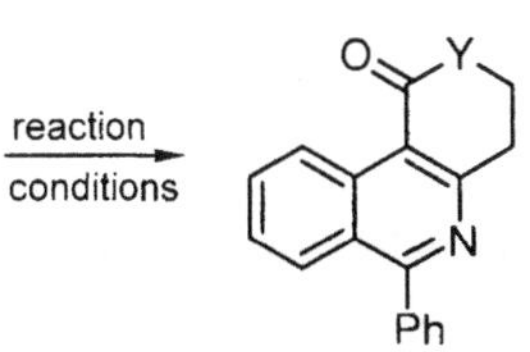

Product	Y	Conditions	Yield (%)
127a	NMe	$MeNH_2$, *p*TsOH 65 °C, 48 h	93
127b	$N(CH_2)_2NMe_2$	$H_2N(CH_2)_2NMe_2$ MeOH, 65 °C, 48 h	92
127c	O	1) Et_3N, H_2O 2) $SOCl_2$, 80 °C, 5 min	96

Scheme 42. Preparation of isoquinoline derivative **126** and its subsequent heterocyclization [7 m, 60 a]

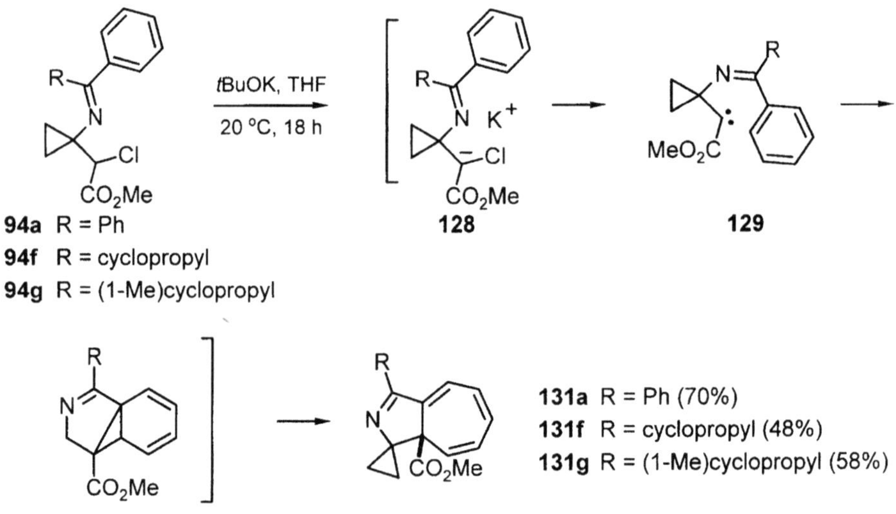

94a R = Ph
94f R = cyclopropyl
94g R = (1-Me)cyclopropyl
128
129
130
131a R = Ph (70%)
131f R = cyclopropyl (48%)
131g R = (1-Me)cyclopropyl (58%)

Scheme 43. Synthesis of the 2-azaazulenes **131** from the Michael adducts **94** of DMPA-H to **1**-Me [60]

The (diphenylmethylene)aminocyclobutenecarboxylates **109** obtained by rearrangement of the DMPA-H adducts of **1**-Me, **2**-Me, contain a 2-azadiene unit and a cyclobutene moiety. Indeed, the parent compound **109a** reacted with 4-phenyl-1,2,4-triazoline-3,5-dione (PTAD, [80]) at room temperature in a [4+2] cycloaddition mode to yield the tricyclic tetraazaundecene **132** in almost quantitative yield (Scheme 44) [8]. As substituted cyclobutenes, compounds **109** should be capable of opening up to the corresponding butadienes [1, 2b, 81]. When compounds **109** were subjected to flash vacuum pyrolysis, the dihydroisoquinolines **135** were obtained, presumably via the expected ring-opened intermediates **133**, which subsequently underwent 6π electrocyclization followed by a 1,5-shift, as is common for other 3-aza-1,3,5-hexatrienes [82].

A very unstable reaction by-product was detected in this reaction, and it showed spectroscopic features in full accordance with the proposed ring-opened intermediates **133**, but attempted purification to get analytically pure material was unsuccessful (Scheme 44) [8].

The presence of the additional cyclopropyl group in the azaazulenes **131f, g** and dihydroisoquinoline **135g** suggested subsequent transformations of these compounds by rearrangement of their cyclopropylketimine moieties. After several unsuccessful attempts applying different conventional methods [83] the cyclopropylketimine moiety in **131f, g** was finally transformed by heating the hydrochlorides **136f, g**, formed by treatment of **131f, g** with hydrogen chloride in ether, with sodium iodide in acetonitrile at 90 °C (Scheme 45) [60b]. Without further purification, the thus formed iminium salts **137f, g** were reduced with sodium borohydride in methanol at –40 °C [84] and gave the tricyclic spirocyclopropane-annelated tertiary amines **138f** and **138g** in 75 and 51% yield, respectively, as mixtures of two and four diastereomers (ratios 1.3:1 for **138f** and 4.7:3.2:2.9:1 for

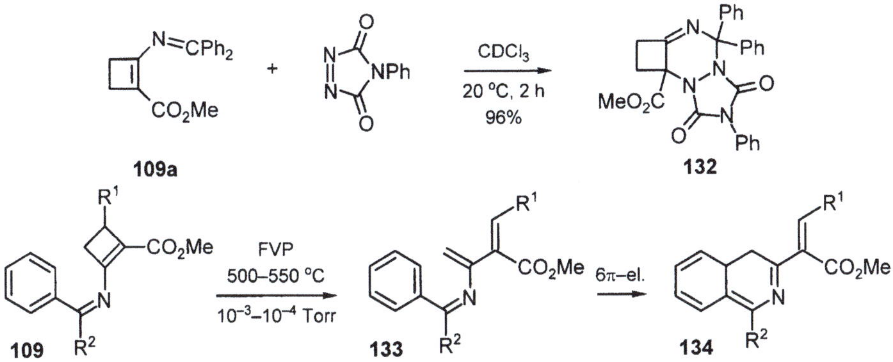

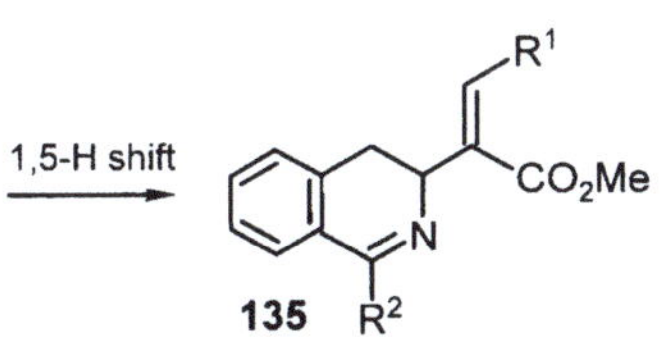

Starting Material	R^1	R^2	Product	Yield (%)
109a	H	Ph	**135a**	68
109e	Me	Ph	**135e**	53
109f	H	*c*Pr	**135f**	50
109g	H	1-Me-*c*Pr	**135g**	68

Scheme 44. Two reaction modes of (diphenylmethylene)aminocyclobutenes **109** [8]

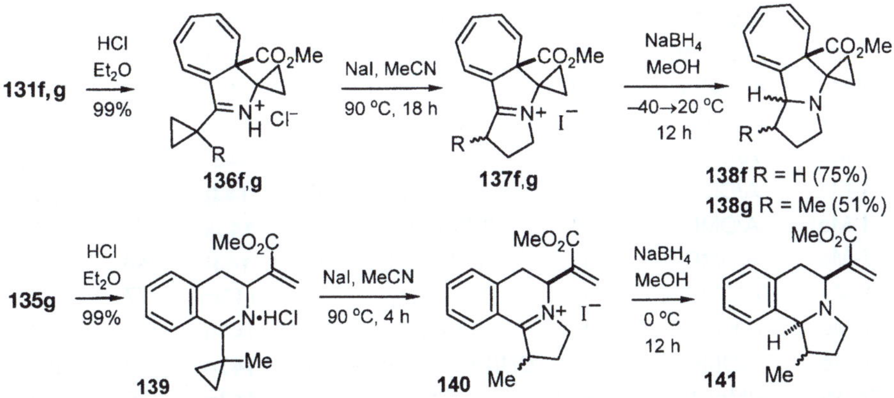

Scheme 45. Preparation of heterocycles **138**, **141** via Cloke rearrangement of the cyclopropylketimine moieties in **131**, **135** [60b]

138g), respectively. The reaction pathway probably involves nucleophilic ring opening of the exocyclic cyclopropyl substituent [44], cyclization, and an enammonium to iminium ion rearrangement [85]. Surprisingly, the spiroannelated cyclopropane moiety remained unchanged, although it should be activated at least in **137f, g** by the adjacent immonium ion moiety [44a]. Essentially, this transformation is an example of the well-known Cloke rearrangement [86].

Eventually, the same method was applied to the dihydroisoquinoline derivative **135g**, and the hexahydropyrroloisoquinoline **141** was isolated in 78% yield (with a 1.8:1 ratio of *cis*- and *trans*-diastereomers) (Scheme 45) [60b].

A short and efficient synthesis of 4-spirocyclopropane-annelated oxa- and thiazoline-5-carboxylates **143** and **144** from chlorocyclopropylideneacetates **1**, **2** and benzamides as well as benzthioamides **142**-X, respectively, has been developed (Scheme 46) [53, 72, 87]. This domino transformation involves a Michael addition of **142**-X across the double bond of **1**-Me or **2**-Me under basic conditions followed by an intramolecular nucleophilic substitution of the α-chlorine substituent.

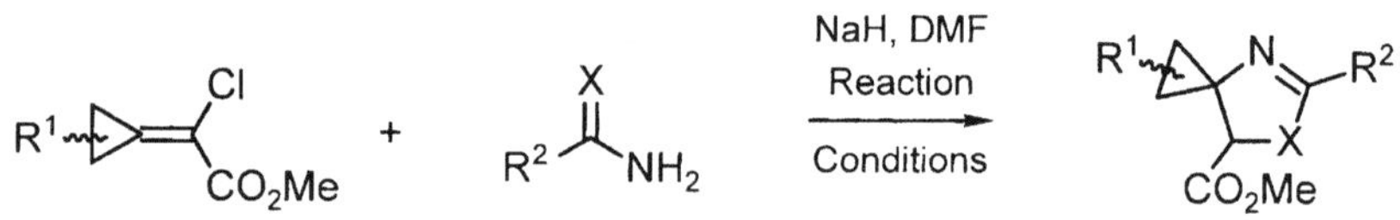

1-Me, **2**-Me   **142**-X (X = O, S)   **143**, **144**

Starting Material	Conditions/ Solvent	R^1	R^2	X	Product	Yield (%)	Ref.
1-Me	A/MeCN	H	Ph	O	**143a**	60	[72]
1-Me	A/MeCN	H	4-Cl-C_6H_4	O	**143b**	59	[72]
1-Me	A/MeCN	H	4-O_2N-C_6H_4	O	**143c**	47	[72]
1-Me	A/MeCN	H	2-MeO-C_6H_4	O	**143d**	75	[53]
1-Me	A/MeCN	H	2-F-C_6H_4	O	**143e**	49	[53]
1-Me	A/MeCN	H	3-I-C_6H_4	O	**143f**	49	[72]
1-Me	A/MeCN	H	3-Me-C_6H_4	O	**143g**	58	[72]
1-Me	A/MeCN	H	2-I-C_6H_4	O	**143h**	72	[72]
1-Me	A/DMF	H	3-NC-C_6H_4	O	**143i**	71	[72]
1-Me	A/DMF	H	3-F-C_6H_4	O	**143j**	75	[72]
1-Me	A/DMF	H	2-Br-C_6H_4	O	**143k**	73	[72]
1-Me	A/DMF	H	2-O_2N-C_6H_4	O	**143l**	48	[72]
1-Me	A/DMF	H	2-Cl-C_6H_4	O	**143m**	77	[72]
1-Me	A/DMF	H	3-Py	O	**143n**	38	[72]
1-Me	A/DMF	H	CH(Bn)(N=CPh_2)	O	**143o**	40	[53]
2b-Me	A/MeCN	Et	Ph	O	**143p**	56	[72]
2b-Me	A/MeCN	Et	2-Me-C_6H_4	O	**143q**	68	[72]
2b-Me	A/MeCN	Et	2-I-C_6H_4	O	**143r**	55	[72]
2b-Me	A/MeCN	Et	3-F-C_6H_4	O	**143s**	46	[72]
1-Me	B/MeCN	H	Ph	S	**144a**	87	[87]
1-Me	B/MeCN	H	Me	S	**144b**	45	[87]
1-Me	B/MeCN	H	4-Br-C_6H_4	S	**144c**	85	[87]
2b-Me	B/MeCN	Et	Ph	S	**144d**	57	[87]
2k-Me	B/MeCN	CH_2CH_2OBn	Ph	S	**144e**	51	[87]

Conditions: A: NaH, −10→20 °C, 12–24 h; B: $NaHCO_3$, 80 °C, 2–5 h

Scheme 46. One-pot preparations of spirocyclopropanated oxazolines **143** and thiazolines **144** from chlorocyclopropylideneacetates **1**, **2** [53, 72, 87]

The structure of the phenyl-substituted oxazoline **143a** was verified by X-ray crystal structure analysis. In most cases the reaction gave better yields of **143** and **144** when performed in acetonitrile than in DMF [87].

Spirocyclopropanated thiazolines **144** could also be prepared in two steps, i.e. by first preparing Michael adducts **145** of thioamides **142** onto chloro esters **1**, **2** under acidic conditions in quantitative yield and then cyclizing the adducts **145** under basic conditions (Scheme 47) [87]. However, only a couple of thiazolines **144a**–**c** were prepared in a pure form along this route, and the yields were lower compared to those by the one-pot method (see Scheme 46). In most cases, a competition with a six-membered ring closure to give the corresponding 5,6-dihydro-1,3-thiazin-4-ones **146** was observed. The latter could not efficiently be separated from the thiazolines **144**. Fortunately, under the action of titanium tetraisopropoxide this cyclization of the intermediates **145** occurred selectively and gave the pure thiazinones **146** in good to very good yields (Scheme 47) [87].

It should also be mentioned that unsubstituted formamide **147** under the same conditions reacted cleanly with two equivalents of the chlorocyclopropylideneacetate **1**-Me by a totally different mode yielding the bisspirocyclopropanated dihydropyrroledicarboxylate **149** (Scheme 47) [53, 72].

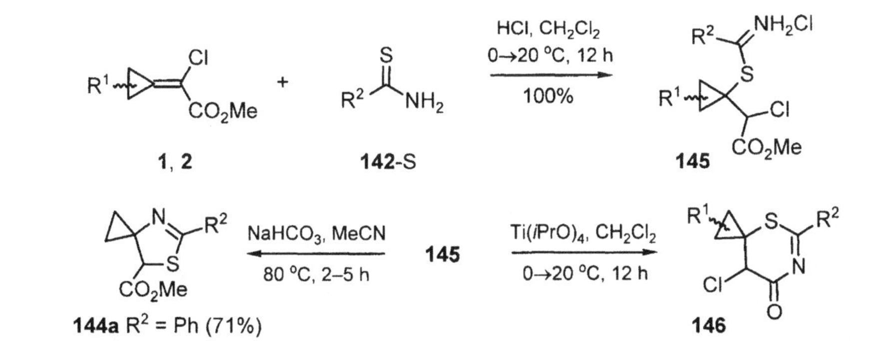

144a R^2 = Ph (71%)
144b R^2 = Me (38%)
144c R^2 = 4-Br-C_6H_4 (82%)

R^1	R^2	Product	Yield (%)
H	Ph	**146a**	71
H	4-Br-C_6H_4	**146b**	51
H	4-*t*Bu-C_6H_4	**146c**	73
Et	Ph	**146d**	52
Et	4-Br-C_6H_4	**146e**	84
Et	4-*t*Bu-C_6H_4	**146f**	88

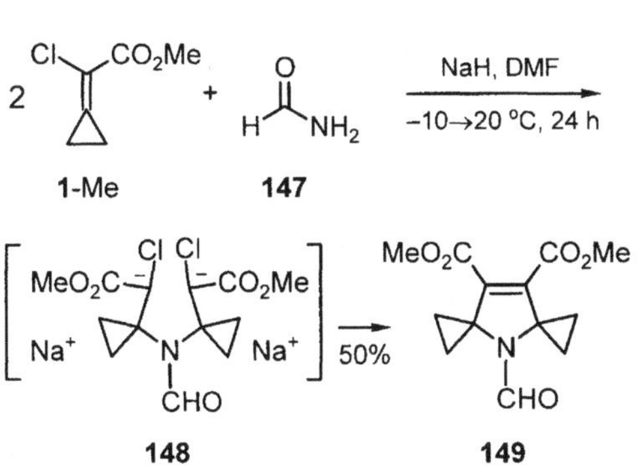

Scheme 47. Different modes of reactivity of bidentate nucleophiles 142, 147 and preparation of thiazolines 144 and thiazinones 146 [53, 72, 87]

The two reaction modes of the Michael adducts **145** demonstrate two general principles for the possible preparation of ordinary size heterocyclic compounds from the chlorocyclopropylideneacetates **1**, **2**. Thus, either the heterocycles **153** can be formed by Michael addition of a bidentate nucleophile **150** onto the chloro ester **1**-Me and subsequent ring closure of the intermediate **151** [26] by nucleophilic substitution of the chlorine atom at the newly formed sp^3 carbon center adjacent to both the carbonyl and the cyclopropyl group (Route B in Scheme 48). Alternatively, the intermediate **151** can cyclize by nucleophilic attack on the ester moiety to give heterocycles of type **152** (Route A in Scheme 48) [26].

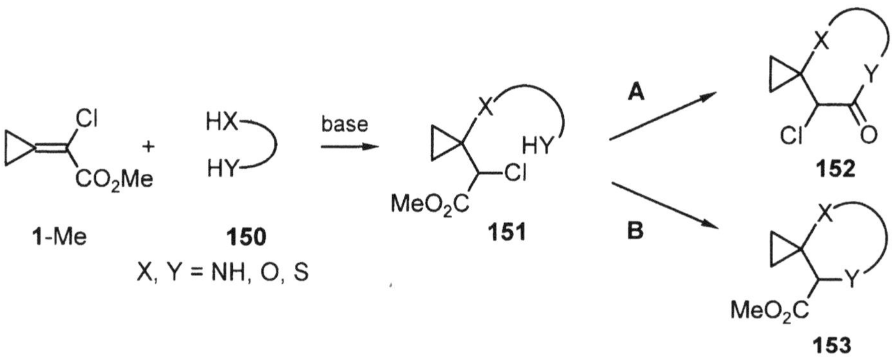

Scheme 48. Two reaction modes of bidentate nucleophiles with the chlorocyclopropylideneacetate **1**-Me

The reaction mode depends mainly on the choice of the base and reaction conditions, but route **B** is usually more common. Thus, tetrahydro-1,4-thiazine derivative **157** the structure of which was proved by X-ray crystal structure analysis (Fig. 7) [22b] was obtained in the reaction of the chloro ester **1**-Me with 2-aminoethanethiol (**156**) using K_2CO_3 or Et_3N as a base in 43 and 20% yield, respectively (Scheme 49) [22b, 26]. In the latter case, the secondary amino group in the primary tetrahydrothiazine product **157** underwent Michael addition to a second molecule of **1**-Me to give 2-[1′-(spirocyclopropanetetrahydrothiazinyl) cyclopropyl]-2-chloroacetate **158** (14% yield). When KOH in the presence of dibenzo-18-crown-6 was employed, however, the seven-membered heterocycles **155** (42%) and **159** (48%) were obtained upon reaction of **1**-Me with 1,3-propanedithiol and 2-aminoethanethiol, respectively (Scheme 49) [26].

3.2.3
Preparation of Oligocondensed Heterocycles and Peptidomimetics

The reaction of the chloro ester **1**-Me with 1,2-dinucleophilic 1,2-disubstituted benzene derivatives **161** in most cases gave the benzene-annelated six-membered heterocycles **162** in moderate yield (Scheme 50) [22b, 26, 73]. The course of the domino-Michael addition-cyclization is also very sensitive to the nature of the base and the reaction conditions applied. Thus, only the primary Michael adducts

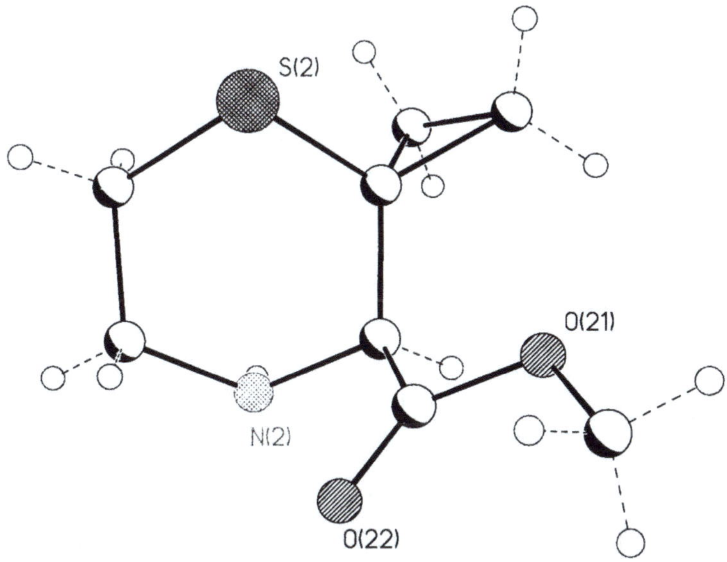

Fig. 7. Structure of the tetrahydro-1,4-thiazine derivative **157** in the crystal [22b]

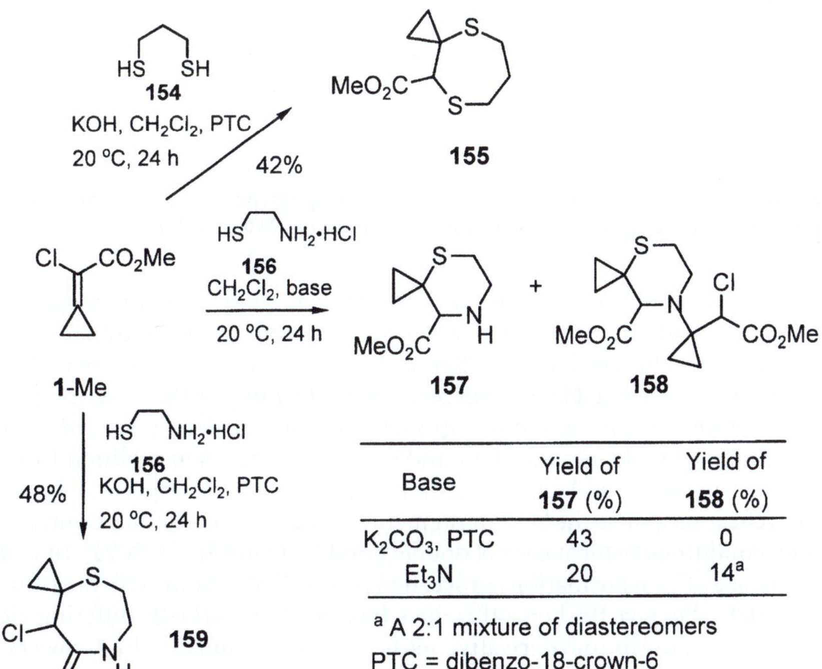

Base	Yield of **157** (%)	Yield of **158** (%)
K_2CO_3, PTC	43	0
Et_3N	20	14[a]

[a] A 2:1 mixture of diastereomers
PTC = dibenzo-18-crown-6

Scheme 49. Heterocyclization of methyl 2-chloro-2-cyclopropylideneacetate (**1**-Me) with non-aromatic bidentate nucleophiles **154**, **156** under various conditions [22b, 26]

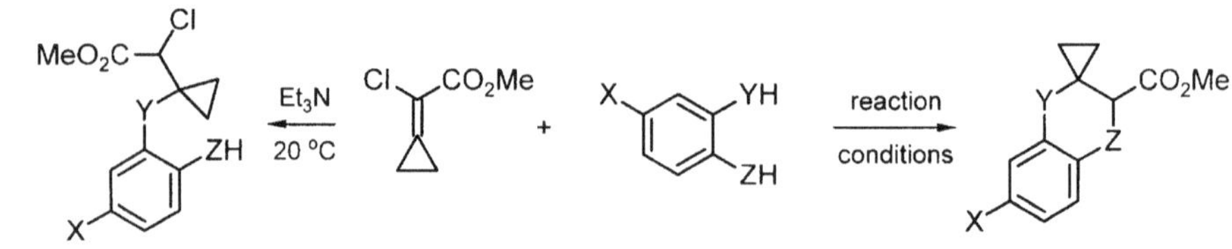

160 **1-Me** **161** **162**

X	Y	Z	Solvent/ t [h]	Product	Yield (%)
OH	S	O	CH_2Cl_2 2	**160a**	86
H	NH	O	DMF 24	**160b**	64
H	S	NH	CH_2Cl_2 24	**160c**	68
Br	S	NH	CH_2Cl_2 24	**160d**	92
NO_2	O	CH_2	CH_2Cl_2 8 d	**160e**	59

X	Y	Z	Reaction Conditions	Product	Yield (%)
OH	S	O	K_2CO_3, KI 20 °C, 12 h	**162a**	87[a]
H	NH	O	DBU 70 °C, 24 h	**162b**	5[a]
H	NH	O	K_2CO_3, PTC 20 °C, 24 h	**162b**	40[b]
H	O	NH	Et_3N 100 °C, 15 h	**162c**	40[a]
H	S	NH	KOH, PTS 20 °C, 24 h	**162d**	46[b]
H	O	O	KOH, PTS 20 °C, 48 h	**162e**	42[b]

[a] Reaction in DMF. – [b] Reaction in CH_2Cl_2, PTC = dibenzo-18-crown-6

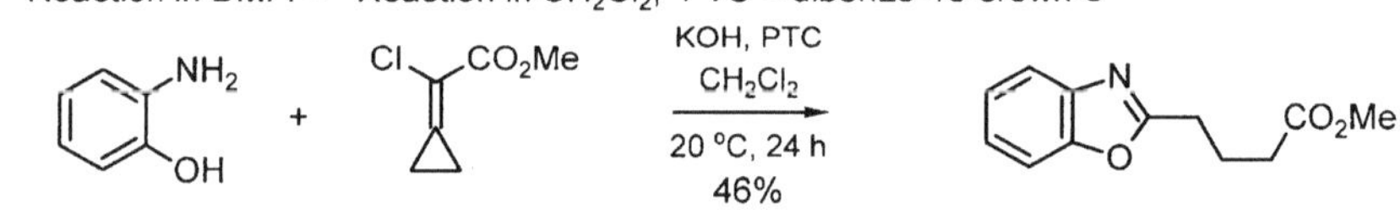

161b **1-Me** **163**

Scheme 50. Heterocyclization of methyl 2-chloro-2-cyclopropylideneacetate (**1**-Me) with aromatic bidentate nucleophiles **161** under various conditions [22b, 26a, 73]

160 were isolated when Et_3N was used as a base at ambient temperature [22b]. The formation of the five-membered ring **163**, without a spirocyclopropane moiety, upon reaction of 2-aminophenol (**161b**) with the chloro ester **1**-Me [26] has later been questioned by an independent performing of this reaction [22b].

Transforming the primary amino groups in the aniline derivatives **161** prior to their addition to **1**-Me into a sulfonamido group or after their addition to **1**-Me, i.e. in the primary adducts **160c, d**, brings some new aspects into these heterocyclizations, since the SO_2Ph fragment is capable to be eliminated under the reaction conditions to form a C=N double bond (Scheme 51) [22b, 73]. In reality, this sequence of transformations proceeded relatively smoothly with the primary adducts **160c, d** to give the benzothiazines **166**, albeit in markedly differing yields (the structure of the bromo derivative **166b**, Fig. 8, was confirmed by X-ray crystal structure analysis, [22b]). The reaction of **1**-Me with the mono(phenylsulfonyl)-phenylene-1,2-diamine **161f** proceeded in a different way and was accompanied by ring opening of the cyclopropane moiety, especially when carried out at elevated temperature [73]. The reaction mixtures obtained from the heterocyclizations of the *N*-(phenylsulfonylamino)phenols **161g–i** with the chloro esters **1**-Me and

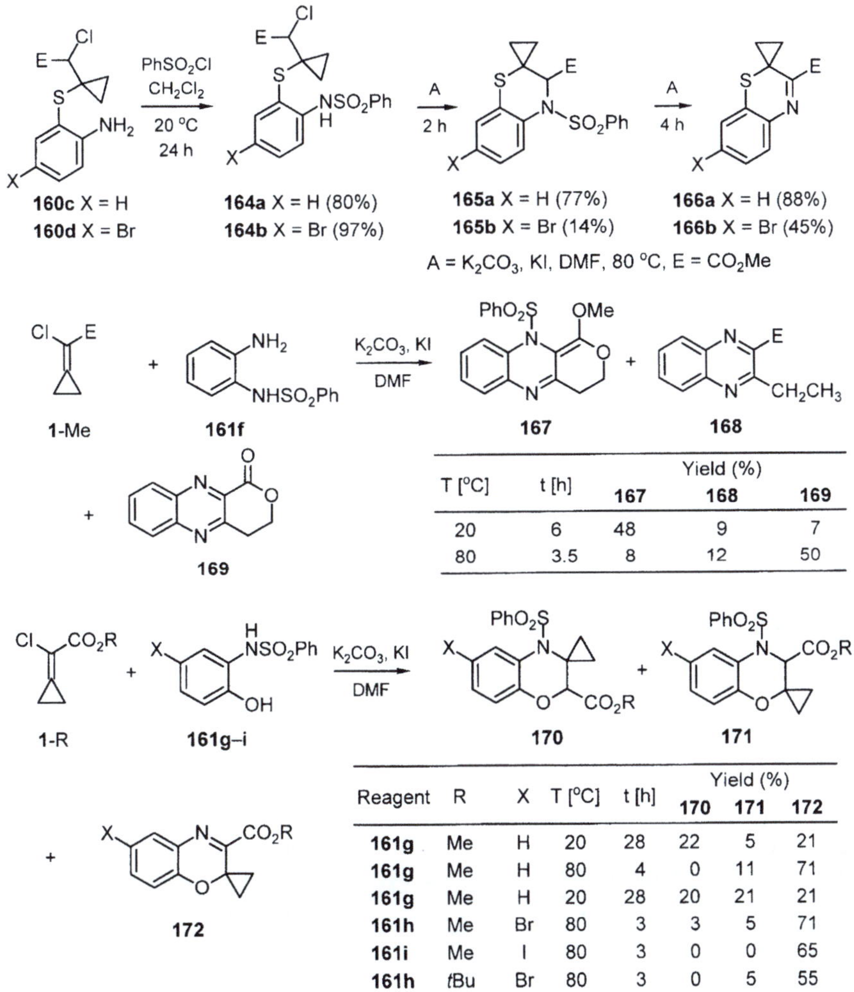

T [°C]	t [h]	Yield (%) 167	168	169
20	6	48	9	7
80	3.5	8	12	50

Reagent	R	X	T [°C]	t [h]	Yield (%) 170	171	172
161g	Me	H	20	28	22	5	21
161g	Me	H	80	4	0	11	71
161g	Me	H	20	28	20	21	21
161h	Me	Br	80	3	3	5	71
161i	Me	I	80	3	0	0	65
161h	*t*Bu	Br	80	3	0	5	55

Scheme 51. Heterocyclizations of 2-[1′-(*o*-*N*-phenylsulfonylamino)arylthiocyclopropyl]-2-chloroacetates **164** and domino heterocyclizations of the chlorocyclopropylideneacetates 1-Me, 1-*t*Bu with various sulfonamides **161f–i** [22b, 73]

1-*t*Bu were also rather complex, but in several cases the desired products of type **172** were isolated in reasonably good yields (Scheme 51) [22b, 73].

Some of these heterocyclic carboxylates easily undergo desmethoxycarboxylation upon heating with morpholine in DMF to yield spirocyclopropane-annelated benzooxazines **173a, b** (Scheme 52) [73]. The latter compounds are capable to undergo Michael additions with thiols or cyanides [88] across the imino group to give the accordingly substituted benzodihydrooxazines **174, 175**.

At last, upon the attempted nucleophilic substitution of chlorine in the Michael adduct **160c** with cyanide an interesting rearrangement was observed which gave the 2-(1′-cyanocyclopropyl)-2-(*o*-aminophenylthio)acetate **176**, ap-

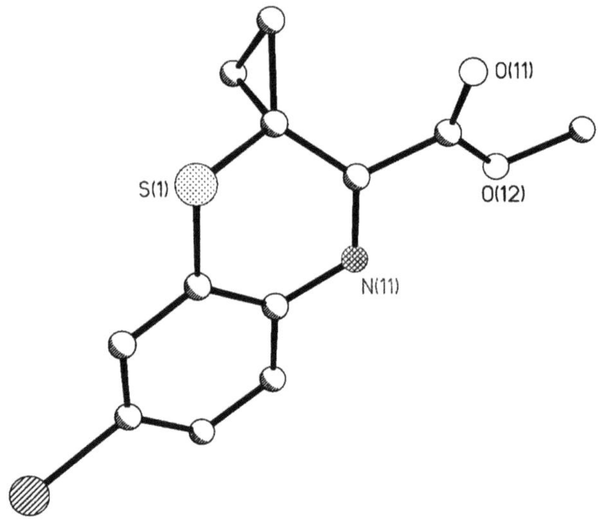

Fig. 8. Structure of benzothiazine **166b** in the crystal [22b]

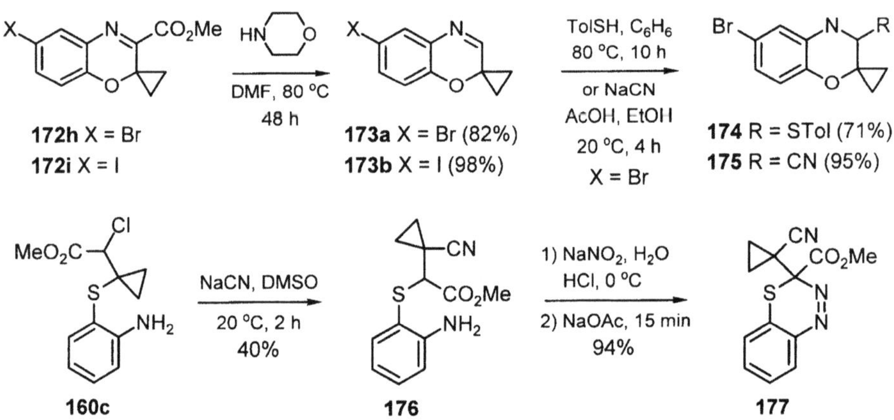

Scheme 52. Transformations of heterocycles **172** and preparation of benzothiadiazine **177** [22b, 73, 88]

parently by an elimination-addition sequence. Subsequent diazotation of **176** and work-up with sodium acetate converted it into the benzothiadiazine derivative **177** (Scheme 52) [22b]. Its structure was proved by an X-ray crystal structure analysis (Fig. 9).

One more useful synthetic application of primary Michael adducts onto 2-chloro-2-cyclopropylideneacetates **1** is the recently reported new approach to octahydrospirocyclopropanepyrazinopyrazines **182**. In the first step, the adducts of primary amines onto **1**-Me (or **1**-*t*Bu) were coupled with *N*-Boc- or *N*-Fmoc-protected glycine derivatives and the resulting products **178**, after deprotection, cyclized under mildly basic conditions to give the spirocyclopropanehexahydropyrazinones **179**, **180** and hexahydrodiazepinediones **181** (Scheme 53) [89].

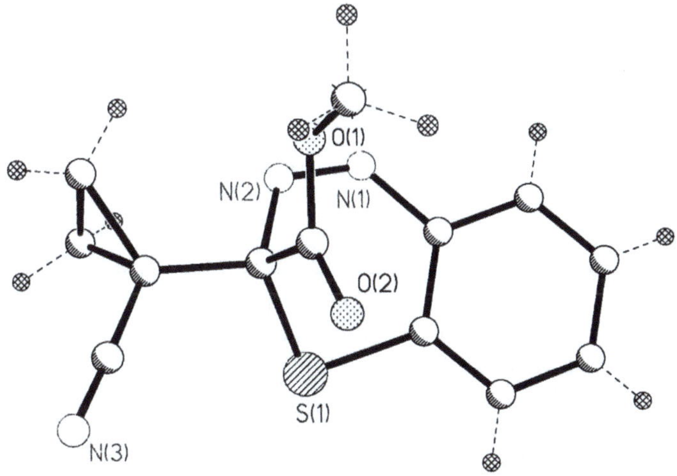

Fig. 9. Structure of the benzothiadiazine derivative **177** in the crystal [22b]

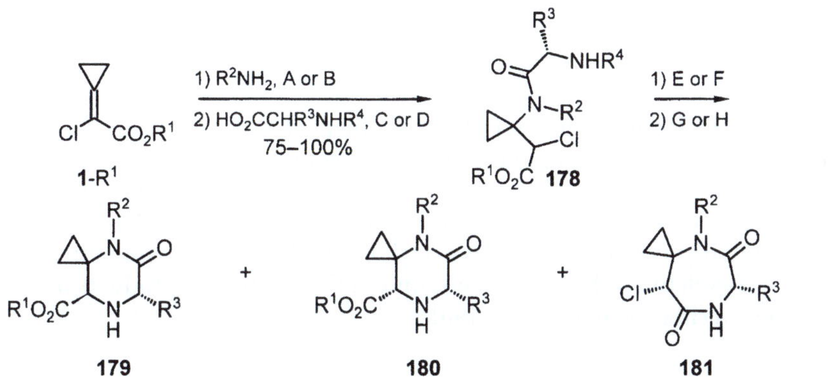

	R^1	R^2	R^3	R^4	Conditions	Yield of **179/180** (%)	Yield of **181** (%)
a	Me	$n\text{C}_5\text{H}_{11}$	H	Boc	A, C, E, G	22/–	60
b	Me	Bn	H	Boc	A, C, E, G	18/–	60
c	Me	$PhCH_2CH_2$	H	Boc	A, C, E, G	18/–	63
d	*t*Bu	$PhCH_2CH_2$	H	Fmoc	A, D, F, H	61/–	0
e	Me	Me	Bn	Boc	B, C, E, G	20/2	19
f	Me	Me	(indol-3'-yl)CH_2	Boc	B, C, E, G	21/2	22

Conditions: A: THF, 0 °C; B: THF, Et_3N, ultrasound; C: DCC, Py or DMAP, THF, 0 °C; D: EDS, Py, 0 °C; E: TFA, CH_2Cl_2, 0 °C; F: aq. NaOH, dioxane; G: aq. $NaHCO_3$, CH_2Cl_2, 0 °C; H: Al_2O_3 basic, MeOH, 0 °C

Scheme 53. Preparation of the spirocyclopropanated hexahydropyrazinones **179**, **180** and hexahydrodiazepinediones **181** from the Michael adducts of primary amines onto 2-chloro-2-cyclopropylideneacetates **1**-Me, **1**-*t*Bu [89]

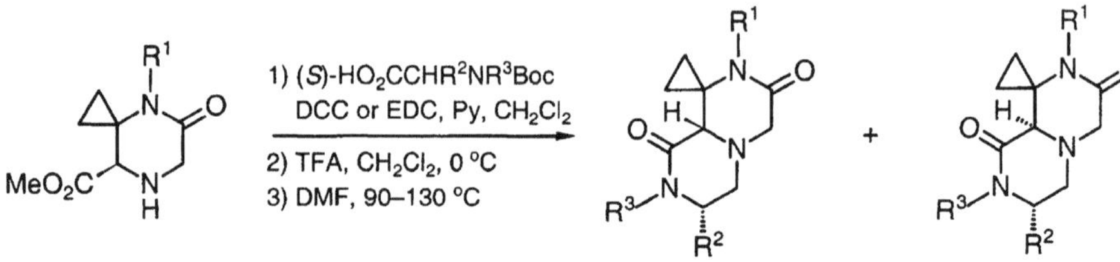

179a–c *trans*-**182** *cis*-**182**

R^1	R^2	R^3		Yield of **182** (%)
nC_5H_{11}	Bn	H	**a**	50
Bn	CH_2CH_2SMe	H	**b**	35
Bn	Bn	H	**c**	59
$PhCH_2CH_2$	(indol-3'-yl)CH_2	H	**d**	56
Bn	$-(CH_2)_3-$		**e**	76

Scheme 54. Cyclization of the spirocyclopropanated hexahydropyrazinones **179** to give spirocyclopropaneoctahydropyrazinopyrazines **182** [89]

The hexahydropyrazinones **179** were subjected to further peptide coupling with various *N*-protected amino acids, deprotection and repeated cyclization to give bicyclic dipeptides with a spirocyclopropaneoctahydropyrazinopyrazine skeleton of type **182** (Scheme 54). Compounds of the types **182** and **179** represent potentially useful classes of geometrically defined peptidomimetics. For example, the skeleton of **179** has been found in the hydrolysis products of the naturally occurring lysomarasmine [90].

3.3 Syntheses of Naturally Occurring and Non-natural Amino Acids

Nature makes use of cyclopropyl groups, not the least as a source for the plant hormone ethylene, since the precursor of ethylene in green plants and their fruits is 1-aminocyclopropanecarboxylic acid (ACC) which was first isolated from cider apples, perry pears and cowberries [91], and is oxidatively degraded by the ethylene forming enzyme (EFE) [92]. Substituted ACC's are generally inhibitors of EFE [93], and most of the other more than two dozen known naturally occurring amino acids containing a cyclopropyl group as well as several of their analogs exhibit interesting biological activities [77, 94]. No wonder that a number of synthetically oriented groups around the world have invested a considerable amount of work into the development of feasible syntheses of such amino acids [95].

A lot of the primary Michael adducts prepared from chlorocyclopropylideneacetates **1**, **2** as described in Sect. 4.1 can serve as convenient synthetic precursors for a large variety of natural and non-natural amino acids (Fig. 10).

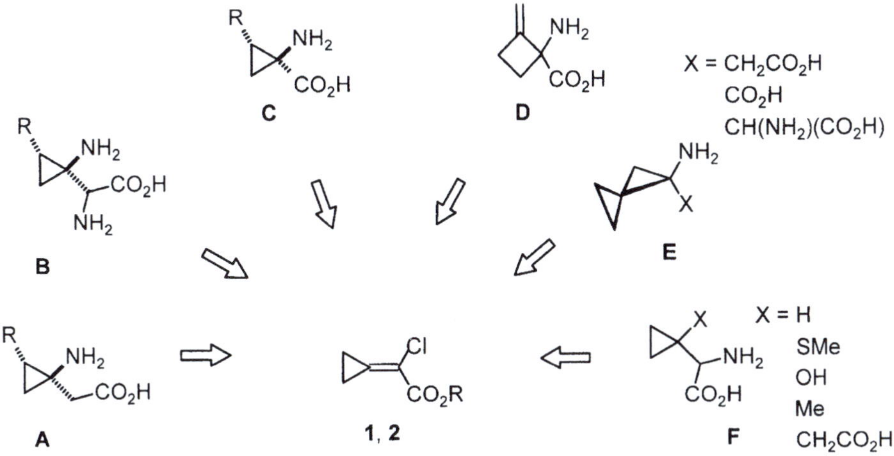

Fig. 10. Different types of cyclopropyl-group containing amino acids prepared from chlorocyclopropylideneacetates **1, 2**

Since three-membered rings have a number of chemical properties in common with C,C double bonds [51], the spiropentyl analogs of some naturally occurring amino acids containing a methylenecyclopropane moiety [96] are also of potential interest.

A very simple synthetic strategy towards racemic 1′-substituted 2-cyclopropylglycines, including the natural products 2-(1′-methylcyclopropyl)glycine (**186b**) [97] and 2-(1′-hydroxycyclopropyl)glycine (cleonin, **186f**) [98] has been elaborated. It consists of a nucleophilic substitution of chlorine with an azide group in the primary Michael adducts of a corresponding nucleophile (methyl anion or equivalent hydroxy group) followed by hydrolysis under basic conditions and, finally, palladium-catalyzed hydrogenolytic removal of all the benzyl protecting groups (Scheme 55).

This convenient three-step synthesis allowed amino acids **186a–e** to be prepared in overall yields ranging from 48 (**186f**, cleonin) to 90% (**186b**) [9]. Since deprotection of benzyl esters may be achieved simultaneously with hydrogenolytic reduction of an azido group, amino acids **186c,f** were prepared from Michael adducts **87c**-Bn and **94b** in only two step reaction sequences (Scheme 55).

According to essentially the same protocol, racemic 2-[2′-(hydroxymethyl)-cyclopropyl]glycine (**189**) and spiropentylglycine (**191**) were also prepared in 31 and 25% overall yield, respectively (Scheme 56) [9, 21]. 3-Methylthio-3,4-methanoproline (**192**) has been synthesized along an analogous route as well [20b].

Racemic β-amino acids, namely 3,3-ethano-β-alanine (**193a**) and its spirocyclopropanated analog **193b**, have been prepared by simple one-step hydrogenolytic reduction and deprotection of the primary Michael adducts **94b** and **94d** (Scheme 57) [9, 21b]. As methyl esters cannot be deprotected under these conditions, only reductive dechlorination and debenzylation was observed for compounds **91t,u** (Scheme 57) [7k]. This looks promising for the construction of

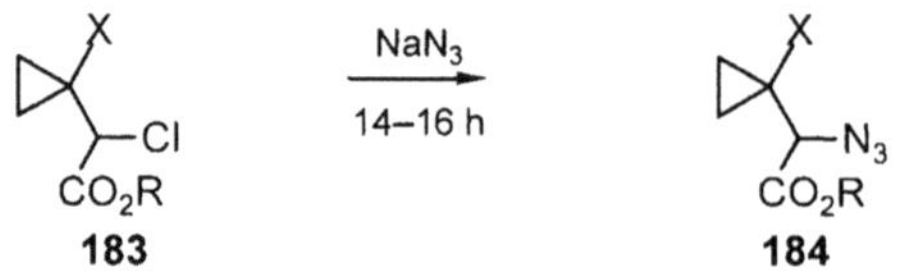

Starting Material	T [°C]	Conditions[a]	R	X	Product	Yield (%)
80a-Me	50	A	Me	H	**184a**	98
88f	55	A	Me	Me	**184b**	98
91a	20	B	Me	NBn_2	**184c**	89
88a	65	A	Me	CH_2CO_2tBu	**184d**	94
85a-Me	50	B	Me	SMe	**184e**	96
87c-Bn	20	B	Bn	OBn	**184f**	91
94b	20	A	Bn	$N{=}CPh_2$	**184g**	88[b]
91h	20	B	Bn	NBn_2	**184h**	91

[a] Conditions: A: DMF, NaI B: H_2O, Aliquat 336®. – [b] Reaction over 14 d.

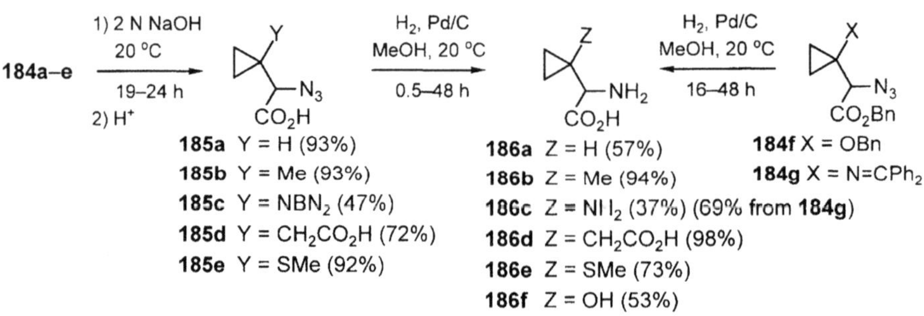

Scheme 55. Three-step preparation of 1′-substituted 2-cyclopropylglycines **186** from Michael adducts onto chloro esters **1**-R [9]

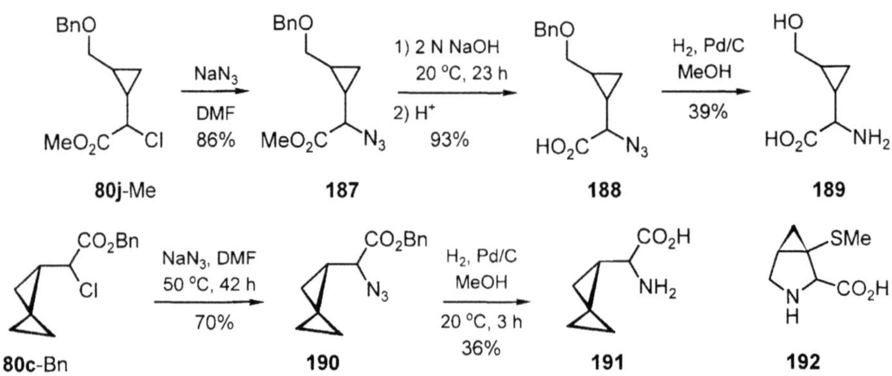

Scheme 56. Preparations of 2-[2′-(hydroxymethyl)cyclopropyl]glycine (**189**) and spiropentylglycine (**191**) [9, 21]

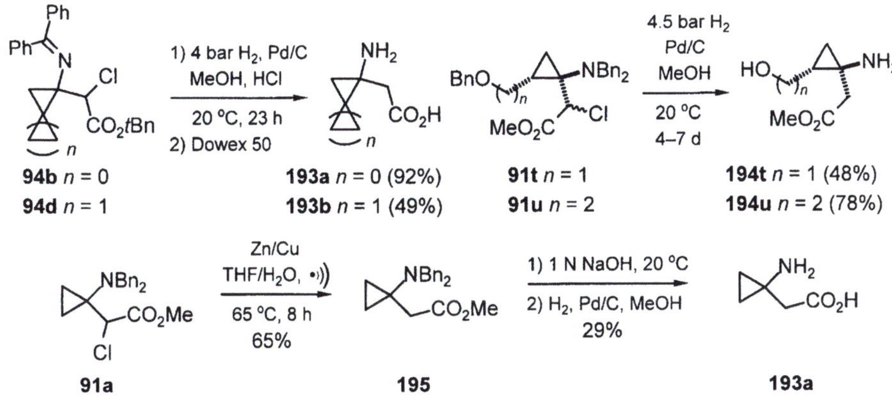

Scheme 57. Preparation of 3,3-ethano-β-alanine (**193a**) and its substituted analogs [7k, 9, 21b]

skeletons like that in the 3,4-methanoprolin derivative **192** from compounds of type **194**.

A less efficient approach to the β-amino acid **193a** included initial reductive removal of the chlorine followed by hydrolysis of the ester and hydrogenolytic deprotection of the amino group (Scheme 57) [9]. Although the overall yield of **193a** was only 19%, this strategy appeared to be appropriate also for application towards the Michael adducts of substituted 2-chloro-2-cyclopropylideneacetates **2**-R (see below).

An essentially analogous set of chemical transformations allowed to prepare enantiomerically pure amino acids of this type with substituents on the cyclopropane moiety. The success strongly depended on the appropriate choice of the starting material, and the best results were obtained using Michael adducts of chloroesters **2**-R with (4*R*,5*S*)-4,5-diphenyloxazolidine-2-one (**100**) which turned out to be a suitable chiral ammonia equivalent permitting to achieve good diastereoselectivities with respect to the stereogenic center α to the alkoxycarbonyl group (Scheme 31). The synthetic strategy shown in Scheme 55 being applied to the preparation of 2′-substituted (1′-aminocyclopropyl)glycines failed, as the nucleophilic substitution in **102** with azide proceeded with total racemization and unsatisfactory yields [10b, c, 62]. However, reductive removal of the chlorine with zinc/copper couple (Scheme 58) followed by electrophilic azidation of the ester enolate with trisyl azide according to a protocol of Evans et al. [99] allowed to introduce the protected α-amino group indirectly with high diastereoselectivity for the newly formed stereogenic center at C-2 with a 2*R*/2*S* ratio of ≥19:1 (the absolute configuration was wrongly depicted in the Schemes of the original publication [62]).

For the unsubstituted compound **197a**-Me the 2*R*/2*S* ratio was only 5.6:1 [10b, c], yet in this case the absolute configuration was checked by X-ray crystal structure analysis. The α-azidoesters **197**-R^2 were converted to the α,β-diamino acids **198**-R^2 as described above, i.e. by catalytic hydrogenation over palladium on charcoal and hydrolysis of the ester group (Scheme 58) [10b, c, 62]. Saponification was not necessary for the benzyl esters **197a, b, i**-Bn.

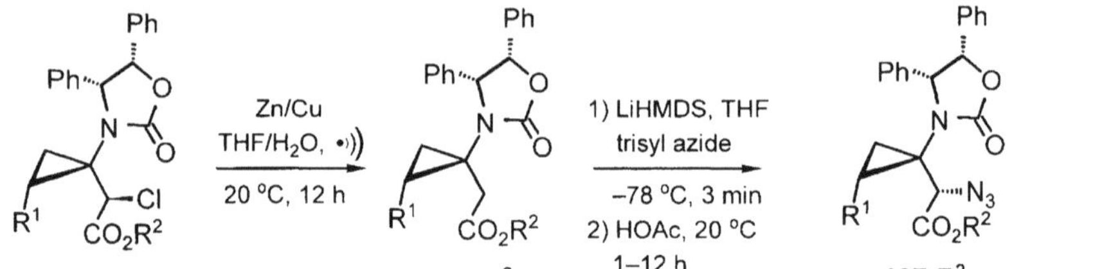

102-R^2 **196**-R^2 **197**-R^2

Starting Material	R^1	R^2	Yield of **196**-R^2 (%)^a	Yield of **197**-R^2 (%)^a	Yield of **198** (%)^a	
102a-Me	H	Me	87	87^b	56	**a**
102b-Bn	Et	Bn	77/86	50/63	73/96	**b**
102b-Me	Et	Me	92/91	74/85	74/85	**b**
102i-Bn	Me	Bn	89/88	79/65	82/30	**i**
102i-Me	Me	Me	83/85	66/71	79/66	**i**
102j-Me	$BnOCH_2$	Me	80/71	84/61	60/–^{c}	**j**
102m-Bn	$MeOCH_2$	Bn	61/84	–	–	**m**
102m-Me	$MeOCH_2$	Me	84/89	83/65	–/83	**m**
102r-Me	$MeSCH_2$	Me	91	–	–	**r**
102t-Me	$CbzNHCH_2$	Me	88	–	–	**t**

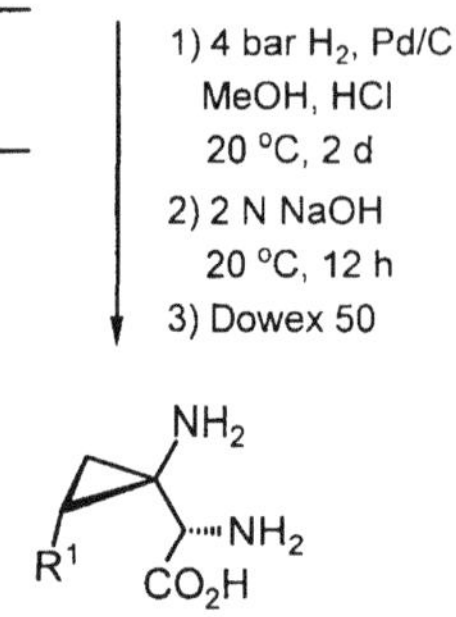

198

^{a}Yields for products obtained from both diastereomers of **102**-R^2. – ^{b}Before azidation, methyl ester **196a**-Me was converted into the benzyl ester **196a**-Bn in 90% yield using a common procedure (Scheme 4). – ^{c}R^1 = CH_2OH

Scheme 58. Preparation of enantiomerically pure 2′-substituted (1′-aminocyclopropyl)glycines **198** from the Michael adducts **102** of (4*R*,5*S*)-4,5-diphenyloxazolidine-2-one (**100**) onto chlorocyclopropylideneacetates **2**-R [10b, c, 62]

The absolute configurations of amino acids **198a** (the racemic counterpart was designated **186c**) and **198i** on the basis of X-ray crystal structure analysis data of the α-azido esters **197a**-Me and **197i**-Bn were (*R*) and (*S*,*S*,*S*), respectively.

Besides the large number of α-amino acids found in nature, β-amino acids are gaining an ever increasing attention [100]. β-Amino acids are e.g. found in the side chain of the cancerostatic taxol [101] as well as the new antibiotics sperabillin and TAN 1057 A/B [102]. The simple deprotection of 2′-substituted cyclopropylacetic acid derivatives **196**-R^2 by hydrogenolysis (and hydrolysis for the methyl esters **196b, i, j, m**-Me) yielded enantiomerically pure 2-(1′-aminocyclopropyl)acetic acids (Scheme 59) [10].

The absolute configuration was determined only for the ethyl- and the methyl-substituted derivatives **200b, i** as being (*S*,*S*) by X-ray crystal structure analysis of their precursors **102b, i**-Bn [10b, c]

The α-azidoesters **197**-Me are not only precursors for α,β-diamino acids **198** but could also be used to prepare 2′-substituted aminocyclopropanecarboxylic acid (ACC) derivatives **203** [10b, c, 62] (Scheme 60). Upon treatment with freshly prepared lithium methoxide in THF and subsequent hydrolysis (HCl), the α-

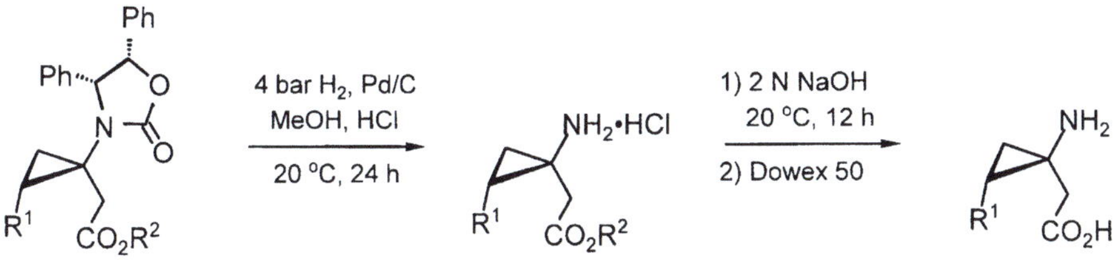

196-R^2 **199**-R^2 **200**

Starting Material	R^1	R^2	Yield of **199**-R^2 (%)[a]	Yield of **200** (%)[a]	Product
196b-Bn	Et	Bn	–	95/94	**200b**
196b-Me	Et	Me	95/95	67	**200c**
196i-Bn	Me	Bn	–	95	**200d**
196i-Me	Me	Me	95	74	**200e**
196j-Me	$BnOCH_2$	Me	95	74[b]	**200f**
196m-Bn	$MeOCH_2$	Bn	–	91	**200g**
196m-Me	$MeOCH_2$	Me	95/95	80	**200h**
196r-Me	$MeSCH_2$	Me	no reaction	–	–
196t-Me	$CbzNHCH_2$	Me	95[c]	–	–

[a]Yields of products obtained from both diastereomers of **196**-R^2. – [b]$R^1 = CH_2OH$. – [c]$R^1 = CH_2NH_2{\cdot}HCl$

Scheme 59. Preparation of enantiomerically pure 2-(1′-aminocyclopropyl)acetic acids **200** by deprotection of 2′-substituted cyclopropylacetic acid derivatives **196** [10]

azido ester **197**-Me yielded the corresponding *α*-keto esters and further the free *α*-keto acids **201** by treatment with LiOH [103]. Oxidative decarboxylation of the *α*-keto acids **201** was achieved with lithium hydroxide/hydrogen peroxide in THF/water to give the ACC derivatives **202**, which were also deprotected by catalytic hydrogenation using 10% palladium on charcoal, to afford 2-substituted 1-aminocyclopropanecarboxylic acids **203**.

Using this approach, the naturally occurring amino acids coronamic acid **203b**, the optical antipode of norcoronamic acid **203i** and the non-naturally occurring, but biologically active 1-amino-2-(hydroxymethyl)cyclopropanecarboxylic acid (**203j**) [77, 93j–m] have been prepared [10b,c, 62], and the ACC (**206**) itself was synthesized by an analogous approach as well (Scheme 60) [10c]. The question, whether (*E*)-methanophenylalanine (**203**, R^1 = Ph), 2,3-methanomethionine (**203**, $R^1 = CH_2SMe$) and oligomethylated ACC's [77] may be prepared along this route, remains to be checked.

Starting from the Michael adducts of (4*R*,5*S*)-4,5-diphenyloxazolidine-2-one (**100**) onto chlorospiropentylideneacetates **2c** (**102c**-Bn and **102c**-Me) and juggling with the set of transformation discussed above, the spiropentane aminocarboxylic acids **207**, **208** and **213** were also prepared either in racemic (**208**) or enantiomerically pure (**207**, **213**) forms, the absolute configurations of which were determined on the basis of X-ray crystal structure analysis of the precursor

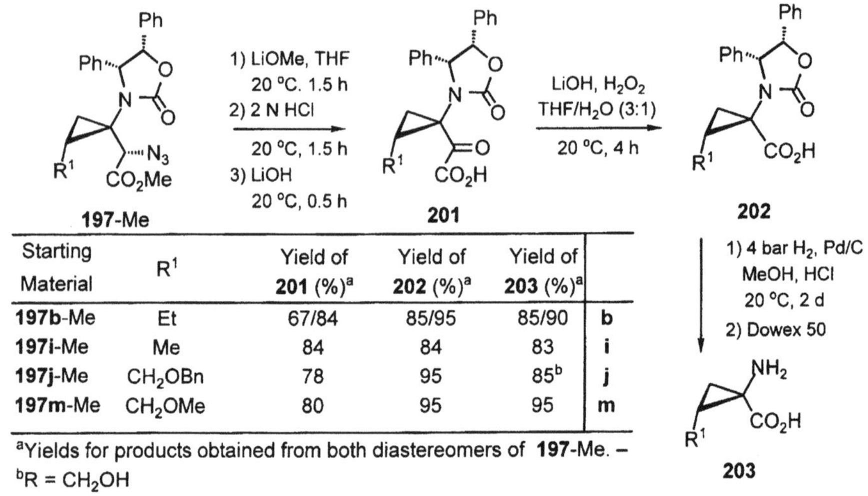

Starting Material	R^1	Yield of **201** (%)[a]	Yield of **202** (%)[a]	Yield of **203** (%)[a]	
197b-Me	Et	67/84	85/95	85/90	**b**
197i-Me	Me	84	84	83	**i**
197j-Me	CH_2OBn	78	95	85[b]	**j**
197m-Me	CH_2OMe	80	95	95	**m**

[a]Yields for products obtained from both diastereomers of **197**-Me. – [b]R = CH_2OH

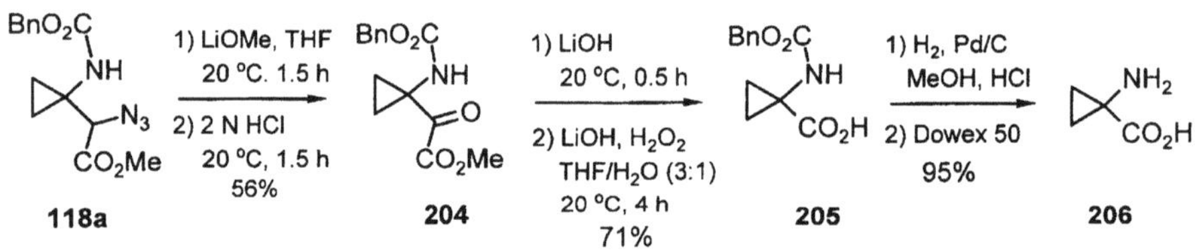

Scheme 60. Preparation of ACC (**206**) and its substituted enantiomerically pure analogs **203** [10b,c,62]

102c of the same absolute configuration or its antipode (Scheme 61) [21]. The two diastereomers of the amino acid **208** were (*R*)-configurated at the stereogenic center of the spiropentane moiety and (*R/S*) at C-2. The precursor **102c** with (*S*)-configuration at the spiropentyl center could not be reduced under the employed conditions to give the correspondingly configurated amino acid ester **209**, which constitutes a certain limitation of this approach. While the hydrochloride of **213** appeared to be rather stable, the free acid **213** when left in an aqueous solution at room temperature slowly isomerized with loss of its optical activity to form the new 1-amino-2-methylenecyclobutanecarboxylic acid **214** in racemic form.

3.4
MIRC and MIMIRC Reactions Leading to Spiropentanes, Tricyclo[3.2.1.0^{2,7}]octanes and Other Tricyclic Skeletons

The Michael addition followed by Intramolecular Ring Closure (MIRC) reactions have been recognized as a general synthetic approach to carbocyclic three-membered ring derivatives [1]. The enhanced Michael reactivity of methyl 2-chloro-2-cyclopropylideneacetate (**1**-Me) towards thiolates, alkoxides, lithiated amides and cyclohexadienolates (see below) allows one to perform highly efficient assemblies of spiropentane, tricyclo[3.2.1.0^{2,7}]octane, bicyclo[2.2.2]octane

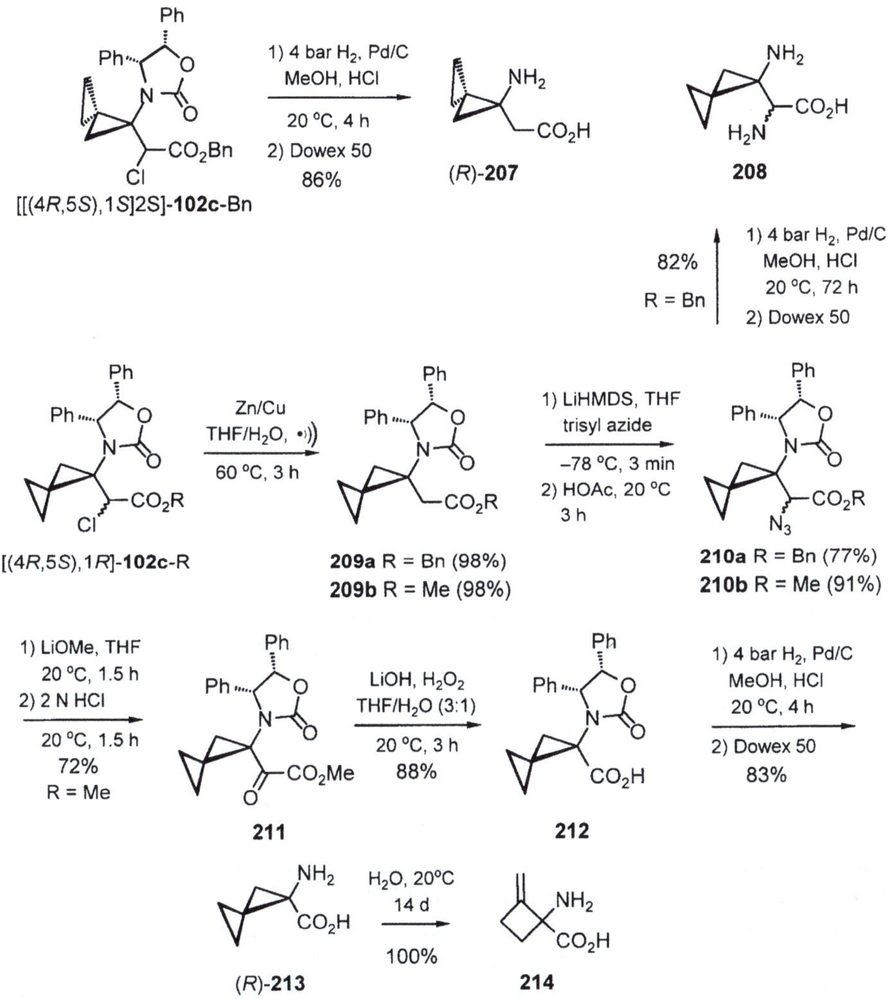

Scheme 61. Preparation of racemic (**208**) and enantiomerically pure (**207**, **213**) spiropentane aminocarboxylic acids from the Michael adducts **102** of (4*R*,5*S*)-4,5-diphenyloxazolidine-2-one (**100**) onto chlorospiropentylideneacetates **2c** [21]

and bicyclo[3.2.1]octane skeletons exploiting this MIRC mechanism [7h, l, 10c, 21a, 53, 104–108]. The typical MIRC reactions were observed in Michael additions of carbon nucleophiles containing acidic protons [109]. In the transformations of **1**-Me discussed here, the intramolecular nucleophilic substitution of chlorine by the newly formed carbon-nucleophilic center led to the formation of a second cyclopropane ring to yield spiropentane derivatives **215** and **216** as single diastereomers (the structure of **215** has been verified by X-ray crystal structure analysis) (Scheme 62) [104].

Albeit the yields were only low to moderate, these results are of principal importance, as they indicate the possibility of a rather general stereoselective synthetic approach to oligosubstituted spiropentanes (corresponding to the smal-

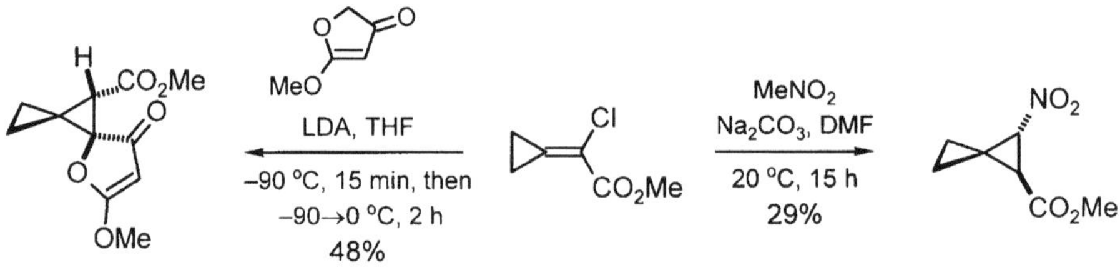

Scheme 62. Two MIRC reactions of methyl 2-chloro-2-cyclopropylideneacetate (1-Me) [104]

lest members of the triangulane family [110]) which are not available along any other synthetic routes.

Under strictly aprotic conditions, the enolate formed upon Michael addition of a nucleophile to the chloro ester **1**-Me, cannot be protonated and will therefore add to another molecule of the chloro ester **1**-Me to give an adduct which can close a three-membered ring by intramolecular nucleophilic substitution. In analogy to the known abbreviation MIRC, such a sequence of transformations is called a Michael Michael Addition Ring Closure (MIMIRC) reaction [107]. Thus, under aprotic conditions, two equivalents of the α-chloro acrylate compound **1**-Me react with one equivalent of a Michael addend to form spiropentane derivatives. Incidentally, when methyl 2-chloro-2-cyclopropylideneacetate (**1**-Me) was treated with half an equivalent of lithium benzylamide (**217a**) generated from benzylamine and butyllithium, the unique 3-chlorospiropentanecarboxylate with an annelated spirocyclopropane-*N*-benzyl-γ-butyrolactam **218** was isolated [21a, 53]. The sequence of events leading to the spiropentane moiety (Scheme 63) apparently starts with a Michael addition of **217a** to **1**-Me leading to the α-chloro enolate **220** which adds again to a molecule of **1**-Me to give the new α-chloro enolate intermediate **221**.

γ-Chlorine elimination from the latter produces a spiropentane derivative **222** in which the benzylamino substituent apparently attacks the appropriately placed methoxycarbonyl substituent in the γ-position to form a γ-lactam. This *N*-benzyl-3-aza-4-oxobicyclo[3.1.0]hexane-1-carboxylate **218** which also has a β-amino acid amide feature, was formed as a single product in 41% yield. Most probably, the other diastereomer of the intermediate **222** is also formed in the domino Michael-Michael-γ-elimination, but subsequently reacts intermolecularly to produce oligomeric amides. The structure of **218** was unequivocally established by an X-ray crystal structure analysis [21a, 53].

With sodium methoxide in THF, **1**-Me reacts analogously, but the sequence stops at the MIMIRC stage corresponding to **222**, i.e. the dimethyl 1-chloro-2-(1′-methoxycyclopropyl)spiropentane-*trans*-1,2-dicarboxylate (**219**).

In contrast to the behavior of unsubstituted formamide (**147**) towards **1**-Me (Scheme 47), its *N*-(phenylethyl)substituted analog **223** reacted with the chloro ester **1**-Me in the MIMIRC mode [10c] to give the spiropentane derivative **224** in 45% yield (Scheme 63).

Upon treatment of the spirocyclopropanated analog of the chloro esters **1**-Me, the 2-chloro-2-spiropentylideneacetate **2c**-Me, with sodium thiophenolate in

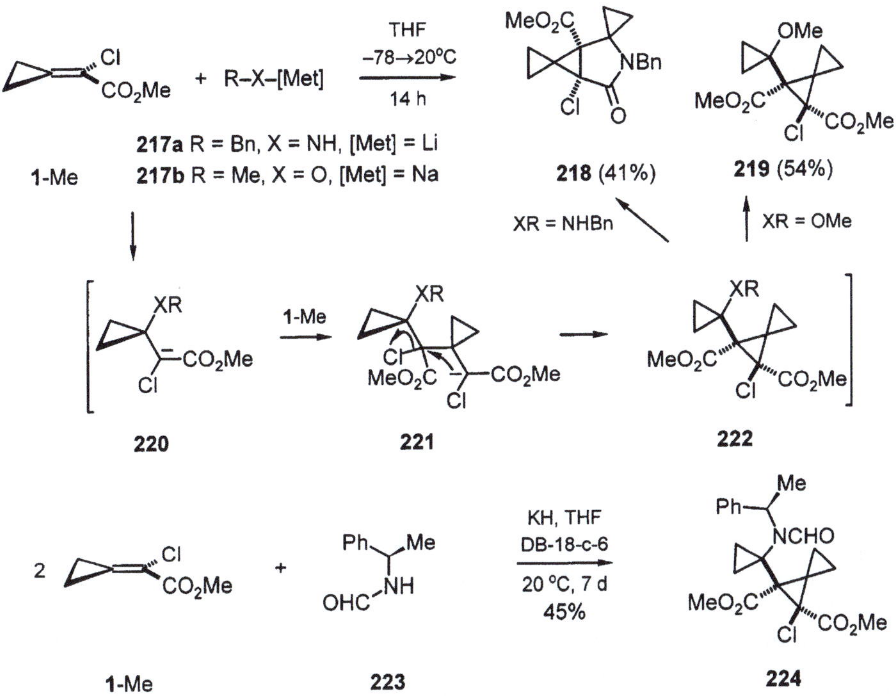

Scheme 63. The formation of spiropentane derivatives **218**, **219**, **224** via MIMIRC reactions of methyl 2-chloro-2-cyclopropylideneacetate (**1**-Me) [10c, 21a, 53]

anhydrous THF (Scheme 64) a reasonable yield (30%) of the spiropentyldispiro[2.0.2.1]heptanedicarboxylate **225b** was obtained as a single diastereomer with (1*S**,2*R**,1′*R**) configuration (according to X-ray crystal structure analysis [7 l]). While a 45% yield was obtained from the reaction of **1**-Me with sodium thiophenolate (Scheme 64) [22c], no MIMIRC product at all could be isolated from the reaction mixture of the chlorodispiroheptylidene acetate **2p**-Me: the normal Michael adduct of thiophenolate was obtained in high yield in this case (93%, Scheme 22).

MIMIRC reactions of an enone moiety generated in the first Michael addition were particularly efficient for the chloro ester **1**-Me. Thus, lithium cyclohexadi-

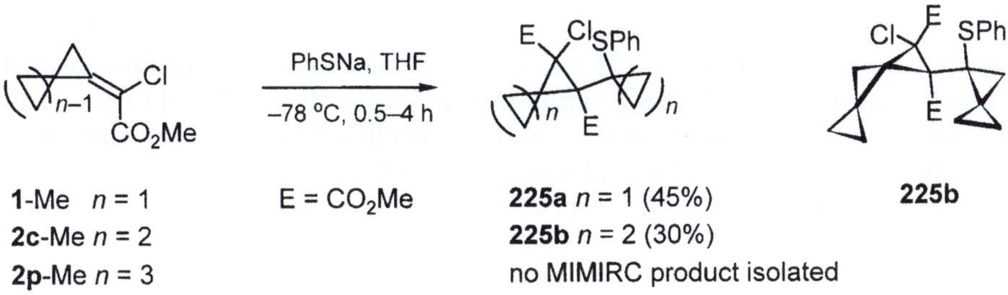

Scheme 64. MIMIRC reactions of the α-chloro acrylate **1**-Me and its spirocyclopropanated analogs **2c**-Me and **2p**-Me with sodium thiophenolate [7h, l, 22c]

enolates **226**, generated from the corresponding cyclohexenones and LDA [111], smoothly added to 2-chloro-2-cyclopropylideneacetate **1**-Me to give spirocyclopropanated tricyclic γ-keto esters **231** [28, 104b–107] in good to excellent yields (Scheme 65).

With lithium (*S*)-methoxymethylpyrrolidinide as a base, the tricyclic esters **231a** and **231e** were obtained in 32 and 85% yield with an enantiomeric excess of 30 and 22%, respectively [104b].

Mechanistically, the formation of the tricyclic γ-keto esters **231** can be rationalized in three different ways. Most probably, an intermolecular Michael addition of **226** to **1**-Me, followed by an intramolecular Michael addition in the enolate-enone **227** leads to **228** with an enolate moiety and a chlorine leaving group in the γ-position; intramolecular nucleophilic substitution then gives **231** (MIMIRC mechanism, see Scheme 65). The first two steps in this sequence would be reversible, the ring closure (γ-elimination) in **228**, which can only occur with the *exo*-chloroenolate shown, makes the overall transformation irreversible. The intermediate **228** could be also formed in an *endo*-selective [2+4] cycloaddition between **226** and **1**-Me. A third possibility would be that the α-chloroenolate intermediate, which is a carbenoid, would α-eliminate a chloride ion to yield carbene **229**, which would subsequently intramolecularly cycloadd onto the double bond of the enone moiety. When R^1 = H, Me as in **231a, b**, reductive cleavage of the C^1–C^7 bond (e.g. with lithium in liquid NH_3) can produce bicyclo[2.2.2]octane derivatives **230**. The 2-benzyloxy derivative **231g**, after reductive removal of the benzyl group by catalytic hydrogenation, undergoes a retro-aldol reaction opening the C^2–C^7 bond to quantitatively give the bicyclo[3.2.1]octanedione derivative **232**. The same reaction occurs when any of the tricyclic γ-keto esters **231** with an oxygen substituent at the bridgehead position, was treated with an acid (i.e. aqueous HCl, CF_3COOH, *p*-TsOH), affording quantitatively the dioxobicyclo[3.2.1]octanecarboxylates **232**.

The very good yields of especially the bridgehead alkoxy derivatives **231** and the excellent regioselectivity with which they can be transformed to other skeletons plus the fact that a spirocyclopropane moiety is a mimic of and can in fact be considered as a "masked" *gem*-dimethyl substituent [28, 112] makes these products versatile precursors of certain natural products. In fact, one may conceive new approaches to the total syntheses of taxol [113–115] and of mediterraneol [116]. Both strategies rely on the MIMIRC reaction of lithium cycloalkadienolates [117] with the α-chloro acrylate **1**-Me to produce a tricyclic precursor to the appropriate bicyclo[*n*.2.1]alkanedione derivative, which are key structural units of several diterpenes and their metabolites.

In spite of the fact that the only α-protons with respect to the carbonyl group in the tricyclic γ-keto esters **231** are at bridgehead positions and thus no real enolates can be formed [118], compounds **231** with R^1 = OR could easily and selectively be deprotonated at C-7, and the resulting lithium derivative then substituted with various electrophiles (Scheme 66).

Thus alkylation and silylation of **231i** could be brought about with methyl iodide and chlorotrimethylsilane in 61 and 66% yield, respectively [108]. In these tricyclic keto esters **231**, the carbonyl group at C–6 exerts its usual electron withdrawing effect and thereby increases the kinetic acidity of the adjacent cyclopropylic proton. Equally important, however, is the effect of the alkoxy group

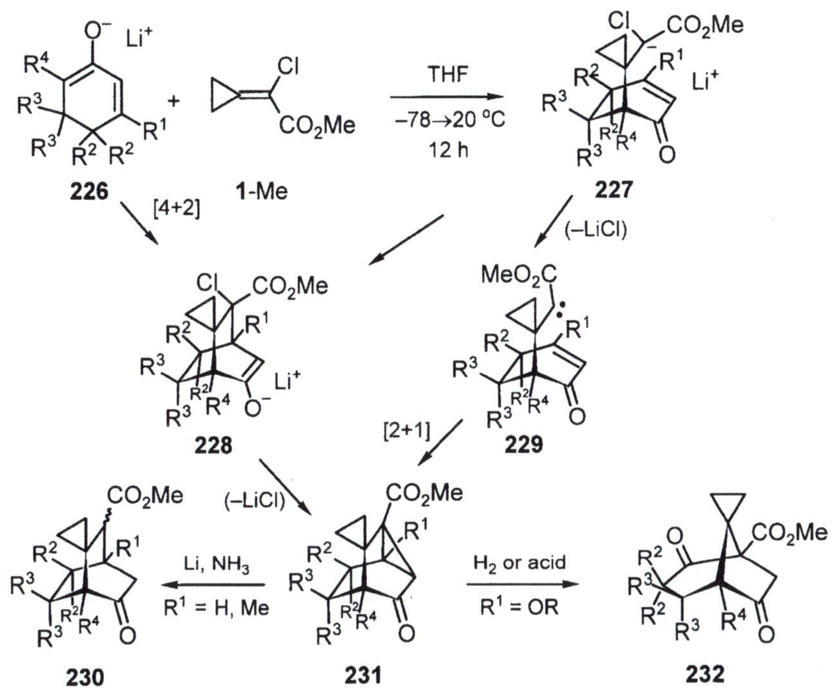

R^1	R^2	R^3	R^4	Product	Yield (%)	Ref.
H	H	H	H	**231a**	34	[107]
Me	H	H	H	**231b**	68	[28]
H	Me	H	H	**231c**	52	[28]
Me	H	Me	H	**231d**	75	[28]
MeO	H	H	H	**231e**	59	[106]
EtO	H	H	H	**231f**	60	[108]
BnO	H	H	H	**231g**	92	[28]
TMSO	H	H	H	**231h**	32	[28]
MeO	H	Me	H	**231i**	83	[108]
EtO	H	Me	H	**231j**	58	[108]
*i*PrO	H	Me	H	**231k**	71	[108]
*n*BuO	H	Me	H	**231l**	72	[108]
Allyl-O	H	Me	H	**231m**	66	[108]
MeO	H	H	Me	**231n**	82	[108]
*n*BuO	H	Me	Allyl	**231o**	61	[108]
H	a	H	H	**231p**	58	[104b]
H	b	H	H	**231r**	55	[104b]

[a] NCO_2Me fragment instead of CR^2_2. – [b] NCO_2Bn fragment instead of CR^2_2.

Scheme 65. MIMIRC reactions of lithium cyclohexadienolates **226** to methyl 2-chloro-2-cyclopropylideneacetate (**1**-Me) [28, 104b, 106–108]

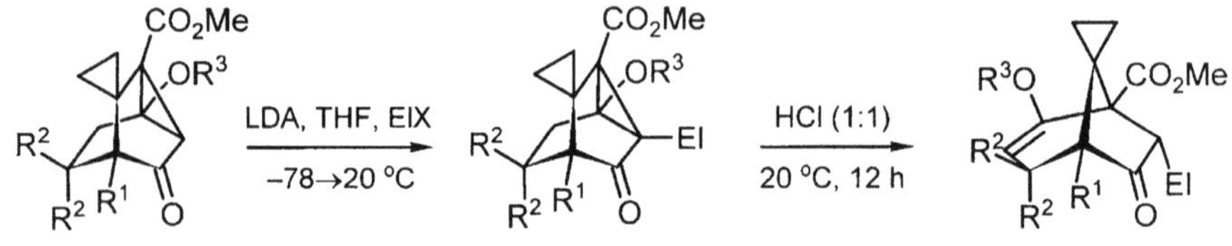

Starting Material	Electrophile	R^1	R^2	El	Product	Yield (%)
231i	MeI	H	Me	Me	**233ia**	61
231i	Me_3SiCl	H	Me	Me_3Si	**233ib**	66
231f	$(PhS)_2$	H	H	PhS	**234fa**	13
231n	$(PhS)_2$	Me	H	PhS	**234na**	73
231i	$(PhS)_2$	H	Me	PhS	**234ia**	69
231i	PhCOCl	H	Me	PhCO	**234ib**	46
231i	$4\text{-}MeOC_6H_4COCl$	H	Me	$4\text{-}MeOC_6H_4CO$	**234ic**	43
231i	$2,4\text{-}(MeO)_2C_6H_3COCl$	H	Me	$2,4\text{-}(MeO)_2C_6H_3CO$	**234id**	65
231i	$2,5\text{-}(MeO)_2C_6H_3COCl$	H	Me	$2,5\text{-}(MeO)_2C_6H_3CO$	**234ie**	35

R^3 = Me except **231f, 234fa** (R^3 = Et)

Scheme 66. Deprotonation of the tricyclic γ-keto esters **231** followed by trapping with an electrophile [108]

on the cyclopropane ring, as it can stabilize the lithiated species by chelation [119]. However, LDA deprotonation of **231**, followed by quenching with diphenyl disulfide or aroyl chloride and acidic workup gave moderate to good yields of *endo*-7-phenylthio- **234fa**, **234na**, **234ia** and *endo*-7-aroylbicyclo[3.2.1]oct-2-en-6-one derivatives **234ib–ie**. Presumably the C^2–C^7 cyclopropyl bond in the highly substituted tricyclic product breaks so easily that it opens up even under the aqueous work-up conditions at ambient temperature. The assignment of the *endo*-configuration for **234** was confirmed by X-ray analysis of **234ia** [108]. LDA deprotonation of the tricyclic γ-keto ester **231i**, followed by quenching with an aldehyde, afforded the expected aldol products **235** with good yields and diastereoselectivities (Scheme 67).

In all cases the reactions proceeded smoothly at –78 °C and the alcohols **235** with an (*R*)-configuration at the newly formed stereogenic center were the major diastereomers in all cases. When the reaction mixtures were allowed to reach room temperature, mixtures of the alcohols **235** and the lactones **236** were isolated instead. The lactones **236** were exclusively obtained when the corresponding alcohols **235** were treated with potassium *tert*-butoxide. On the other hand, treatment of the alcohols **235** with dilute hydrochloric acid afforded mixtures of the alkenes **237** and the tetracyclic compound **238**, the structure of which was confirmed by X-ray analysis.

Organolithium compounds add to the tricyclic γ-keto ester **231** from the *exo* side to give the *endo*-alcohols **239** in moderate to excellent yields (Scheme 68) [108]. Treatment of the alcohols **239** with concentrated hydrochloric acid afford-

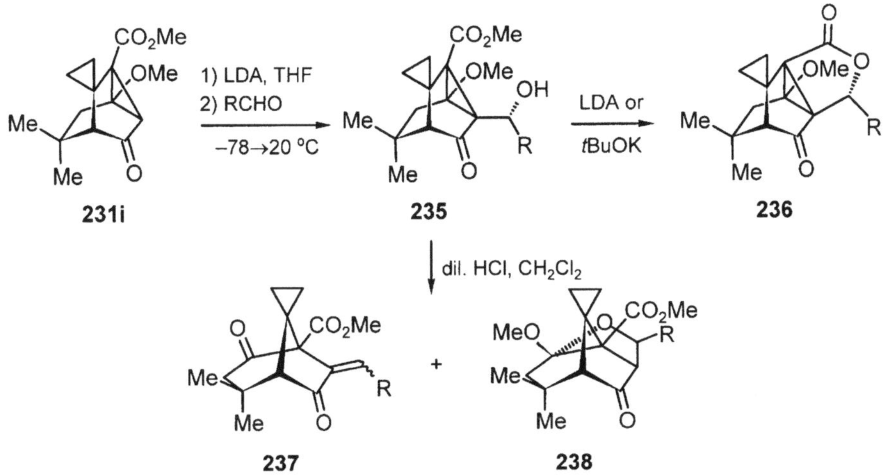

Aldehyde RCHO R	Product	Conversion (%)	Yield (%)	Diastereomer (1'-*R*/1'-*S*)
*c*Hex	**235a**	75	58	86:14
2-Furyl	**235b**	92	67	87:13
Ph	**235c**	80	64	90:10
2-Me-C_6H_4	**235d**	88	71	95:5
3-Me-C_6H_4	**235e**	92	71	90:10
2-MeO-C_6H_4	**235f**	82	64	93:7
3-MeO-C_6H_4	**235g**	92	72	96:4
4-Me-C_6H_4	**235h**	93	81	96:4
4-MeO-C_6H_4	**235i**	92	73	93:7
4-NO_2-C_6H_4	**235j**	93	64	96:4
2,5-$(MeO)_2$-C_6H_3	**235k**	87	58	86:14

Scheme 67. Additions of the lithiated tricyclic γ-keto ester **231 i** to aldehydes and subsequent transformations of the products [108]

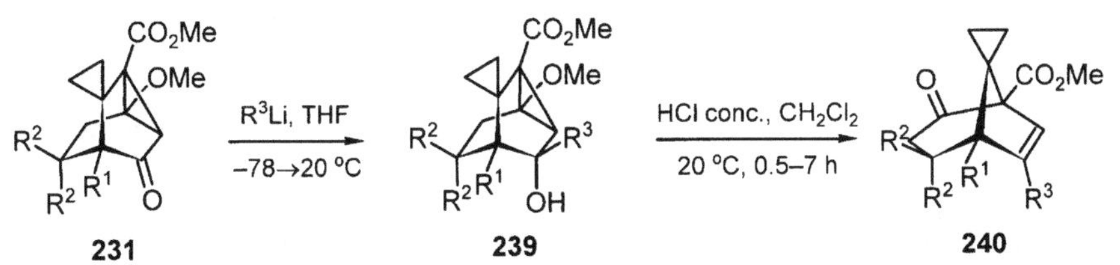

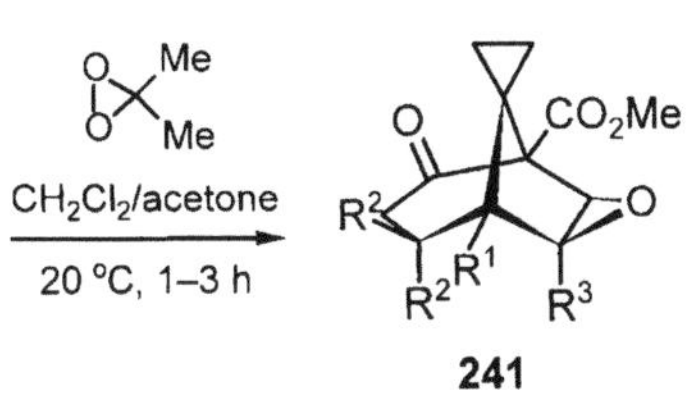

Starting Material	R^1	R^2	R^3	Yield of **239** (%)	Yield of **240** (%)	Yield of **241** (%)
231i	H	Me	Ph	66	86	98
231n	Me	H	Me	95	71	92

Scheme 68. Transformations of the tricyclic γ-keto esters **231** [108]

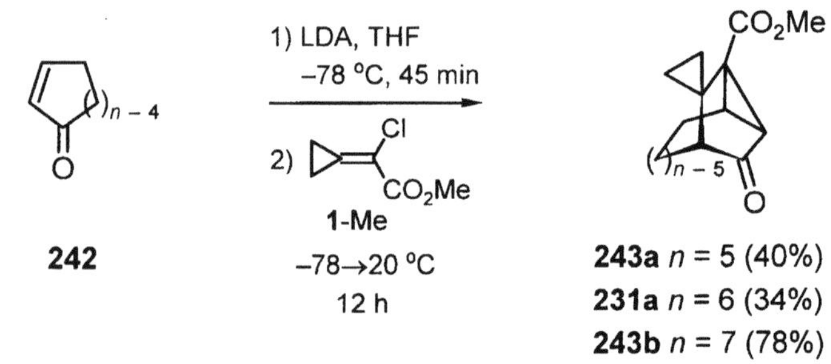

Scheme 69. MIMIRC reactions of cyclic dienolates of different ring size with the chloro ester **1**-Me [104b, 107]

ed the bicyclo[3.2.1]oct-6-en-2-one derivatives **240**, in contrast to expectations based on literature precedence [117]. As could be anticipated, epoxidation of the bicyclooctenones **240** with dimethyldioxirane gave the *exo*-epoxides **241** in excellent yields. However, treatment of **240** with *m*-chloroperbenzoic acid, even under buffered conditions, led to complete decomposition.

Cycloalkadienolates with other than six-membered ring sizes reacted with the chloro ester **1**-Me in the same manner (Scheme 69) [104b, 107], and the yields of the resulting unsubstituted tricyclic esters **243 a, b** were even better than that of **231 a** from the unsubstituted cyclohexadienolate [107].

The Wittig olefination of the tricyclic γ-oxo ester **231f** affords the 6-methylene tricycle **244**. Treatment of the latter with diluted hydrochloric acid results in the formation of the 6-methyl-2-oxobicyclo[3.2.1]oct-6-ene derivative **240f** in 94% yield (Scheme 70) [106, 108]. This transformation may be realized also in a one-pot procedure giving compound **240n** in 70% yield (Scheme 70) [104b]. The overall yield in the step-by-step preparation of this compound, as shown in Schemes 65 and 68, was only 55% starting from the chloro ester **1**-Me.

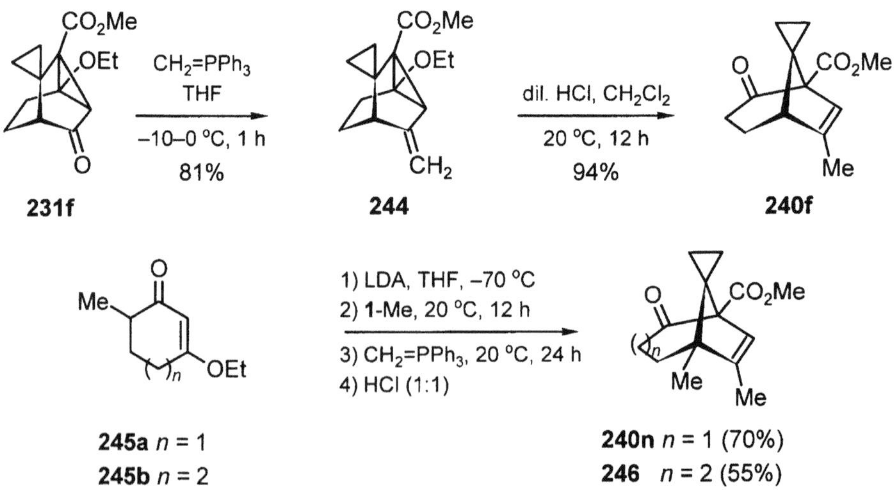

Scheme 70. Step-by-step and one pot preparation of the enones **240f, n** and **246** via Wittig olefination [104b, 106, 108]

As the frames of this review do not allow the coverage of the rich and attractive chemistry of polycycles like **231**, **243** exhaustively, the original publications [104b–108] may be recommended for the first-hand acquaintance. The last achievement in this field was a recently reported preparation of the dione **232a** and the enone **240f** from methyl 2-chloro-2-cyclopropylideneacetate **1**-Me in 79 and 58% yield, respectively, performed on polymer support [120].

3.5 Miscellaneous

So-called domino or cascade reactions have become more and more important for the efficient synthesis of complex organic molecules [121]. In this respect methyl 2-chloro-2-cyclopropylideneacetate (**1**-Me) has been used as a dienophile to trap cyclic dienes which were produced by intramolecular Heck reactions in Diels-Alder cycloadditions. Thus, the spirocyclopropanated functionalized bicyclo[4.3.0]nonenes **248**, **250** (Fig. 11) were obtained from the bromodiene **247** or enynes **249** in 56–83% yield (Scheme 71) [122, 123].

The application of this synthetic approach to bromodialkenyl amines **251** results in a novel one-pot synthesis of tetrahydroisoindolines **252** (Scheme 72) [124].

Methyl 2-bromo-2-cyclopropylideneacetate (**11a**) has never been tested in these reactions, but has been used as a starting material for the stepwise construction of 1,6-heptadienes with methylenecyclopropane units for intramolecular Heck reactions. Thus, bromo ester **11a**, after reduction, subsequent conversion of the resulting alcohol to the bromide and coupling with enolates of substituted malonates, was transformed into dienes of the type **254** (Scheme 73) – versatile synthetic blocks for the preparation of functionally substituted spirocyclopropanated bicyclo[4.3.0]nonenes **255a–d** by a domino Heck-Diels-Alder reaction [122a].

More examples of palladium-catalyzed reactions applied to compounds prepared from the chloroester **1**-Me are shown in Scheme 74.

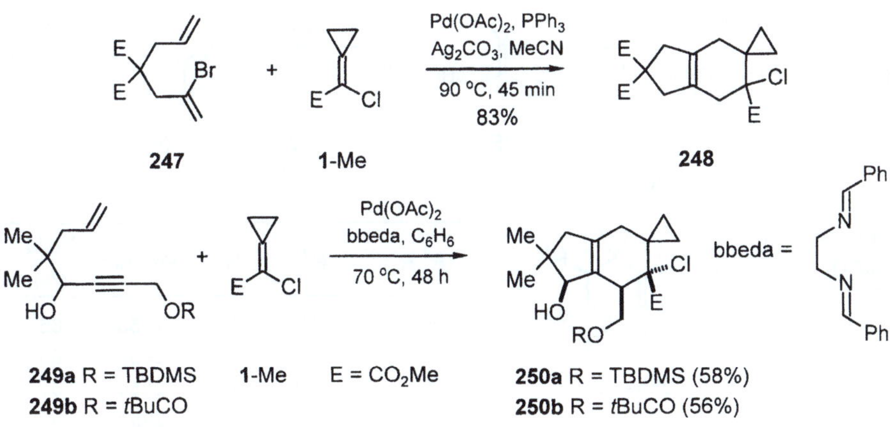

Scheme 71. The application of methyl 2-chloro-2-cyclopropylideneacetate (**1**-Me) in domino Heck-Diels-Alder chemistry [122, 123]

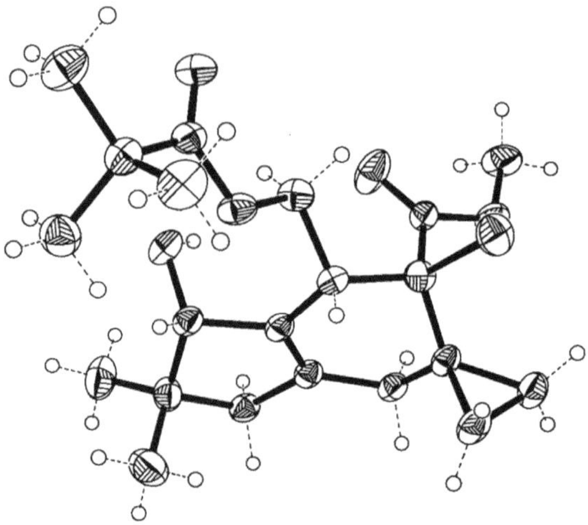

Fig. 11. Structure of the spirocyclopropanated bicyclo[4.3.0]nonene derivative **250b** in the crystal [123]

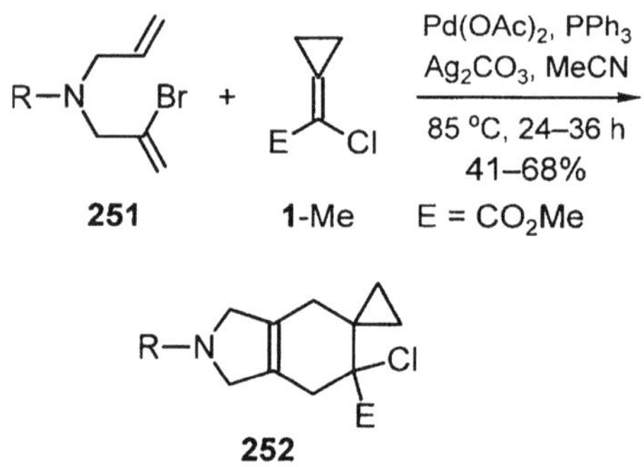

Starting Material	R	Product	Yield (%)
251a	CHO	**252a**	55
251b	MeCO	**252b**	68
251c	Boc	**252c**	43
251d	Ts	**252d**	44

Scheme 72. Methyl 2-chloro-2-cyclopropylideneacetate (**1**-Me) in a Heck-Diels-Alder reaction of bromodialkenyl amines **251** [124]

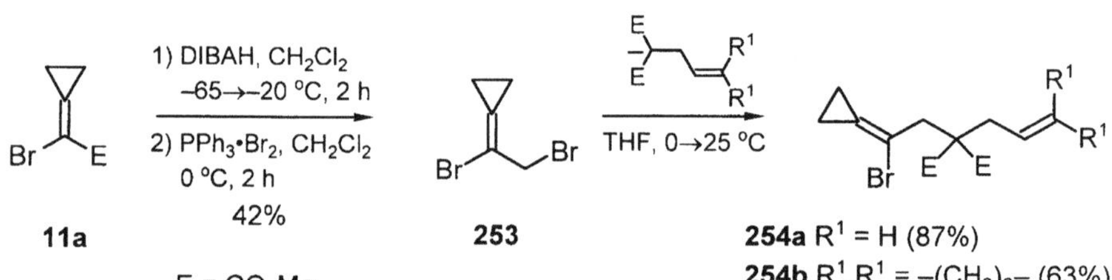

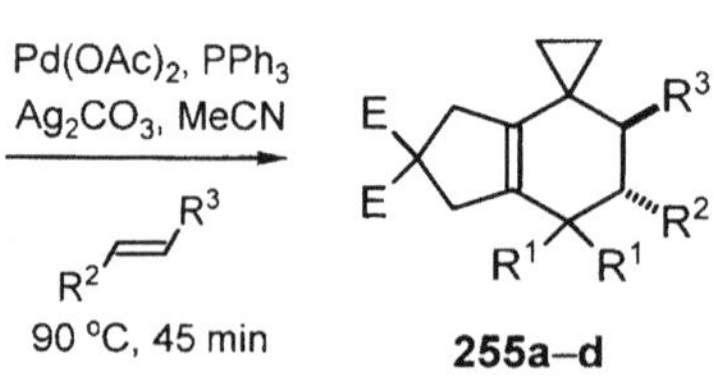

R^1 (R^1,R^1)	R^2	R^3	Product	Yield (%)
H	COMe	H	**255a**	86
H	CO_2Me	H	**255b**	86
H	CO_2tBu	H	**255c**	68
–$(CH_2)_2$–	CO_2Et	CO_2Et	**255d**	30

Scheme 73. Intramolecular Heck reaction of the dienes **254** prepared from methyl 2-bromo-2-cyclopropylideneacetate (**11a**) [122]

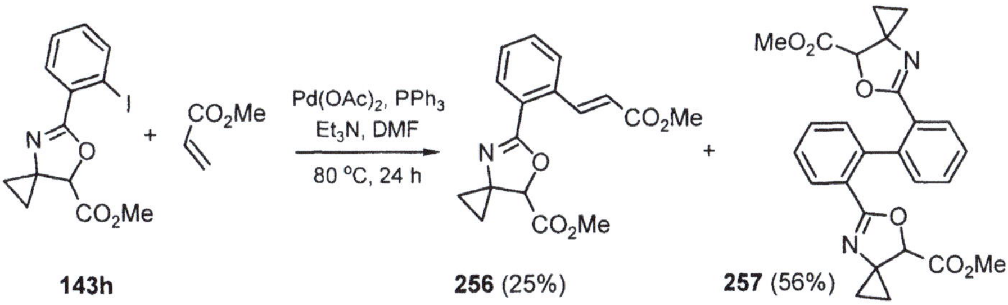

143h **256** (25%) **257** (56%)

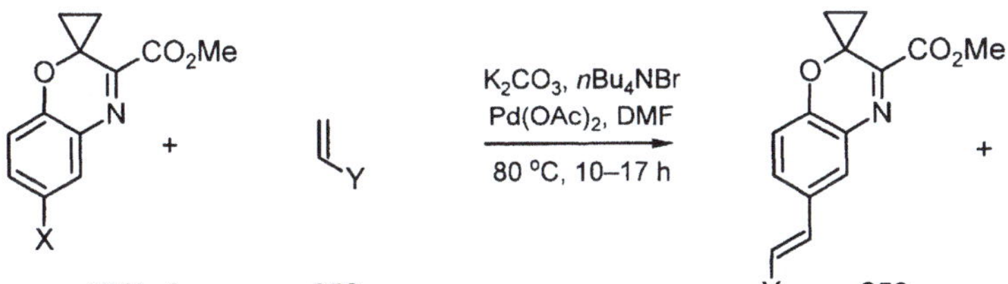

172h, i **258** **259**

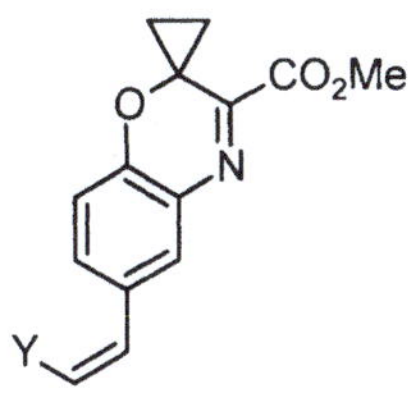

260

X	Y	Product	Yield (%)	Yield of **260** (%)
Br	CO_2Me	**259a**	20	0[a]
I	CO_2Me	**259a**	86	0
I	CHO	**259b**	89	0
I	C(O)Me	**259c**	72	0
I	CN	**259d**	63	24
I	Ph[b]	**259e**	72	0[a]
I	OEt[b]	**259f**	45	15

[a]In the presence of nBu$_3$N/PPh$_3$, the three-membered ring in **260** underwent ring opening to the propene fragment [73]

[b]8–35% of different products were also formed

Scheme 74. Palladium-catalyzed further transformations of heterocycles prepared from methyl 2-chloro-2-cyclopropylideneacetate (**1**-Me) [53, 73]

In an attempt to perform a Heck coupling of the iodophenyloxazoline **143h** with methyl acrylate, the bis(phenyloxazolinecarboxylate) **257** was isolated as the main product along with 25% of the target product **256** [53]. On the other hand, Heck couplings of the halobenzoxazines **172h,i**, particularly the iodo derivative **172i**, with a number of alkenes **258** went well and gave the corresponding products **259** in good to very good yields. In a few cases (Y=CN, OEt), the (*Z*)-isomers **260** were isolated along with the (*E*)-isomers (Scheme 74) [73].

Upon heating at 500°C or irradiation of a solution in acetonitrile with a low pressure mercury lamp, the chloro ester **1**-Me underwent a well-known [4] methylenecyclopropane-to-methylenecyclopropane rearrangement to give the interesting methylenecyclopropane derivative **261** along with an unidentified ring-enlarged product, presumably 3-chlorodihydrofuranone derivative **262** (Scheme 75) [7 m].

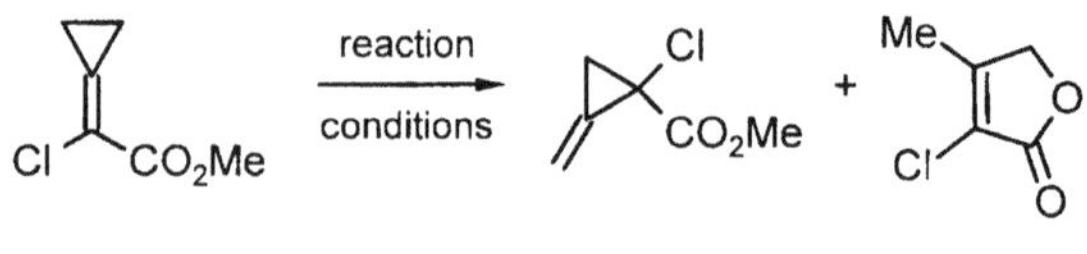

1-Me 261 262

Reaction Conditions	Yield of **261** (%)	Yield of **262** (%)
Pyrolysis at 500 °C	13	7
MeCN, hν (254 nm)	10	8

Scheme 75. Thermal and photochemical rearrangements of methyl 2-chloro-2-cyclopropylideneacetate (1-Me) [7m]

Two remarkable intramolecular reorganizations of the Diels-Alder adducts of methyl 2-chloro-2-cyclopropylideneacetate (**1**-Me) onto furans should be mentioned here. In a reaction of **1**-Me with 2,5-bis(trimethylsilyl)furan (**263**) the unstable adduct **264** underwent a spontaneous deprotection followed by ring opening to give the dioxospiro[2.5]octene derivative **265** (Scheme 76) [7 m]. The Pauson-Khand reaction of the transformed Diels-Alder adducts **70, 71** of **1**-Me have been discussed above (Scheme 18), however, when the compounds *endo*,-*exo*-**62e** were treated under Pauson-Khand conditions, but at higher temperature, the interesting $Co_2(CO)_8$-promoted stereoselective rearrangement, in the presence as well as in the absence of an alkyne component, has been observed. The *cis*- and *trans*-substituted 6-methylenespiro[2.4]hexanes **266** were isolated as main products in these reactions (Scheme 76) [19b].

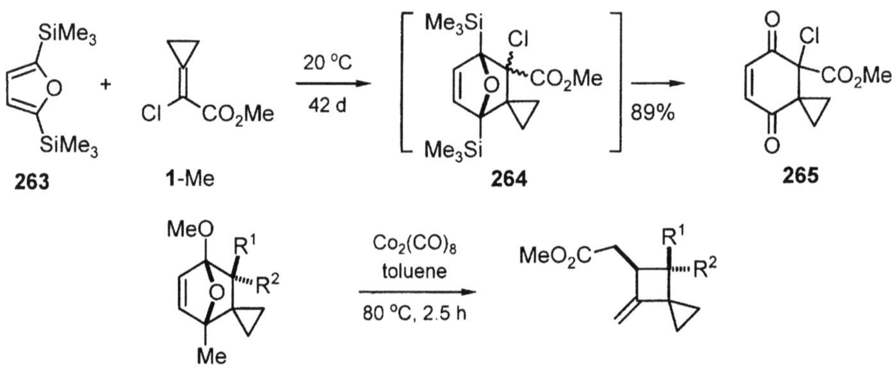

exo-**62e** R^1 = CO_2Me, R^2 = Cl
endo-**62e** R^1 = Cl, R^2 = CO_2Me

trans-**266** R^1 = CO_2Me, R^2 = Cl (41%)
cis-**266** R^1 = Cl, R^2 = CO_2Me (63%)

Scheme 76. Skeletal transformations of the Diels-Alder adducts **62e, 264** of methyl 2-chloro-2-cyclopropylideneacetate (**1**-Me) [7m, 19b]

4
Combinatorial Approaches to Libraries of Compounds Using Polymer-bound 2-Chloro-2-cyclopropylideneacetates

The construction of small molecule libraries by combinatorial chemistry carried out with polyfunctional reactive building blocks on a polymer support as a convenient way of rapidly generating a large number of related compounds has become a favorite methodology in the area of drug discovery [125]. Whereas peptide synthesis on polymer support has been comprehensively elaborated so that it may even be called routine synthetic methodology, many traditional chemical transformations still require further development before they can find general application in combinatorial chemistry both on a polymeric support or in solution [126]. One of the most important conditions for a synthetic sequence to be applicable in a combinatorial approach is an almost quantitative or at least very high yield in every step. In this respect the alkyl 2-chloro-2-cyclopropylideneacetates are predestined to be applied for the construction of small molecule libraries, since several types of their reactions have been demonstrated to proceed with very high yields, and there is ample reason to do so, since many cyclopropane-containing molecules exhibit various biological activities [77].

There are three straightforward possibilities of binding a chlorocyclopropylideneacetate molecule onto a polymer resin (Fig. 12): a) binding using an ester functionality as in **267**, b) binding by a functionality in a substituent on the three-membered ring as in **268** and c) binding by way of a Michael addition of a polymer-bound nucleophile onto the chloro esters **1**, **2** as in **269**.

Exploiting the known reactivities of chlorocyclopropylideneacetates **1**–**3**, the synthesis of a compound library from polymer-bound chlorocyclopropylideneacetic acid **1**-H has been realized in a sequence of transformations as shown in Scheme 77 [127], following attachment of the acid **1**-H to a resin (**A**), a Michael addition of nucleophile (**B**), nucleophilic substitution of the chlorine (**C**), modification of one of the substituents (**D**) and cleavage of the substrate to polymer bond (*E*).

Since the polymer resin should be resistant to basic conditions as applied for steps **B**–**D** and, on the other hand, permit to cleave off the final product under mild conditions, the acid **1**-H was bound to the acid sensitive resin TentaGel S AC **275** – a crosslinked polyethyleneglycol resin – which fulfilled all the requirements, using a Mitsunobu reaction [128] (Scheme 78).

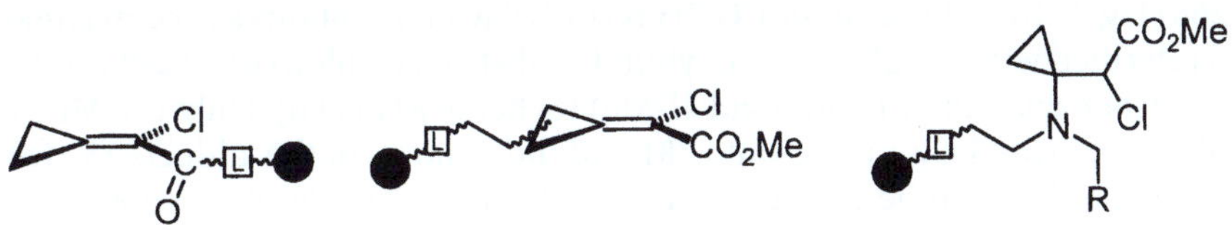

267 **268** **269**

Fig. 12. Three possibilities to bind the cyclopropylideneacetate building block to polymer support

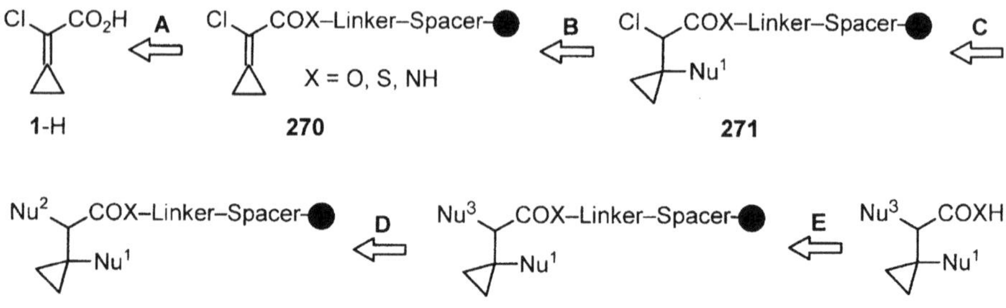

Scheme 77. Synthetic sequence to obtain a combinatorial library of cyclopropane derivatives from 2-chloro-2-cyclopropylideneacetic acid 1-H [11b, 127]

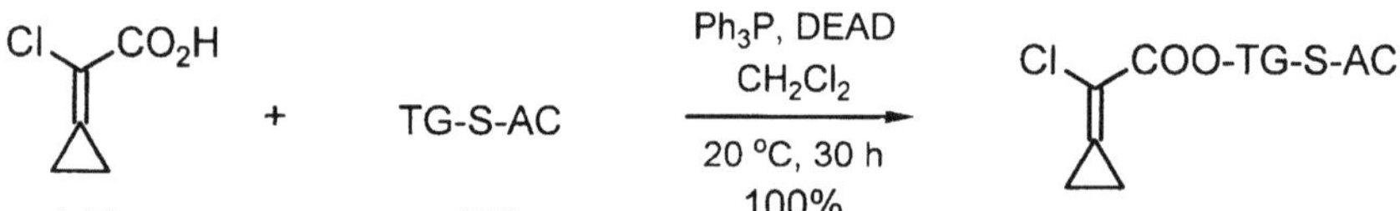

Scheme 78. Attachment of chlorocyclopropylideneacetic acid (1-H) to TentaGel S AC resin 275 [11b, 127]

The preparation of combinatorial libraries from the tentagel-bound chloro ester **276** was performed according to the generally accepted methodology (Scheme 79).

As synthetic steps, the Michael additions of nitrogen nucleophiles were followed by nucleophilic substitutions of the chlorine atom with a primary amine and, finally, alkylations of the then secondary amino group with various alkyl bromides were performed just as previously developed for the chloro ester **1**-Me in solution (see, e.g. Schemes 25, 27, 36 etc.). With differently substituted pyrazoles as Michael addends, different primary amines and alkyl bromides, combinatorial libraries consisting of 8, 24 and 84 compounds were thus successfully prepared in ca. 60% yield and proved by the LC-MS technique to contain all the individual compounds in about equal amounts (Scheme 80) [127].

According to essentially the same protocol, a bundle of cyclopropylidenepiperidine derivatives **278** consisting of 6 individual compounds was prepared as well [127].

To examine the second way of binding the chloro ester **1**-Me to a polymer as in **268** (Fig. 12), the TentaGel-S-COOH resin **279** which contains a succinyl linker, was esterified with the 2′-(4-hydroxybutyl)-substituted chloro ester **280** (prepared by debenzylation of compound **21**-Me) (Scheme 81) [11b]. Only the Michael addition of secondary amines onto **281** and the removal of the adducts **283** from the resin has been probed so far, any further transformations of polymer-bound compounds **281** need yet to be developed.

Another variant of a polymer-bound methyl 2′-(4-hydroxybutyl)-2-chloro-2-cyclopropylideneacetate **280** was prepared using a Merrifield resin containing a hydroxyethyl end group with a hydroxymethyldihydropyranyl (DHP) linker **284**

276

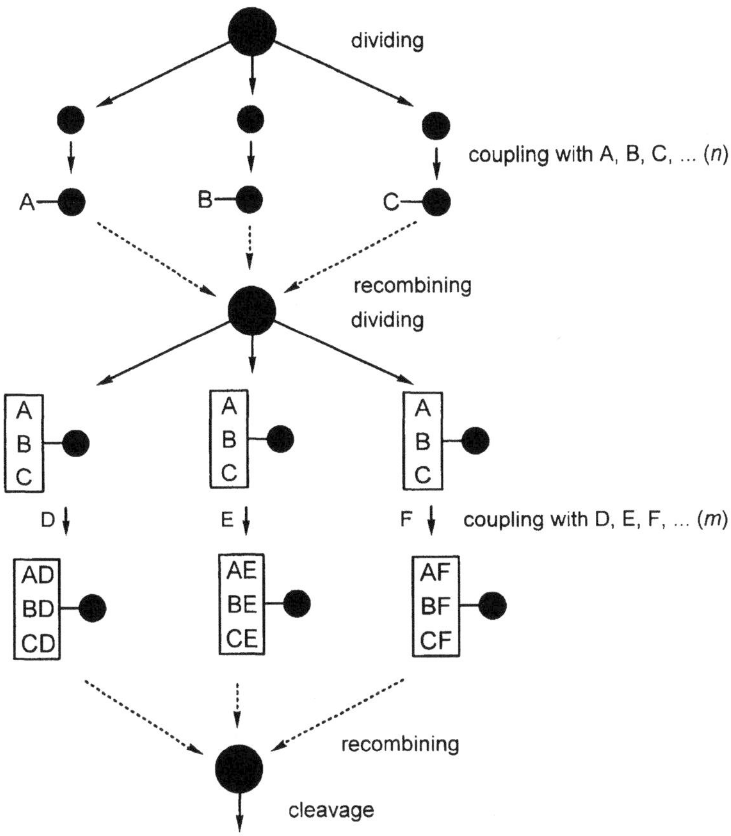

274 (*n* × *m* compounds)

Scheme 79. The split and combine method of preparing a combinatorial small molecule library

(Scheme 81) [11b]. Using the same type of resin **284**, the third approach as in **269** (Fig. 12) in which the binding to the polymer is achieved by help of an ethanolamino linker that adds to the 2-chloro-2-cyclopropylideneacetate (**1**-Me) in the first step of the reaction sequence has also recently been realized (Scheme 81) [11b].

5 Conclusion

In conclusion, the chemistry presented here illustrates the broad synthetic applicability of methyl 1-chloro-1-cyclopropylideneacetate (**1**-Me) and its derivatives. The remarkably efficient method for the syntheses of such compounds presented here, can undoubtedly be applied to generate a large variety of similar methylenecyclopropane derivatives with different substituents. Such chloro esters demonstrate unique reactivities and thereby have a wide potential as

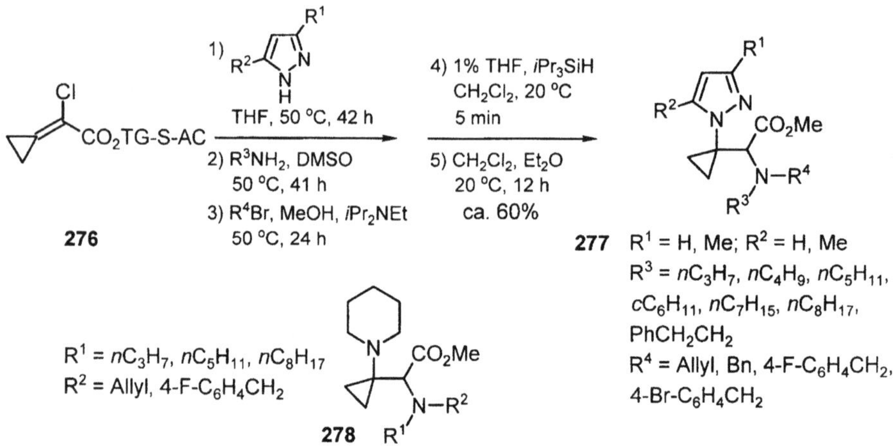

Scheme 80. Preparation of combinatorial libraries of new pyrazole **277** and piperidine **278** derivatives [127]

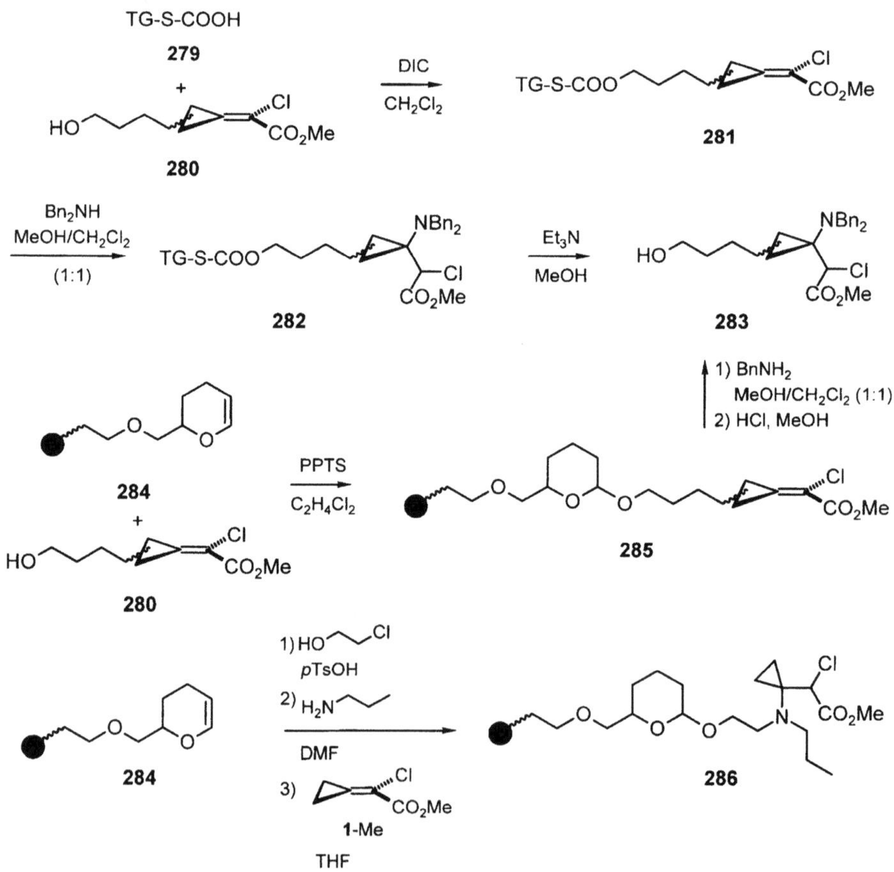

Scheme 81. Examinations of other possibilities to bind the chlorocyclopropylideneacetate building block onto a polymer support [11b]

building blocks for organic synthesis. Their chemical versatility illustrated above allows to conclude that further progress in this chemistry can and will be achieved towards the preparation of the naturally occurring and biologically active derivatives, especially applying them as the polyfunctional reactive building blocks in the modern strategy of preparations on a polymer support.

Acknowledgements. The work of our own group described herein has been supported by the Deutsche Forschungsgemeinschaft, the Volkswagen-Stiftung and the Fonds der Chemischen Industrie as well as BAYER, BASF AG, Chemetall GmbH, Degussa, Hoechst and Hüls AG through generous gifts of chemicals. A.d.M. is grateful to the group of dedicated and enthusiastic young chemists who - over the years - have made this research flourish. The authors are indebted Dr. B. Knieriem, Göttingen, for his careful reading of the manuscript and Mr. S. Beußhausen for his technical assistance.

References

1. de Meijere A (ed) (1997) Houben-Weyl, Thieme, Stuttgart, vol E 17a–c
2. a) de Meijere A (ed) (1986) Small ring compounds in organic synthesis I. Top Curr Chem 133. Springer, Berlin Heidelberg New York; b) de Meijere A (ed) (1987) Small ring compounds in organic synthesis II. Top Curr Chem 135. Springer, Berlin Heidelberg New York; c) Rappoport Z (ed) (1987) The chemistry of the cyclopropyl group. Wiley, Chichester; d) de Meijere A (ed) (1988) Small ring compounds in organic synthesis III. Top Curr Chem 144. Springer, Berlin Heidelberg New York; e) de Meijere A, Blechert S (eds) (1989) Strain and its implications in organic chemistry. NATO-ASI Series C, Kluwer, Dordrecht, vol 273; f) de Meijere A (ed) (1990) Small ring compounds in organic synthesis IV. Top Curr Chem 155. Springer, Berlin Heidelberg New York; g) Wong HNC, Hon M-Y, Tse C-W, Yip Y-C, Tanko J, Hudlicky T (1989) Chem Rev 89:165; h) de Meijere A (ed) (1996) Small ring compounds in organic synthesis V. Top Curr Chem 178. Springer, Berlin Heidelberg New York
3. Trost BM (1989) In: de Meijere A, Blechert S (eds) Strain and its implications in organic chemistry. NATO-ASI Series C, Kluwer, Dordrecht, vol 273, p 1 ff
4. a) Binger P, Büch HM, Top Curr Chem (1987) 135:77; b) Binger P, Schmidt T (1997) In: de Meijere A (ed) Houben-Weyl, Thieme, Stuttgart, vol E 17c, p 2217; c) Brandi A, Goti A (1998) Chem Rev 98:589; d) de Meijere A, Kozhushkov SI, Khlebnikov AF (1999) Top Curr Chem, this volume
5. For previous reviews of our own work in this area see: a) de Meijere A (1987) Chemistry in Britain 23:865; b) de Meijere A, Wessjohann L (1990) Synlett 20; c) Mißlitz U, de Meijere A (1989) In: Regitz M, Hanack M (eds) Houben-Weyl, Thieme, Stuttgart, vol 4, p 664 and 769; d) de Meijere A (1992) In: Yoshida Z, Ohshiro Y (eds) New Aspects of Organic Chemistry II. Kodansha-VCH, Tokyo-Weinheim, p 181
6. a) Stohlmeier M (1988) Dissertation, Universität Hamburg; b) Banwell MG, Corbett M, Gulbis J, MacKay MF, Reum ME (1993) J Chem Soc Perkin Trans 1:945
7. a) Weber A, de Meijere A (1980) Angew Chem 92:135 Angew Chem Int Ed Engl 19:138; b) Liese T, Splettstößer G, de Meijere A (1982) Tetrahedron Lett 23:3341; c) Liese T, de Meijere A (1982) Angew Chem 94:65 Angew Chem Int Ed Engl 21:65; d) Weber W, de Meijere A (1985) Chem Ber 118:2450; e) Liese T, de Meijere A (1986) Chem Ber 119:2995; f) Liese T, Jaekel F, de Meijere A (1990) Org Synth 69:144; g) Liese T (1983) Dissertation, Universität Hamburg; h) Seyed-Mahdavi F (1986), Dissertation, Universität Hamburg; i) Wenck H (1986) Dissertation, Universität Hamburg; j) de Meijere A, Kozhushkov SI (unpublished results); k) Kordes M (1999) Dissertation, Universität Göttingen; l) Michaelsen H (1994) Dissertation, Universität Göttingen; m) Wessjohann L (1990) Dissertation, Universität Hamburg
8. Wessjohann L, Giller K, Zuck B, Skattebøl L, de Meijere A (1993) J Org Chem 58:6442
9. Wessjohann L, Krass N, Yu D, de Meijere A (1992) Chem Ber 125:867

10. a) Es-Sayed M, Gratkowski C, Krass N, Meyers AI, de Meijere A (1993) Tetrahedron Lett 34:289; b) Gratkowski C (1994) Dissertation, Universität Göttingen; c) Es-Sayed M (1992) Dissertation, Universität Hamburg
11. a) Ricker CB (1996) Diplomarbeit, Universität Göttingen; b) Ricker CB (1999) Dissertation, Universität Göttingen; c) Labahn T, Noltemeyer M, Heiner T, Ricker CB, de Meijere A (1999) Acta Cryst C (to be submitted)
12. a) Glück C, Poignée V, Schwager H (1987) Synthesis: 260; b) Sepiol J, Soulen RL (1975) J Org Chem 40:3791
13. Kostikov RR, de Meijere A (1984) J Chem Soc Chem Commun 1528
14. In the few known cases of intermolecular reactions of thermally or photochemically generated vinylcarbenes, the yields are markedly lower: a) Franck-Neumann M, Lohmann JJ (1977) Angew Chem 89:331 Angew. Chem Int Ed Engl 16:323 Angew Chem Suppl 1715; b) Stang PJ (1978) Chem Rev 78:383; c) Franck-Neumann M, Lohmann JJ (1979) Tetrahedron Lett 2397; d) Bruce MI (1991) Chem Rev 91:197
15. a) Liese T, Splettstößer G, de Meijere A (1982) Angew Chem 94:799 Angew Chem Int Ed Engl 21:790; b) Liese T, Teichmann S, de Meijere A (1988) Synthesis 25
16. Liese T, Seyed-Mahdavi F, de Meijere A (1990) Org Synth 69:148
17. Our recent experience allows one to increase the yield of this transformation up to 85% if this first S_N2'-type replacement has been performed at 20 °C for 24 h followed by heating under reflux. Ricker CR, Rauch K (unpublished results)
18. a) Salaün J, Bennani F, Compain J-C, Fadel A, Ollivier J (1980) J Org Chem 45:4129; b) Spitzner D, Swoboda H (1986) Tetrahedron Lett 27:1281; c) Osborne NF (1982) J Chem Soc Perkin Trans 1:1435
19. a) de Meijere A, Teichmann S, Seyed-Mahdavi F, Kohlstruk S (1996) Liebigs Ann 1989; b) Kohlstruk S (1992) Dissertation, Universität Göttingen
20. Seebach D, Hungerbühler E, Naef R, Schnurrenberger P, Weidmann B, Züger M (1982) Synthesis: 138; b) Krass N, Wessjohann L, Yu D, de Meijere A (1989) In: de Meijere A, Blechert S (eds) Strain and its Implications in Organic Chemistry. NATO-ASI Series C, Kluwer Academic Publ, Dordrecht, p 509
21. a) de Meijere A, Ernst K, Zuck B, Brandl M, Kozhushkov SI, Tamm M, Yufit DS, Howard JAK, Labahn T (1999) Eur J Org Chem in press; b) Ernst K (1994) Dissertation, Universität Göttingen
22. a) Brandl M, de Meijere A (unpublished results); b) Sokolov VI, de Meijere A (unpublished results); c) Rauch K, de Meijere A (unpublished results)
23. de Meijere A (1984) Bull Soc Chim Belg 93:241
24. Bengtson G (1985) Dissertation, Universität Hamburg
25. a) Michalski SA (1989) Diplomarbeit, Universität Hamburg; b) Michalski SA (1992) Dissertation, Universität Hamburg
26. a) de Meijere A, Teichmann S, Yu D, Kopf J, Oly M, von Thienen N (1989) Tetrahedron 45:2957; b) Teichmann S (1988) Dissertation, Universität Hamburg
27. Seyed-Mahdavi F, Teichmann S, de Meijere A (1986) Tetrahedron Lett 27:6185
28. Spitzner D, Engler A, Wagner P, de Meijere A, Bengtson G, Simon A, Peters K, Peters E-M (1987) Tetrahedron 43:3213
29. de Meijere A, Wenck H, Seyed-Mahdavi F, Viehe HG, Gallez V, Erden I (1986) Tetrahedron 42:1291
30. Heiner T (1996) Dissertation, Universität Göttingen
31. Dolbier WR Jr, Lomas D, Garza T, Harmon C, Tarrant P (1972) Tetrahedron 28:3185
32. Stella L, Janousek Z, Merényi R, Viehe HG (1978) Angew Chem 90:741 Angew Chem Int Ed Engl 17:691
33. Yu D (1989) Dissertation, Universität Hamburg
34. Narasaka K, Soai K, Mukaiyama T (1974) Chem Lett 1223
35. Conditions as optimized by Paulsen et al: Paulsen H, Hölck J-P (1982) Liebigs Ann Chem 1121. Cf. also [74]
36. a) Staudinger H, Meyer J (1919) Helv Chim Acta 2:635; b) Review: Gololobov YuG, Kasukhin LF (1992) Tetrahedron 48:1353; c) Mitchell D, Liebeskind LS (1990) J Am Chem Soc 112:291

37. Reviews: a) Subramanian LR, Zeller K-P (1997) In: de Meijere A (ed) Houben-Weyl, vol E 17a, Thieme, Stuttgart, p 256; b) Simmons HE, Cairns TL, Vladuchik SA, Hoiness CM (1973) Org React 20:1
38. a) Corey EJ, Chaykovsky M (1962) J Am Chem Soc 84:3782; b) Corey EJ, Chaykovsky M (1965) J Am Chem Soc 87:1353; c) Asunskis J, Shechter H (1968) J Org Chem 33:1164; d) Landor SR, Punja N (1967) J Chem Soc C 2495
39. de Meijere A, Kozhushkov SI, Yufit DS, Boese R, Haumann T, Pole DL, Sharma PK, Warkentin J (1996) Liebigs Ann 601
40. Regitz M, Heydt H (1984) In: Padwa A (ed) 1,3-Dipolar cycloaddition chemistry, Wiley, New York, vol 1, p 393
41. Aspartt-Pascot L, Bastide J (1971) Comptes Rendus Acad Sci Ser C 273:1772
42. Moffat JB (1978) J Phys Chem 82:1083
43. a) Zorn C, Goti A, Brandi A, Johnsen K, Kozhushkov SI, de Meijere A (1998) Chem Commun: 903; b) Zorn C, Goti A, Brandi A, Johnsen K, Noltemeyer M, Kozhushkov SI, de Meijere A (1999) J Org Chem 64:755
44. a) Danishefsky S (1979) Acc Chem Res 12:66; b) Loreto MA, Pellacani L, Tardella PA (1979) J Heterocyclic Chem 16:1233
45. a) Caserio MC, Graham WH, Roberts JD (1960) Tetrahedron 11:171; b) Roberts JD, Mazur RH (1951) J Am Chem Soc 73:2509; c) Renk E, Roberts JD (1961) J Am Chem Soc 83:878; Review: Klunder AJH, Zwanenburg B (1997) In: de Meijere A (ed) Houben-Weyl, vol E 17c, Thieme, Stuttgart, p 2419 ff
46. a) Primke H, Sarin GS, Kohlstruk S, Adiwidjaja G, de Meijere A (1994) Chem Ber 127:1051; b) Primke H (1989) Dissertation, Universität Hamburg
47. a) Anchel M, Hervey A, Robbins WJ (1950) Proc Nat Acad Sci USA 36:300; b) Anchel M, Hervey A, Robbins WJ (1952) Proc Nat Acad Sci USA 38:927; c) McMorris TC, Anchel M (1965) J Am Chem Soc 87:1594; d) McMorris TC, Kelner MJ, Chadha RK, Siegel JS, Moon S-s, Moya MM (1989) Tetrahedron 45:5433; e) Niwa H, Ojika M, Wakamatsu K, Yamada K, Ohba S, Saito Y, Hirono I, Matsushita K (1983) Tetrahedron Lett 24:5371; f) Ohba S, Saito Y, Hirono I, Niwa H, Ojika M, Wakamatsu K, Yamada K (1984) Acta Cryst C 40:1877; g) Ojika M, Wakamatsu K, Niwa H, Yamada K (1987) Tetrahedron 43:5261; h) Kelner MJ, McMorris TC, Beck WT, Zamora JM, Taetle R (1987) Cancer Med Research 47:3186; i) Kelner MJ, McMorris TC, Taetle R (1990) J Natl Cancer Inst 82:1562; j) McMorris T C, Kelner MJ, Wang W, Moon S, Taetle R (1990) Chem Res Toxicol 3:574; k) McMorris TC, Kelner MJ, Wang W, Estes LA, Montoya MA, Taetle R (1992) J Org Chem 57:6876; l) Review: Yamada K, Ojika M, Kigoshi H (1998) Angew Chem 110:1918 Angew Chem Int Ed Engl 37:1818
48. Ptaquilosin is the aglycon of the potent carcinogen ptaquiloside, which has the same skeleton as illudin M: a) Kigoshi H, Imamura Y, Mizuta K, Niwa H, Yamada K (1973) J Am Chem Soc 115:3056; b) Kigoshi H, Sawada A, Nakayama Y, Niwa H, Yamada K (1989) Tetrahedron Lett 30:1983; c) Kigoshi H, Imamura Y, Mizuta K, Niwa H, Yamada K (1993) J Am Chem Soc 115:3056. Ptaquiloside: d) Kushida T, Uesugi M, Sugiura Y, Kigoshi H, Tanaka H, Hirokawa J, Ojika M, Yamada K (1994) J Am Chem Soc 116:479; e) Kigoshi H, Imamura Y, Mizuta K, Niwa H, Yamada K (1993) J Am Chem Soc 115:3056; f) Kigoshi H, Tanaka H, Hirokawa J, Mizuta K, Yamada K (1992) Tetrahedron Lett 33:6647
49. a) Schore NE (1991) Org React 40:1; b) Schore NE (1991) In: Trost BM, Fleming I (eds) Comprehensive Organic Synthesis, Pergamon Press, Oxford, vol 5, p 1037 ff; c) Schore NE (1988) Chem Rev 88:1081; d) Pauson PL (1985) Tetrahedron 41:5855; e) Pauson PL (1987) In: de Meijere A, tom Dieck H (eds) Organometallics in Organic Chemistry. Springer, Berlin Heidelberg New York, vol 1, p 233; f) Krafft ME, Chirico X (1994) Tetrahedron Lett 35:4511; g) Oppolzer W, Robyr C (1994) Tetrahedron 50:415
50. a) Ager DJ, East MB (1993) Tetrahedron 49:5683; b) Vogel P, Fattori D, Gasparini F, Drian CL (1990) Synlett 173
51. The central carbon atom in methylenecyclopropane is almost sp-hybridized, like that in allene; the carbon atoms in cyclopropane are almost sp^2-hybridized, cf de Meijere A (1979) Angew Chem 91:867 Angew Chem Int Ed Engl 18:809

52. a) Vill V (1992) Mol Cryst Liq Cryst 213:67; b) Tabohasi T, Sakurai T, Higuchi R, Mikami N, Yamamoto E, Takeuchi K (1987) Eur Pat Appl EP 244,129CA (1988) 108:P66076j; c) McGaffin G (1993) Dissertation, Universität Hamburg; d) Byron DJ, Matharu AS, Rees M, Wilson RC (1995) Mol Cryst Liq Cryst A258:217
53. Tamm M (1997) Dissertation, Universität Göttingen
54. a) O'Donnell MJ, Polt RL (1982) J Org Chem 47:2663; b) O'Donnell MJ, Bennett WD, Polt RL (1985) Tetrahedron Lett 26:695; c) O'Donnell MJ, Falmagne J-B (1985) Tetrahedron Lett 26:699; d) O'Donnell MJ, Zhou C, Mi A, Chen N, Kyle JA, Andersson PG (1995) Tetrahedron Lett 36:4205; e) O'Donnell MJ, Yang X, Li M, Tetrahedron Lett (1990) 31:5135; f) O'Donnell MJ, Bennett WD (1988) Tetrahedron 44:5389; g) O'Donnell MJ, Zhou C, Chen N (1996) Tetrahedron: Asymmetry 7:621
55. Review: Lipshutz BH, Sengupta S (1992) In: Paquette LA (ed) Org React, Wiley, New York, vol 41 p 135
56. Tamm M, Thutewohl M, Ricker CB, Bes MT, de Meijere A (1999) Eur J Org Chem:2017
57. Blandov A, de Meijere A (unpublished results)
58. Wessjohann L, McGaffin G, de Meijere A (1989) Synthesis 359
59. Es-Sayed M, Heiner T, de Meijere A (1993) Synlett 57
60. a) Wessjohann L, Skattebøl L, de Meijere A (1990) J Chem Soc Chem Commun: 574; b) Giller K, Baird MS, de Meijere A (1992) Synlett: 524
61. a) O'Donnell MJ, Wojciechowski K, Ghosez L, Navarro M, Sainte F, Antoine J-P (1984) Synthesis: 313; b) Stork G, Leong AYW, Touzin AM (1976) J Org Chem 41:3491
62. Es-Sayed M, Gratkowski C, Krass N, Meyers AI, de Meijere A (1992) Synlett: 962
63. a) Matsui S, Hashimoto Y, Saito K (1998) Synthesis: 1161; b) Badone D, Bernassau J-M, Cardamone R, Guzzi U (1996) Angew Chem 108:575 Angew Chem Int Ed Engl 35:535; c) Tamura O, Hashimoto M, Kobayashi Y, Katoh T, Nakatani K, Kamada M, Hayakawa I, Akiba T, Terashima S (1992) Tetrahedron Lett 33:3487; d) Newman MS, Kutner A (1951) J Am Chem Soc 73:4199; e) Stefanovsky JN, Spassov SL, Kurtev BJ, Balla M, Ötvös L (1969) Chem Ber 102:717. Other methods of preparation: f) Söderbaum HG (1896) Chem Ber 29:1210; g) Auerbach RA, Kingsbury CA (1971) Tetrahedron 27:2069; h) Hassner A, Burke SS (1974) Tetrahedron 30:2613; i) Okuda M, Tomioka K (1994) Tetrahedron Lett 35:4585; j) Foglia TA, Swern D (1969) J Org Chem 34:1680; k) Pirkle WH, Simmons KA (1983) J Org Chem 48:2520; l) Hoppe D, Follmann R (1976) Chem Ber 109:3047; m) Murthy KS, Keshava DDN (1984) J Heterocycl Chem 21:1721; n) Banks MR, Cadogan JIG, Gosney I, Hodgson PKG, Thomson DE (1991) J Chem Soc Perkin Trans 1:961; o) Murthy KS, Keshava DDN (1984) Synth Commun 14:687; p) Park, YS, Boys ML, Beak P (1996) J Am Chem Soc 118:3757
64. (*S*)-4-Phenyloxazolidine-2-one (**98**), (*S*)-4-phenyloxazolidine-2-thione (**99**) and (4*R*,5*S*)-4,5-diphenyloxazolidine-2-one (**100**) are very poor nucleophiles. A substoichiometric amount (10 mol%) of potassium hydride is necessary to generate a fraction of the potassium salt of **13** which has a sufficient nucleophilicity. This holds for the anions generated by deprotonation at the carbamate position and activation by addition of dibenzo-18-crown-6 to the reaction mixture, and causes the Michael addition to be very slow at –78 °C. The best temperature is $-20 \rightarrow 0$ °C. The higher the temperature, the more decomposition product is formed during the Michael addition.
65. a) Duhamel L, Duhamel P, Launay JC, Plaquevent JC (1984) Bull Soc Chim Fr II: 421; b) Potin D, Williams K, Rebek Jr J (1990) Angew Chem 102:1485 Angew Chem Int Ed Engl 29:1420; c) Fehr C, Galindo J (1988) J Am Chem Soc 110:6909
66. For the analogous reactions which did not get any further development see also ref. [9]
67. a) Salaün J (1987) In: Rappoport Z (ed) The chemistry of the cyclopropyl group, Wiley, New York, p 809ff; b) Gajewsky JJ, Oberdier JP (1972) J Am Chem Soc 94:6053; c) Erden I, de Meijere A, Rousseau G, Conia JM (1980) Tetrahedron Lett 21:2501; d) Crandall JK, Conower WW (1978) J Org Chem 43:3533; e) Aue DH, Meshishneck MJ, Shellhamer DF (1973) Tetrahedron Lett 4799
68. a) Aue DH, Lorens RB, Helwig GS (1973) Tetrahedron Lett: 4795; b) Aue DH, Lorens RB, Helwig GS (1979) J Org Chem 44:1202; c) Crandall JK, Conover WW (1973) J Chem Soc

Chem Commun 33; d) Crandall JK, Conover WW (1974) J Org Chem 39:63; e) Tsuge O, Ohnishi T, Watanabe H (1981) Heterocycles 16:2085; f) Tsuge O, Watanabe H (1976) Heterocycles 4:1905; g) Tsuge O, Watanabe H (1977) Heterocycles 7:709; h) Tsuge O, Watanabe H, Kiryu Y (1979) Bull Chem Soc Jpn 52:3387; i) Tlili T (1988) J Soc Chim Tunis 2:3
69. Dauban P, Dubois L, Dau METH, Dodd RH (1995) J Org Chem 60:2035; b) Tanner D (1994) Angew Chem 106:525 Angew Chem Int Ed Engl 33:599 and references cited therein; c) Goodman M (1994) Chemistry and Biology 231
70. a) Amice P, Conia JM (1974) Bull Soc Chim Fr: 1015; b) Amice P, Conia JM (1974) Tetrahedron Lett: 479; c) Snider BB, Roush DM, Rodini DJ, Gonzalez D, Spindell D (1980) J Org Chem 45:2773; d) Ficini J, Krief A (1970) Tetrahedron Lett: 885; e) Ficini J, Dureault A (1977) Tetrahedron Lett 809
71. The ring enlargement of a cyclopropylmethylene to a cyclobutene is well known: a) Wiberg KB, Burgmaier GJ, Warner P (1971) J Am Chem Soc 93:246; b) de Meijere A, Wenck H, Kopf J (1988) Tetrahedron 44:2427; c) Ho G-J, Krogh-Jespersen K, Moss RA, Shen S, Sheridan RS, Subramanian R (1989) Am Chem Soc 111:6875
72. Tamm M, Nötzel M, Noltemeyer M, de Meijere A (2000) J Org Chem (submitted)
73. Kuchuk I, de Meijere A (unpublished results)
74. Adachi T, Yamada Y, Inoue I (1977) Synthesis 45
75. a) Dürckheimer W, Blumbach J, Lattrell R, Scheunemann KH (1985) Angew Chem 97:183 Int Ed Engl 24:180; b) Bell MR, Carlson JA, Oesterlin R (1972) J Org Chem 37:2733; c) Jungheim LN, Sigmund SK, Jones ND, Swartzendruber JK, Lilly E (1987) Tetrahedron Lett 28:289; d) Campbell MM, Connarty BP (1982) Heterocycles 19:1853; e) Campbell MM, Carruthers NI, Mickel SJ, Winton PM (1984) J Chem Soc Chem Commun: 200; f) Barrett AGM, Cheng MC, Spilling CD, Taylor SJ (1989) J Org Chem 54:992
76. a) Johnson G, Ross BC (1981) J Chem Soc Chem Commun 1269; b) Branch CL, Pearson MJ (1986) J Chem Soc Perkin Trans I: 1077
77. As for the biological activities of cyclopropane derivatives see: a) Salaün J, S. Baird MS (1995) Curr Med Chem 2:511; b) Salaün J (1999) Top Curr Chem, this volume
78. Hegedus LS, Montgomery J, Narukawa Y, Snustad DC (1991) J Am Chem Soc 113: 5784
79. Kirmse W (1965) Angew Chem 77:1 Angew Chem Int Ed Engl 4:1
80. Review: Korobitsyna IK, Khalikova AV, Rodina LL, Shusherina NP (1983) Khim Geterotsikl Soed 147 Chem Heterocycl Comp (Engl Transl) 117
81. Breslow R (1963) In: de Mayo P (ed) Molecular rearrangements. Wiley, New York p 233 ff
82. a) Govindan CK, Taylor G (1983) J Org Chem 48:5348; b) Wendling LA, Bergman RG (1976) J Org Chem 41:831; c) Rossi E, Celentano G, Stradi R, Strada A (1990) Tetrahedron Lett 31:903
83. a) Boeckman RK Jr, Jackson PF, Sabatucci JP (1985) J Am Chem Soc 107:2191; b) Wasserman HH, Dion RP, Fukuyama J (1989) Tetrahedron 45:3203
84. a) Málek J (1988) Org React 36:310; b) Hutchins RO, Su W-Y, Sivakumar R, Cistone F, Stercho YP (1983) J Org Chem 48:3412
85. a) Stevens RV (1977) Acc Chem Res 10:193; b) Boeckman RK Jr, Sabatucci JP, Goldstein SW, Springer DM, Jackson PF (1986) J Org Chem 51:3740; c) Wasserman HH, Fukuyama JM (1991) Tetrahedron Lett 32:7127; d) Maryanoff BE, McComsey DF, Mutter MS, Sorgi KL, Maryanoff CA (1988) Tetrahedron Lett 29:5073
86. a) Cloke JB (1929) J Am Chem Soc 51:1174; b) Gotkis D, Cloke JB (1934) J Am Chem Soc 56:2710; c) Stevens RV, Ellis MC, Wentland MP (1968) J Am Chem Soc 90:5576
87. Nötzel M (1999) Part of a forthcoming Dissertation, Universität Göttingen
88. a) Stacy GW, Day RI, Morath RJ (1955) J Am Chem Soc 77:3869; b) Stacy GW, Craig PA, DayRI (1958) J Org Chem 23:1760; c) Sandhu JS, Sethi PS, Mohan S (1971) J Indian Chem Soc 48:89
89. Belov VN, Funke C, Labahn T, Es-Sayed M, de Meijere A (1999) Eur J Org Chem 1345
90. a) Robert M, Barbier M, Lederer E, Roux L, Biemann K, Vetter W (1962) Bull Soc Chim Fr 187; b) Hardegger E, Liechti P, Jackman LM, Boller A, Plattner PA (1963) Helv Chim Acta 46(1):60

91. a) Burrouhs LF (1957) Nature 179:360; b) Vähätalo M-L, Virtanen AI (1957) Acta Chem Scand 11:741
92. a) Rosenthal GA (1982) Plant Nonproteinogenic Amino and Imino Acids. Biological and Toxicological Properties, Academic Press, New York; b) Barrett GC (1985) Chemistry and Biochemistry of the Amino Acids, Chapman and Hall, New York, p 64; c) Boller T, Kende H (1980) Nature 286:259; d) Pirrung MC (1987) J Org Chem 52:4179; d) Hill RK, Prakash SR, Wiesendanger R, Angst W, Martinoni B, Arigoni D, Liu H-W, Walsh CT (1984) J Am Chem Soc 106:795; e) Hoffman NE, Yang SF, Ichihara A, Sakamura S (1982) Plant Physiol 70:195
93. a) Pirrung MC, McGeehan GM (1985) Angew Chem 97:1074 Angew Chem Int Ed Engl 24:1044; b) Mitchel RE, Pirrung MC, McGeehan GM (1987) Phytochemistry 26:2695; c) Ichihara A, Shiraishi K, Sato H, Sakamura S, Nishiyama K, Sakai R, Furusaki A, Matsumoto T (1977) J Am Chem Soc 99:636; d) Bernabe M, Fernadez OC, Fernadez-Alvarez E (1979) Annales de Quimica 75:977; e) Harding WM, DeShazo ML (1967) Arch Biochem Biophys 118:23; f) Johnson RL, Koerner JF (1988) J Med Chem 31:2057; g) Pellicciari R, Natalini B, Marinozzi M, Monahan JB, Snyder JP (1990) Tetrahedron Lett 31:139; h) Stewart FHC (1981) Aust J Chem 34:2431; i) Tsang JW, Schmied B, Nyfeler R, Goodman M (1984) J Med Chem 27:1663; j) Pirrung MC, McGeehan GM (1986) J Org Chem 51:2103; k) Baldwin JE, Adlington RM, Rawlings BJ (1985) Tetrahedron Lett 26:481; l) Pirrung MC, Dunlap SE, Trinks UP (1989) Helv Chim Acta 72:1301; m) Aitken DJ, Royer J, Husson H-P (1990) J Org Chem 55:2814
94. Stammer CH (1990) Tetrahedron 46:2231; b) Pirrung MC, Cao J, Chen J (1995) J Org Chem 60:5790; c) Burgess K, Lim D, Ho K-K, Ke C-Y (1994) J Org Chem 59:2179
95. For several recent communications on this topic see: a) Zindel J, de Meijere A (1995) J Org Chem 60:2968; b) Burgess K, Ke C-Y (1996) Synthesis 1463; c) Hercouet A, Bessières B, Le Corre M (1996) Tetrahedron Asymmetry 7:1267; d) Franzone G, Carle S, Dorizon P, Ollivier J, Salaün J (1996) Synlett 1067; e) Hercouet A, Godbert N, Le Corre M (1998) Tetrahedron: Asymmetry 9:2233; f) Fadel A, Khesrani A (1998) Tetrahedron: Asymmetry 9:305; g) Chinchilla R, Falvello LR, Galindo N, Nájera C (1998) Tetrahedron: Asymmetry 9:2223; h) Atlan V, Racouchot S, Rubin M, Bremer C, Ollivier J, de Meijere A, Salaün J (1998) Tetrahedron: Asymmetry 9:1131
96. a) Baldwin JE, Adlington RM, Bebbington D, Russell AT (1994) Tetrahedron 50:12015 and ref. [2] therein; b) Lai M-t, Liu M-d, Liu H-w (1991) J Am Chem Soc 113:7388; c) Lai M-t, Liu H-w (1992) J Am Chem Soc 114:3160; d) Kurokawa N, Ohfune Y (1985) Tetrahedron Lett 26:83; e) Li K, Du W, Que NLS, Liu H-w (1996) J Am Chem Soc 118:8763
97. Shoji J-I, Sakazaki R, Kato T, Tori K, Yoshimura Y, Matsuura S (1981) J Antibiotics 34:370
98. Kato K, Takita T, Umezawa H (1980) Tetrahedron Lett 21:4925
99. a) Evans DA, Britton TC (1987) J Am Chem Soc 109:6881; b) Evans DA, Ellman JA (1989) J Am Chem Soc 111:1063
100. a) Aoyagi T, Yoshida S, Matsuda N, Ikeda T, Hamada M, Takeuchi T (1991) J Antibiotics 44:573; b) Angelastro MR, Mehdi S, Burkhart JP, Peet NP, Bey P (1990) J Med Chem 33:13
101. a) Bunnage ME, Davies SG, Goodwin CJ (1994) J Chem Soc Perkin Trans I: 2385; b) Wang Z-M, Kolb HC, Sharpless KB (1994) J Org Chem 59:5104
102. a) Katayama N, Nozaki Y, Tsubotani S, Kondo M, Harada S, Ono H (1992) J Antibiotics 45:10; b) Funabashi T, Tsubotani S, Koyama K, Katayama N, Harada S (1993) Tetrahedron 49:13; c) Sokolov VV, Kozhushkov SI, Nikolskaya S, Belov VN, Es-Sayed M, de Meijere A (1998) Eur J Org Chem: 777
103. Manis PA, Rathke MW (1980) J Org Chem 45:4952
104. a) Zindel J (1993) Dissertation, Universität Göttingen; b) Gutke H-J (1998) Dissertation, Universität Göttingen
105. Spitzner D, Engler A, Liese T, Splettstößer G, de Meijere A (1982) Angew Chem 94:799 Angew Chem Int Ed Engl 21:790
106. Hadjiarapoglou L, de Meijere A, Seitz H-J, Klein I, Spitzner D (1994) Tetrahedron Lett 35:3269
107. Hadjiarapoglou L, Klein I, Spitzner D, de Meijere A (1996) Synthesis 525

108. de Meijere A, Hadjiarapoglou LP, Noltemeyer M, Gutke H-J, Klein I, Spitzner D (1998) Eur J Org Chem 441
109. From a formal point of view, the one-pot heterocyclizations discussed in Sect. 4.2 may be also called MIRC reactions. In the present section this term implies the formation of a cyclopropane ring in accordance with the generally accepted MIRC mechanism [1]
110. a) de Meijere A, Kozhushkov SI (1995) In: Halton B (ed) Advances in Strain in Organic Chemistry, JAI Press Ltd., London, vol 4, p 225ff; b) Zefirov NS, Kuznetsova TS, Zefirov AN (1995) Izv Akad Nauk Ser Khim 1613 Russian Chem Bull (Engl Trans) 44:1543
111. a) Hagiwara H, Kodama T, Kosugi H, Uda H (1976) J Chem Soc Chem Commun 413; b) Gröger C, Musso H (1976) Angew Chem 88:415 Angew Chem Int Ed Engl 15:373
112. Tsuji T, Kikuchi R, Nishida S (1985) Bull Chem Soc Jpn 58:1603
113. a) Wani MC, Taylor HL, Wall ME, Coggon P, McPhail AT (1971) J Am Chem Soc 93:2325; b) Vidensek N, Lim P, Campell A, Carlson C (1990) J Nat Prod 53:1609; c) Witherup KM, Look SA, Stasko MW, Ghiorzi TJ, Muschik GM, Cragg GM (1990) J Nat Prod 53:1249
114. a) Swindell CS (1991) Org Prep Procedures Int 23:465; b) Kingston DGI, Molinero AA, Rimoldi JM (1993) Prog Chem Org Nat Prod 61:1; c) Nicolaou KC, Dai W-M, Guy RK (1994) Angew Chem 106:38, Angew Chem Int Ed Engl 33:15; d) Boa AN, Jenkins PR, Lawrence NJ (1994) Contemporary Organic Synthesis 1:47
115. a) Holton RA, Somoza C, Kim H-B, Liang F, Biediger RJ, Boatman PD, Shindo M, Smith CC, Kim S, Nadizadeh H, Suzuki Y, Tao C, Vu P, Tang S, Zhang P, Murthi KK, Gentile LN, Liu JH (1994) J Am Chem Soc 116:1597; b) Holton RA, Kim H-B, Somoza C, Liang F, Biediger RJ, Boatman PD, Shindo M, Smith CC, Kim S, Nadizadeh H, Suzuki Y, Tao C, Vu P, Tang S, Zhang P, Murthi KK, Gentile LN, Liu JH (1994) J Am Chem Soc 116:1599; c) Nicolaou KC, Yang Z, Liu JJ, Ueno H, Nantermet PG, Guy RK, Claiborne CF, Renaud J, Couladouros EA, Paulvannan K, Sorensen EJ (1994) Nature 367:630
116. a) Francisco C, Banaigs B, Teste J, Cave A (1986) J Org Chem 51:1115; b) Kakiuchi K, Fukunaga K, Matsuo F, Ohnishi Y, Tobe Y (1991) J Org Chem 56:6742
117. a) Nagaoka H, Shibuya K, Yamada Y (1993) Tetrahedron Lett 34:1501; b) Nagaoka H, Baba A, Yamada Y (1991) Tetrahedron Lett 32:6741
118. Wrobel J, Takahashi K, Honkan V, Lannoye G, Cook JM, Bertz SH (1983) J Org Chem 48:139
119. a) Militzer H-C, Schömenauer S, Otte C, Puls C, Hain J, Bräse S, de Meijere A (1993) Synthesis 998; b) Padwa A, Wannamaker MW (1986) Tetrahedron Lett 27:2555
120 Gutke H-J, Spitzner D (1999) Tetrahedron 55:3193
121. For reviews see a) Trost BM (1977) Tetrahedron 33:2615; b) Trost BM (1980) Acc Chem Res 13:385; c) Tsuji J (1980) Organic Synthesis with Palladium Compounds, Springer, Berlin; d) Tsuji J (1986) Tetrahedron 42:4361; e) Tsuji J, Minami I (1987) Acc Chem Res 20:140; f) Consiglio G, Waymouth RM (1989) Chem Rev 89:257; g) Trost BM (1978) Angew Chem 101:1199 Angew Chem Int Ed Engl 28:1173; h) Frost CG, Howarth J, Williams JMJ (1992) Tetrahedron: Asymmetry 3:1089; i) de Meijere A, Meyer FE (1994) Angew Chem 106:2473 Angew Chem Int Ed Engl 33:3279; j) Bräse S, de Meijere A (1998) In: Diederich F, Stang PJ (eds) Metal-catalyzed Cross-coupling Reactions, Wiley-VCH, Weinheim, p 99ff; k) de Meijere A, Bräse S (1991) J Organomet Chem 576:88
122. a) Ang KH, Bräse S, Steinig AG, Meyer FE, Llebaria A, Voigt K, de Meijere A (1996) Tetrahedron 52:11503; b) Meyer FE, Ang KH, Steinig AG, de Meijere A (1994) Synlett 191
123 Körbe S, de Meijere A (unpublished results)
124 Bhat A, Steinig A, Appelbe R, de Meijere A (2000) Eur J Org Chem (to be submitted)
125. a) Terrett NK, Gardener M, Gordon DW, Kobylecki RJ, Steele J (1995) Tetrahedron 51:8135; b) Balkenhohl F, Bussche-Hünnefeld C, Lansky A, Zechel C (1996) Angew Chem 108:2436 Angew Chem Int Ed Engl 35:2288; c) Nefzi A, Ostresh JM, Houghten RA (1997) Chem Rev 97:449; d) Szardenings AK, Burkoth TS, Lu HH, Tien DW, Campbell DA (1997) Tetrahedron 53:6573; e) Dömling A (1998) Comb Chem & High Troughput Screen 1:1
126. Terrett NK (1998) Combinatorial Chemistry, Oxford University Press, Oxford
127. Khlebnikov AF, de Meijere A (unpublished results)
128. Hughes DL (1992) In: Paquette L (ed) Org React, Wiley, New York, vol 42, p 335

Author Index Volume 201 – 207

Author Index Vols. 26 – 50 see Vol. 50
Author Index Vols. 51 – 100 see Vol. 100
Author Index Vols. 101 – 150 see Vol. 150
Author Index Vols. 151 – 200 see Vol. 200

The volume numbers are printed in italics

MIX
Papier aus verantwortungsvollen Quellen
Paper from responsible sources
FSC® C105338

If you have any concerns about our products,
you can contact us on
ProductSafety@springernature.com

In case Publisher is established outside the EU,
the EU authorized representative is:
Springer Nature Customer Service Center GmbH
Europaplatz 3, 69115 Heidelberg, Germany

Printed by Libri Plureos GmbH
in Hamburg, Germany